795=

265
316
385
419

35-1
17-2,3
13
5.8, 29
23
39
14.3
29

HINDUISM

AND

ITS MILITARY ETHOS

HINDUISM

AND

ITS MILITARY ETHOS

AIR MARSHAL (Retd) RK NEHRA

Lancer • New Delhi • Frankfort, IL

www.lancerpublishers.com

LANCER

Published in the United States by

The Lancer International Inc
19558 S. Harlem Ave., Suite 1,
Frankfort, IL. 60423.

First published in India by

Lancer Publishers & Distributors
2/42 (B) Sarvapriya Vihar,
New Delhi-110016

ISBN-13: 978-1-935501-23-7 • ISBN-10: 1-935501-23-2

Online Military Bookshop
www.lancerpublishers.com

IDR Net Edition
www.indiandefencereview.com

Dedication

The book is dedicated to my loving and loved wife Sneh Nehra.
It was her encouragement and unstinted support,
which enabled me to write this book.

Raj Nehra

CONTENTS

PREFACE

Hinduism is the most ancient and complex religion of humankind. Hindus like to believe that their religion is timeless; they call it *anadi* – without beginning, and *anant* – without end. Hinduism is a fascinating religion with its earthy smell, limitless horizons and bewildering variety; it has universal appeal.

There is no way we can present a philosophical view of Hinduism in a book of this nature. All we do is to try to tell the story of Hinduism in simple language of the layman, taking care to avoid myriad complexities of the religion. This does involve some compromises; hopefully, the reader would bear with that.

We cover the story about Hinduism in three broad headings:

- Hindu Religion and its Godhood
- Hindu History – Ancient to Modern
- Hindu Military Value System

Hinduism is basically a mythology based religion; the reader will get acquainted with many mystifying details of Hindu mythology. A broad view of Hindu literature, starting with the Rig Veda, the first religious book of humankind, is given. The fascinating story of development of Hindu Godhood is covered; the tale starts with 33 gods of the Rig Veda, covers the One Single God of the Upanishads, ending with *Trimurti* (Hindu Triad) and other gods of the *Puranas* and the epics. In the ancient times, it was the Vedic *dharma* that prevailed in Bharat-varsha; the present-day India has the Puranic *dharma*.

Hindus, whilst having the most voluminous religious literature, have been very economical in writing about their history. The Hindu history is mostly known from the pen of foreigners who visited India from time to time. The early Hindu history is actually more mythology, and less history. The recorded Hindu history starts from about 6th century BC, but mostly with patchy details. Some foreign tribes came to rule North West India, starting as early as 2nd century BC and going up to 3rd century AD. The Muslim rule started in 1200 AD, which was replaced by the British rule around 1750 AD.

Out of the recorded Hindu history of around 2300 years, Bharat was under jackboots of slavery for some 1300 years – a dubious record. The book attempts to analyze the reasons that were responsible for prolonged periods of Hindu slavery. It is baffling to see the great Hindu civilization going under with such extraordinary ease. It would appear that reasons for Hindu slavery lay in their mind, rather than in their muscle. The ancient Hindus were a set of martial people who lived by the sword. Somewhere along the line, Hindus lost their way and their martial spirit. The book concludes that Hindus developed a deluded sense of *dharma* under influence of Buddhism; that was the main reason for their downfall. This aspect is discussed in some detail in last part of the book.

Section A
World Religions

1
ABOUT THE CONTENTS

The story in this book has three separate, but inter-related elements:

- Hindu Civilization, Religion and Godhood
- Hindu (Indian) History — Ancient to Modern
- Hindu Military Value System

This book about Hinduism starts by giving a brief account of other major religions of the world including Buddhism, Christianity and Islam. Chapters 4 to 15 explore the Hindu religion in all its diversity, glory and complexity. An effort has been made to present a simplified view of Hinduism to which, hopefully, an *aam admi* (layman) can easily relate.

Hindu Civilization, Religion and Godhood

Man felt the need of religion at a very early stage of development of human society. As soon as man's food and shelter needs were met, he started looking for some sort of a supernatural power. Throughout human history, no society *so far* has been able to live without a belief system in supernatural power, later called God. Such a belief system came to be called religion. Even in the present materialistic world (*Kali Yuga* for Hindus), the vast majority believes in God and religion of one's choice. Self-proclaimed atheists are a microscopic minority.

Hinduism is the first formal religion developed by mankind. No one has an idea as to when Hinduism evolved, and who were its founders (or founder). What is known for sure is that this religion got evolved in the holy and ancient land of Bharatvarsh, later called Hindustan, and then India. Widely different dates have been suggested for the start of the Hindu religion. These vary from 2000–1500 BC as put forward by the western historians, to hundreds of thousands of years as per the Yuga theory believed by the orthodox Hindus. There is also an intermediate view, which talks in terms of 4000–3000 BC.

Compared to other religions, Hinduism is a very complex religion, difficult to understand, especially by laymen. That is due to:

- Very ancient nature of the religion *appealing to the intellect*
- Prevalence of large number of intricate philosophies, some of these in conflict with each other
- Innumerable number of gods and goddesses, and conflict between polytheism and monotheism

Hinduism presents some of the most original and sublime thought processes that the human mind could evolve. The philosophies contained in the *Bhagwad Gita* and the *Upanishads* are unparalleled. The *Rig Veda* is the first religious book of mankind. Then there are the *Puranas* — a storehouse of ancient wisdom, legends and mythology. The two epics of the *Ramayana* and *Mahabharta* play a central and crucial role in defining contours of Hinduism. In addition, there is a vast range and variety of other religious Hindu literature.

The journey of development of Hindu Godhood makes a fascinating subject. The *Rig Veda* introduces the very concept of God, which is further developed in the *Upanishads*. That is followed by a period of almost 'no-god' of Buddhism. Thereafter, in the closing centuries before the birth of Christ emerge a new series of Hindu gods in the *Puranas* and the epics.

Hindu (Indian) History

We have to understand the past so that we can analyze the present and plan for the future. The present does not lend itself to any realistic analysis; it is too close to events, and we get overtaken by our prejudices and pre-formed views. To know the past, we have to study history, which is full of pitfalls. Generally, there is always more than one version and view of history. It may not be always possible to find objective narratives of history.

Especially, Hindu history presents innumerable challenges. Ancient Hindu history is an artful mix of mythology and history; so much that it is almost impossible to distinguish between the two. Hindus possibly have the most extensive ancient literature of religious nature; it is not always possible to tell what part is history, and what part is mythology.

Even the later recorded Hindu history is known mostly from chronicles of foreigners who visited India (Bharat) from time to time. These include:

- One Greek: Megasthenes — around 320 BC, during the rule of Chandragupta Maurya
- Three Chinese:
  Fa-Hien: 400 to 411 AD — Gupta Empire
  Hiuen Tsang: 630 to 644 AD — Harshvardhan Rule
  I Tsing: 671 to 695 AD — No central rule in India

In addition, there were a number of Muslim historians in the second millennium AD; the most important of them was Al-Beruni, who accompanied Mahmud of Ghazni around 1010 AD. Then, during the 18th and 19th century AD, there were an avalanche of British historians; they really dug up the Hindu past. In spite of what some hardcore analyst may express, British historians present a largely unbiased view of Hindu history, with perhaps an odd undeserved emphasis or deviation here and there. The indigenous Indian (read Hindu) sources are very economical on listing and description of Hindu historical events. For example, the invasion of India by Alexander in 326 BC was a major event in Indian history — the Hindu sources make very little reference to it, and give very few details. Very scant details are available for a period of 500 years (200 BC to 300 AD) when foreign tribes ruled over North West India, and again from 700 AD to 1000 AD, i.e. before coming in of the Muslims.

Hindu Military Value System

The very first human requirements on earth were food and shelter. As supplies of these were limited, there was need for competition.

Gradually, requirements of man increased from food and shelter to include items like cattle, land, etc. Over time, need was also felt for the intangibles like self-respect, freedom, glory and dominance (over others); these are a part of human DNA. In other words, the situation was ideal for disagreement and conflict. As the intensity of conflict increased, it acquired contours of what we now call war.

War is a natural part of the development of human society. It is as integral to the human psyche as love, spirituality, religion and other such emotive issues. We may not love war, but we cannot avoid it. So, we must not only learn to live with war, but be comfortable with it. Societies which want to live with dignity have to be always prepared for war. People, who could learn this basic lesson, dominated the world; those who did not, were dominated. This is one lesson that cries out loud and clear from the pages of history.

In keeping with the above principle, the early Vedic (Hindu) people were martial and warlike. They valued and loved their freedom, roamed the jungles and lived by hunting. They propitiated their gods by animal sacrifices; the horse-sacrifice had special significance. The Hindu holy book, the incomparable *Bhagwad Gita* emphasizes the duty to engage in (righteous) war — to kill, or be killed.

However, over time, due to a number of factors, things started to change. Emphasis in Hinduism shifted to spirituality and other-worldly matters, especially under the influence of Buddhism and Jainism. Self-defeating concepts like *ahimsa* (non-violence), *shanti* (peace), *satya* (truth) started making inroads into the Hindu psyche. At some stage Hindus got obsessed with the 'next life'. War and issues connected with it were relegated to the background. The Hindus had to pay very dearly for that mindset. That is the central theme of this book and we will examine the issue in detail, as our narrative develops.

Contents of the Book

The book is about Hindus, and their military value system. As such, the Muslims get mentioned only in passing, especially as their military value system has been generally in order, and needs no specific changes. The author holds the Indian Muslim community in high esteem.

The book is divided into 6 Sections, designed according to the following pattern:

Section A
World Religions (Chapters 1 to 3)

The section dwells on the nature of religion and covers in brief major religions of the world, other than Hinduism, i.e. —

- India Born Religions: Buddhism, Jainism and Sikhism
- Judaic Religions : Judaism, Christianity and Islam

Section B
Hinduism (Chapters 4 to 9)

This is the main section of the book, covering Hindu religion in all its diversity and complexity. It has six chapters, under the following headings:

- An Overview of Hinduism: A Short History
- Hindu Religious Literature: From Vedas to Tantras
- Hindu Godhood: From Vedic gods to Brahman and Ganesha
- Hindu Creation Concepts: An Overview
- Manu — the Hindu Adam: Origin and Progeny
- Hindu Miscellaneous: Ocean Churning and Others Issues

Section C
Some Issues of Hindus (Chapters 10 to 15)

The section covers some contemporary issues of the Hindus like Polytheism, Untouchability, *Ahimsa* (non-violence), Hindu Tolerance, etc. It also analyses some Hindu sayings like:

- *Satya Mev Jayate* (Truth always triumphs)
- *Sarv Dharam Sambhav* (Equal respect for all religions).

These sayings have been counter-productive for the Hindus.

Section D
History of India (Chapters 16 to 26)

This section gives an overview of (recorded) Hindu history from the very early period (7th century BC) to 20th century AD. The period from the Mauryan Empire to the Muslim period, followed by the

British rule is covered. There are chapters on Rajputs, Marathas and South India.

Section E
Free India and Its Armed Forces (Chapters 27 to 37)

Current events starting with the Partition of India are covered in this section. There are chapters on Indian Armed Forces, the Indian Atom Bomb and the Ayodhya Temple. The battles fought by the Indian Forces post 1947 are analyzed with a degree of candor, perhaps not seen earlier; the author may get a lot of flak for being frank, and not being discreet, which is the norm as far as military issues are concerned.

General issues connected with the conduct of war are covered in this section; these include 'Role of Generals' and 'Shedding of Blood' in war.

Section F
Hindu Military Reverses — Analysis (Chapter 38 to 43)

It analyses the causes of Hindu military defeats over the centuries, and suggests measures to remedy the situation.

2
BUDDHISM, JAINISM AND SIKHISM
India Born Religions

2.1 Background

The two earliest formal religions to emerge in the world were Hinduism and Judaism, in that order. Hinduism can be considered to have evolved around 4000–3000 BC and Judaism around 2000 BC. Hinduism emerged in the fertile land of Bharat, served by many river systems. Judaism emerged in the sandy wastes of West Asia. In the rest of the world, largely localized pagan belief systems (perhaps not religions) prevailed. We may recall that both the Greek and Roman civilizations prospered without the base of any formalized religion — as we understand religion today.

Hinduism, largely, remained confined to Bharat. Judaism first spread to the surrounding areas, and then beyond. The first religion to spread far and wide was Christianity. Islam out-did Christianity in its pace of spread. In Bharat, two other religions, Buddhism and Jainism were born in the 6th Century BC. At present, the world has two families or streams of religions, i.e. —

- India Born Religions (IBRs): Hinduism, Buddhism, Jainism, Sikhism
- The Judaic family of religions: Judaism, Christianity, Islam

First, we take the IBRs for our study. As Hinduism is the main topic of our study, we will take it up later, starting with Chapter 4. We start with our study of Buddhism, Jainism and Sikhism

2.2 Buddhism

The founder of Buddhism was Gautama Budha. He was born as a Kshatriya prince Siddharta, belonging to the Sakya tribe. The tribe ruled around an area near the Indo–Nepal border, with its headquarters at Kapilavastu. Budha is believed to have lived from 563 to 486 BC, or so. Budha means 'The Enlightened One'; he is also called Tathagat and Sakya Muni.

Buddhism is based on the basic principle that the world is full of misery and sorrow, and that the object of religion is to find *Nirvana* or *Moksha*, i.e. deliverance from the cycle of rebirth. Budha did not believe in God as creator of the universe; he neither accepted nor rejected the existence of God. Budha was not for an immortal soul, though he believed in rebirth (Rebirth without the concept of an immortal soul may appear somewhat strange to most people). Budha rejected the authority of the Vedas and supremacy of the priestly class, i.e. the Brahmins.

Buddhism advocated virtuous and holy life, and avoidance of sinful thought. He expressed that no injury, by word or deed, be done to any living being. Budha strongly emphasized *ahimsa* (non-violence). He denounced animal sacrifice, which was part of the Vedic rituals, though there was perhaps no injunction against meat eating. Budha was against the caste system, but did not particularly work for its abolition.

Buddhist canons revolve around the four Noble Truths and the Eight Fold Path. The four Noble Truths are:

- Life is full of Suffering and Sorrow
- Suffering is caused by Desire
- Only elimination of Desire can lead to the end of Suffering
- Desire can be eliminated by following the Eight-Fold path

The Eight Fold path is also called the Middle Path. Adherence to the Middle Path suggests neither a life of ease, nor of religious asceticism, which was advocated by Jainism also prevalent at that time. The Middle Path would eventually lead to renunciation of desire and detachment from worldly things, and thus lead to *Nirvana*.

During their worldly sojourn, some beings may accumulate enough merit to escape rebirth. But, they may choose to stay in this world of sorrow and pass on some of their earned merit to other people; in other words, take on the sins of others on themselves. These people are potential Budhas, called Bodhisattvas. All of us have the potential of becoming the Budha through the Bodhisattva route, provided we can earn enough merit. Tradition has it that Gautama Budha had lived earlier through many lives as Bodhisattva, doing acts of kindness and mercy; that brought him the reward of *Nirvana* in his life as Gautama Budha. There is also talk of emergence of another Budha after the death of Gautama. Around the beginning of the Christian era the concept of Maitreya, the future Budha, became quite widespread.

No supernatural status was claimed for the Budha. Budha did not want any worship of himself, or his images to be made. However, within a few centuries of his death, Budha's images started appearing; his worship was also started.

Jataka tales are stories of the Budha in his earlier births, when he lived as Bodhisattva. These are quite voluminous and have plenty of information about contemporary life of those times.

Buddhist canons were reduced to writing some 5 or 6 centuries after his death. Thus, it is very difficult to say what are the original sayings of the Budha and what are later interpolations. Not much after the death of the Budha, Buddhism got split into two sects:

Hinayana (Thervada)   — the Lesser Vehicle
Mahayana   — the Greater Vehicle

Hinayana is more orthodox and less tolerant of other viewpoints. Hinayana first traveled to Ceylon and from there to South East Asia, i.e. Burma, Thailand, Cambodia, etc. Mahayana is more receptive to other viewpoints and modified itself to fit in with local concepts. It traveled to Tibet, China, Japan, etc. Bodhisattva is a concept of the Mahayana sect.

Buddhism spread in India, especially after the push given to it by Emperor Ashoka. Some kings of the barbarian tribes who ruled over

North West India from the 2nd Century BC to the 3rd Century AD converted to Buddhism. As such, Buddhism became prevalent even in areas now in Pakistan and Afghanistan. The Indo–Greek king Milinda or Menander (160–140 BC) became a Buddhist. Milinda had sessions of questions and answers on Buddhism with a Buddhist monk Nagasena; these are published under the title *Milinda Panha* — Questions of Milinda. Another famous king to convert to Buddhism was Kanishka of the Kushan tribe; he ruled over North India, probably from 78 to 114 AD, or may be some 30 years later. He is reputed to have called a council of Buddhist Sangha to sort out the differences between the Buddhist sects.

There is some divergence of opinion as to the extent that Buddhism spread in India. One view is that Buddhism got a major hold over the people of India, and Hinduism might have been gasping for breath under the onslaught of Buddhism. Other scholars opine that the spread of Buddhism was in patches, both area-wise and time-wise.

King Ashoka of the Mauryan Empire helped the spread of Buddhism. Around 180 BC, the Mauryan Empire was replaced by the Brahmin Shunga Dynasty which ruled over a shrunken empire in the East. The Shunga dynasty worked for the restoration of Hinduism. So did the Kanva dynasty, which replaced the Shungas. Later, the Gupta dynasty (320–520 AD) strove for the revival of Hinduism.

At some stage, the Mahayana branch of Buddhism introduced Hindu gods as agents to whom people could pray. Consequently, images of the Budha started appearing, and he began to be deified. By about 4th century AD, a religion having ingredients of both Hinduism and Buddhism started emerging in Bharat. Around the 5th Century AD, Budha was incorporated in Hinduism as the ninth *avatar* (incarnation) of Vishnu. In 9th century AD, Adi Shankaracharya worked tirelessly for the revival of Hinduism; he was highly successful. Turks came in the 13th century AD; they burnt up Buddhist Viharas (monasteries) in Afghanistan, Punjab and later, in the East of Bharat. So, Buddhism gradually got enfeebled and finally, almost extinct in India. It may not be an exaggeration to say that Hinduism presently prevailing in Bharat has a (thick) coating of Buddhism. Concepts like *ahimsa*, renunciation, etc are largely taken from Buddhism.

2.3 Jainism

The founder of Jainism was Vardhaman Mahavira (a Kshatriya) who lived from 599 to 527 BC, or around that period. Jains believe that there were in all 24 Tirthankars, out of which Vardhaman Mahavira was the last and Parasnath, the last but one. The first Tirthankar was Rishaba Deva, also called Adinath.

Mahavira believed in the dualistic philosophy that matter and soul are the only two ever existing elements; of these, matter is inert, and the soul is eternal and evolutionary. By severe austerities and self-mortification, the bondage of *karma* (action) can be shaken off and *moksha* (freedom from rebirth) achieved. Jainism emphasizes *ahimsa* (non-violence) and renunciation of pleasure and worldly goods. Its commitment to *ahimsa* is obsessive; even unconscious killing of ants and insects during walking is a sin. Agriculture was ruled out for Jains as that might result in unintended killing of insects. Existence of Vedic authority was not accceptable to the Jains, nor was the claim that knowledge was revealed only to the Brahmins (Hindu priests). Existence of an all powerful God was not central to the Jain doctrine. Purification of the soul was the goal of living. The way to deliverance is through Triratna, the three levels of Right Path, Right Knowledge and Right Conduct.

In time, Jainism was split into two sects:

Digambara — Sky-clad or Naked
Shatembara — Clad in White

Unlike Buddhism, Jainism did not travel out of India. Presently, Jains are, more or less, part of the Hindu society. Jainism is well known for its austerity, vegetarianism, medievalism and *ahimsa* (non-violence). All of that affected Hinduism to a substantial extent; *ahimsa* got lodged in the Hindu psyche.

2.4 Sikhism

Sikhism was founded by Guru Nanak towards the end of 15th century AD. Nanak's teachings combined much wisdom from the

Vedas, Buddhism and the Muslim Koran. It was started as a sort of reform movement in Hinduism to take people away from wrong ritualistic practices and undesirable social customs. Guru Nanak was against the caste system and untouchability practiced by the Hindus.

Sikhism is a strictly monotheist religion — there is only one Wahe-Guru, or Sat-Guru; Sikh Gurus are his emissaries. Guru Nanak taught that God should be experienced through personal devotion, through '*naam jaap*'.

The Sikhs had ten gurus; the last guru Govind Singh founded the Khalsa in 1699 AD and gave Sikhs distinctive physical identity, by prescribing five articles beginning with 'k' — *kesh* (hair), *kangha* (comb), *kachha* (underwear), *kirpan* (sword) and *kara* (iron bangle). Guru Govind made Sikhs a martial race to fight against the oppression of the Mughal rulers. He decried that after him, the Adi Granth (the Sikh holy book) would be the Guru.

Maharaja Ranjit Singh was a great Sikh ruler who ruled over major parts of Punjab, Kashmir and some parts of Afghanistan from 1801 to 1839 AD. Ranjit Singh's empire broke up after his death.

3
THE JUDAIC RELIGIONS

3.1 Background

Before considering the three Judaic religions individually, let us see some of their common features.

In ancient Mesopotamia (modern day Iraq), around 2,000 BC, there was a man named Abraham. Abraham had two sons; the elder Ishmael through an Egyptian maid Hagar; the younger Isaac through his wife Sarah. The descendents of Isaac were Jews. Some 2,000 years after Abraham, out of the Jews grew the Christians, who in due course overtook the Jews in popularity and numbers. Ishmael and his mother were banished by Abraham to the desert sands of Mesopotamia/Arabia. Arabs are considered to be the descendents of Ishmael. Out of the Arabs arose Islam, some 2,600 years after Abraham. Thus, Abraham is the common ancestor of Judaism, Christianity and Islam. Due to this, the three religions are called Judaic (from Judaism) religions.

Now, Abraham was descended from Shem, one of the three sons of Noah (of Noah's Ark fame) who had descended from Adam. So, the three religions are also called semetic (derived from Shem) religions. However, sometimes the word semetic is used only for Judaism.

The three religions are also called Religions of the Book, as each has one Book of Final Adjudication, namely:

- Old Testament (Torah, Talmud, Old Bible) for Jews
- New Testament (New Bible) for Christians
- Koran (Quran) for Muslims

Here, we may relate an interesting story. At one time, God asked ‹ Abraham to give in sacrifice his most prized possession. Abraham took one of his two sons for the sacrifice, which was to be in the form of 'burnt offering', i.e. through fire. As Abraham was on the point of killing his son, a ram appeared nearby. God's angel asked Abraham to sacrifice the ram, instead of the son; Abraham complied. As per the Jews and Christians, that son was Isaac. However, Muslims say that that the son was Ishmael; Muslims celebrate their annual festival of Eid-ul-zuha (Bakr-Eid) in commemoration of that event. It is believed that the place of that sacrifice was the present Temple Mount in Jerusalem. Around 950 BC, King Solomon built the famous Jewish Temple at that site. This temple was later demolished. Presently, the Muslim Shrine called Dome of the Rock, built in 688 AD stands on the site; it is sort of out of bounds for the Jews.

The three Judaic religions emerged out of the desert sands of West Asia. Compared to that, the three 'India born religions' emerged in the fertile land of Bharat. All the three founders of Judaic religions were born in ordinary middle class families. But, the two *avatars* (incarnations) of Hinduism and founders of Buddhism and Jainism were scions of royal families.

The common lineage of the three Judaic religions is as follows:

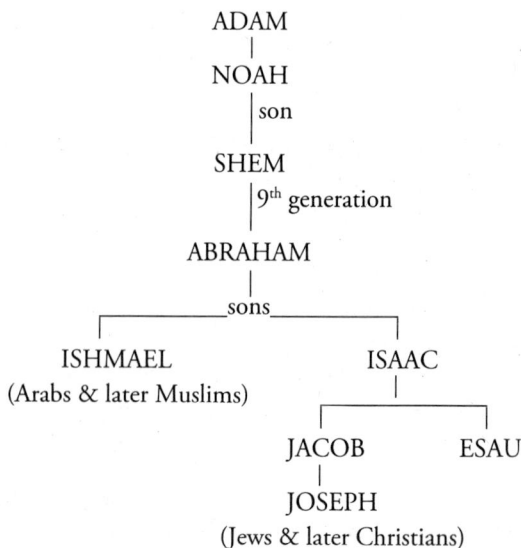

ADAM
|
NOAH
| son
SHEM
| 9th generation
ABRAHAM
|
_____sons_____
| |
ISHMAEL   ISAAC
(Arabs & later Muslims) |
                          _____|_____
                          | |
                       JACOB   ESAU
                          |
                       JOSEPH
              (Jews & later Christians)

Following comments are offered on the above table:

- Adam was the first man on earth. His grandson was Noah, in whose lifetime, the famous flood occurred and who made the Ark. Noah had three sons, out of whom one was named Shem. The 9th generation from Shem was Abraham who had two sons, Ishmael and Isaac.
- Abraham's son Isaac had two sons Jacob and Esau. Jacob had 12 sons, out of whom Joseph was, by far, the most famous. He lived in Egypt, as a ruler. Jews are the descendents of Joseph and his 11 brothers — 12 tribes of Israel.
- Christians emerged out of the Jews after about 2,000 years of the advent of Judaism. Christ was born a Jew.
- Arabs are considered to be the descendents of Ishmael. Out of the Arabs emerged Islam some 2,600 years after the advent of Judaism.

3.2 Judaism

Judaism, religion of the Jews and the first religion of the Judaic family, was founded around 2000 BC, by Abraham at Urs in Chaldea, South Iraq. The God of Jews is called Yahweh (YHWH). Yahweh entered into a covenant with the Jews that he would look after them, and give them their promised land of Cannan (Israel and Judea), provided the Jews —

- Obey His laws, and
- Worship no God other than Yahweh.

At some stage, Jews were taken in slavery to Egypt. Around 1200 BC, a child, later named Moses, was born to Jewish parents in Egypt; he was abandoned at birth. Moses was brought up as a prince in the court of the Egyptian Pharaoh Ramses I. After Moses grew up, he had his first encounter with God (Yahweh), whom Moses failed to recognize. He also had spats with the Pharaoh Ramses II. Finally, Moses was successful in taking Jews out of the Egyptian slavery. Yahweh parted the Red Sea for the fleeing Jews, who were being pursued by the forces of the Pharaoh.

It was Moses who gave flesh and blood to the Jews and made them what they are today. Moses spoke to Yahweh (who was behind a bush) for the second time on Mount Sinai. Yahweh gave Moses the Ten Commandments written on clay tablets by the finger of Yahweh. Jews could get to the promised land of Israel and Judea, only after roaming around in the Sinai desert for 40 years; Moses died during those wanderings. David and Solomon were two great Jewish kings, who ruled over Israel and Judea during 900 to 1000 BC. Around 950 BC, Solomon built the famous Jewish temple in Jerusalem.

In spite of the covenant between Yahweh and the Jews as referred above, the Jews had to suffer unfathomable atrocities and suffering during most part of their history. They were expelled from their land of Israel and Judea, time and again. This could be explained, perhaps, by the fact that Jews may have violated the conditions laid down by Yahweh. It is reported that sometimes in the distant past, the Jews worshipped another god called Baal. They might have violated other laws, as well.

Christians hold the Jews responsible for the crucification of Jesus Christ, though the sentence was formally ordered by the Roman governor of the area, Pontius Pilate. Christians would lose no opportunity to pile murder and mayhem on the Jews. Though the Christian Crusades of the 12th and 13th century AD were against the Muslims, Christians indulged in mass slaughter of the Jews, whenever an opportunity presented itself. Then, there were the Spanish Inquisitions in which Jews were liquidated on a mass scale. In our own times, millions of Jews were slaughtered by Hitler; a figure of six million is often quoted. Then, there is intense enmity between the Muslims and the Jews, primarily centered on the locations of their respective holy places in Jerusalem. Presently, the conflict between the Jews and Muslims threatens World peace.

In spite of the untold sufferings of the Jews, they showed unparalleled tenacity in holding on to their faith. It was due to that that Jews could return to their promised land of Israel after almost 2,000 years. There is no parallel of that in world history.

3.3 Christianity

Christianity was established around 30 AD, in Israel by Jesus Christ, who was born a Jew. It is indeed extra-ordinary that Jesus could establish the base of a major religion like Christianity, in about 3 years' time. Founders of all other religions have had to struggle almost their entire lifetime for the same. Very few details of the life of Christ before the age of 30 years are available, except some information about his birth. Christ was crucified by the age of about 33 years, under orders of the Roman Governor of the Area, Pontius Pilate, on charges leveled by the Jewish clergy. The crucification took place on a Friday, now called 'The Good Friday'. Christians believe that Christ arose from the dead on the third day — a Sunday, and ascended to heaven.

Christ's mother was the 'virgin' Mary; his father was Joseph. Next to Christ, the two most important persons in Christianity were Saint Peter and Saint Paul. Peter was sort of a senior among the associates or followers of Christ, who came to be called the apostles. Paul is the one who provided flesh and blood to Christianity, and made it as we see it today. Paul had no opportunity to meet Jesus in his lifetime. Paul and Peter, in that order, went to Rome and propagated Christianity there. Both were executed under orders of the Roman emperor Nero. Another important person was Mary Magdalene, a constant companion of Jesus. At some stage, the Church associated her name with a prostitute. Lately, a version has emerged that Christ was married to Mary Magdalene and they had a child, born after Christ's demise. The Church totally rejects this version.

After the death of Christ, a lot of debate and conflict arose about the Godhood of Christ. One section held the view that Christ was 'God-Like' (Similar in Being/Essence, or Consubstantial), but not God. They expressed that Christ had himself said that the Father (God) was greater than him. This section was called Arians (after a priest called Arius), or Unitarians — meaning that there was only one God (The Father). The other section expressed that Christ was also God (Same in Being/Essence) in the form of the Trinity of 'the Father, the Son and the Holy Spirit (or Ghost)'; they were called Trinitarians. Names of the two factions in Greek, the popular language at that time, were:

HOMOOUSIOS — Same in Being; Trinitarians
HOMOIOUSIOS — Similar in Being; Unitarians

Note: There is a difference of only an 'i' between the two.

The two factions even indulged in bloodshed. The issue was decided in favor of the Trinitarians at the Council of Nicaea in 325 AD, called by the Roman Emperor Constantine, who at that time was perhaps not a formally converted Christian. He is reported to have converted to Christianity at his death-bed around 335 AD. Even after the Council of Nicaea, the conflict situation continued for a century or two.

Christianity spread fairly fast, though they had to face a lot of persecution during the first few centuries, especially at the hands of Roman Emperors, of whom Nero was one. Presently, Christianity is the largest religion (some 2.4 billion) of the world, spread over all the continents.

3.4 Islam

Islam was founded in Mecca/Medina, Arabia around 622 AD (Hizri) by Prophet Muhammad (pbuh). It had a relatively easy passage. Within about 100 years of the Prophet's death in 632 AD, Islam had put its flag over vast lands in West Asia and beyond. Islam was spread partly by persuasion, partly by the sword.

The five pillars of Islam are:

- Belief in One Allah and His Prophet Muhammad (pbuh)
- *Namaz* (prayer) five times a day
- Zakat — Alms giving
- Fasting during Ramadan (a Muslim month)
- Haj (pilgrimage) to Mecca at least once in a life-time

After the death of the Prophet, Islam split into two sects, i.e. Sunnis and Shias. Shias believe that the Caliphate should have stayed within the family of the Prophet, i.e. his daughter Fatima and her husband Ali, and their progeny. Sunnis believe that that need not be, and the Caliph should be elected. Koran (Quran) is the holy book of Islam. Sayings of the Prophet are called Hadit.

3.5 Jerusalem

Jerusalem is a city most holy to all the three Judaic religions. The history of Jews revolves around Jerusalem and its Solomon Temple. The temple was built by King Solomon around 950 BC at the site, presently called the Temple Mount; this is the most holy site in the world. Tradition has it that Abraham had planned to sacrifice his son for God, on a rock below the Temple Mount. It is also believed that in the ancient past the 'Ark of the Covenant' (carrying the Ten Commandments) lay in or under the Solomon's Temple. One version says that that Ark was taken to Ethopia, by or on behalf of Queen of Sheba (an Ethiopian).

The Solomon Temple was destroyed by Babylonians in 579 BC, and they took the Jews into Babylonian slavery (581–538 BC). The temple was rebuilt. It was again destroyed by the Romans in 70 AD, when the Jews rebelled against the Roman occupation. The temple could not be rebuilt. Presently, a part of the wall of the temple stands at the site. This portion of the wall is called the 'Wailing Wall', which is most holy to the Jews. Jews from all over the world come to pray at this Wall.

For Christians, Jerusalem is associated with the last days, crucifixion and ascension to heaven of Jesus Christ; as such, it is also most holy to them. It has the Church of the Holy Sepulcher built over or near the cave where the body of Christ lay for two days before he ascended to heaven. In 625 AD, Byzantine Christian Emperor Heraclius captured Jerusalem from the Persians (non-Muslims at that time) and restored the 'True-Cross' there.

The Muslims lay their claim to Jerusalem in their belief that Prophet Muhammad (pbuh) ascended to heaven from the Temple Mount. Muslims captured Jerusalem from the Christians in 637 AD, i.e. within 5 years of the Prophet's death. In 688 AD, Muslims built the Dome of Rock at the site of ascension of the Prophet, which is the exact spot (the Temple Mount) on which Solomon's Temple once stood. The famous Al Aqsa mosque was also built nearby. The 'Wailing Wall' of the Jews stands abutting the Muslim 'Dome of Rock'; these two are the most disputed pieces of real estates, which threaten world peace.

637 AD onwards Jerusalem was under control of the Muslims. It was captured by the Christians in 1099 AD as a result of the First Crusade; it stayed with the Christians for 90 years. In 1187 AD, the Great Muslim General Saladin (a Kurd) captured Jerusalem and dragged the Christian True-Cross on the ground. Thereafter, Christians mounted six more

Crusades with a view to capture Jerusalem, but were not successful. Presently, Jerusalem is a great bone of contention between the Jews and the Muslims, with the Christians watching from the sidelines.

3.6 Spread of Religions

Hinduism is the first formal religion of mankind; yet, it remained confined to Bharat. Christianity and Islam emerged on the world scene thousands of years after Hinduism. Yet, these two religions could catch the imagination of the world and spread to various nooks and corners of the world. There must be complex reasons for the same; we try giving a rather simple reason below:

It would appear that the primary reason for the fast spread of Christianity and Islam was the stark simplicity of the creeds of the two faiths. These are easy to understand by laymen with average (or even below average) intelligence. Christianity has the simple principle of 'faith in Christ', who would look after you and intercede on your behalf with God, of whom he is a part. There are no complex philosophies to be understood and no gods to choose from. Islam is even simpler than Christianity. There is but one God Allah, and Muhammad is His Prophet (pbuh). Then, there are few injunctions of Islam, which we have listed earlier. By comparison, Hindu philosophy is highly complex and their view of life difficult to understand. The number of Hindu gods is almost endless. Hindu religion will be covered in detail later.

We may mention one more general aspect of religion. It is fashionable to say for the record that religion teaches love, brotherhood and peace; and that it is against hatred, violence and conflict. But history shows that there is no dearth of hatred, violence and conflict in the name of religion. The hatred and violence take place not only inter-religion, but also intra-religion. Such conflicts have occurred in the classical ages, in the medieval ages, and in the modern age, which is generally considered to be an enlightened era. The Muslim Al Qaeda is all charged up against the Christian USA, and to some extent, against the Hindu India. The Shias and Sunnis (both Muslim) are determined to eliminate each other in Iraq. Then, there was the slaughter of millions of Jews during the Second World War. In the domain of religion, the few things that appear to be presently missing are love, brotherhood and peace.

Section B
Hinduism

4
HINDUISM — AN OVERVIEW

4.1 Hinduism — Introduction

Hinduism is the most ancient religion of humankind; it is a religion of bewildering complexity and extraordinary diversity. It is a great folly to try to understand such a religion in a few pages. But we have no alternative but to commit this folly; so, we go ahead.

With a bit of hyperbole, it could be stated that no one has the least idea as to when and how Hinduism started. There are mythological claims, learned conjectures and considered opinions. There are generally more than one version of events, and differing opinions on the very nature of Hinduism. Especially, the question of Hindu godhood is very complex; different interpretations are prevalent. No claim is made that the versions given in the following pages are the only versions; there could be other versions, equally valid.

Most religions have a mix of mythology and history; some more, some less. In this time and age of enlightenment, the Vatican is looking for (non-existing) miracles of Mother Teresa, with a view to bestow sainthood on her. In her own lifetime, the saintly mother claimed no such powers. Due to its very ancient nature, the Hindu religious history is overloaded with a thick layer of mythology. There is such an artful mix of history and mythology that it is difficult to tell one from the other. Mythology has this unique feature — whilst one may not believe in it, it is not easy to reject mythology outright; a lingering doubt remains that there may be an element of truth in mythology.

In ancient times, at the very beginning, Bharat (India) was called Jambudvipa — literally 'island' of the jambolan (*jamun* — the Indian blue berry) tree; but meaning 'continent' of the jambolan tree. Tradition has it that the primordial Mountain Meru was all covered by the *jamun* tree. It was around Mount Meru that lay the Jambudvipa in concentric circles, separated by oceans (hence the use of word *dvipa* or Island). The fruit *jamun* appeared to be the favorite of both gods and men; its berries gave a sort of an intoxicating drink. Even today, the *jamun* berry is considered to have lot of medicinal value. We may note in passing that the British planted *jamun* trees in large numbers on the boulevards of New Delhi which they built.

The word Hindu and therefore, Hinduism, are foreign to Bharat. In the North-West of Bharat was the vast river Sindhu, which formed a boundary. The ancient Iranians, pronounced 's' as 'h'; thus, 'Sindhu' became 'Hindu'. The Greeks were not very found of 'h'; they converted Hindu into Indu or Indus. Later on, from the 7th century AD onwards, Arab traders picked up the Iranian given word Hindu; its usage increased so much that even the local inhabitants started using it. At some stage, the religion practiced by the people of this region came to be called Hinduism. Thus, the word Hinduism has only geographical connotation; it has no religious meaning.

The original name of the religion of the people of Jambudvipa (Bharat/India) is Vedic *dharma*; some may call it Arya *dharma*. Later, it came to be called Sanatan *dharma* — Eternal Faith, or Universal Truth. It incorporates not only religious thought processes, but also the way of life and living, including culture, philosophy and mystic thought. Unlike all other religions, Vedic or Sanatan *dharma* has no single founder, no known approximate date of its start, and no single book of adjudication. Orthodox Hindus say that their religion is *anadi* (no beginning) and *anant* (no end). The religion incorporates thousands of years of accumulated wisdom of sages and *rishis* (seers). These seers must have done *tapasya* (asceticism/penance) for eons on the mighty Himalayas to come up with the sublime and original thought processes contained in Hinduism

As mythology is Hinduism's strength, history is its weakness. Hindus have been very economical in describing and recording events relating to Hindu history. A significant portion of Hindu history is known from pens of foreign historians who visited India from time to time. During major part of the second millennium AD, Muslims ruled over India; they had no interest in the Hindu past. Hindus themselves were engaged in their struggle for survival and existence, under jackboots of slavery. As such, a lot of literature about ancient Hindu history was lost or misplaced during the Muslim rule.

The British came to India in the 18th century. They showed an uncanny interest in the Hindu past and took to unraveling it with unbridled enthusiasm and rare passion. It was the British who brought to notice the two jewels in Hindu history, i.e. the Indus Valley Civilization and Ajanta/Ellora Caves. It was a Britisher, James Princep who deciphered the Brahmi script in 1837 AD. That was the key to understanding the Rock and Pillar Edicts of Emperor Ashoka. The Vedas were given a worldwide luster by Max Muller. The market is awash with books on Hindu history and Hindu religion, written by British and European authors. In quality of contents, these compare very favorably with books by Indian authors. Notwithstanding a bias here and there, British authors have projected an objective view of events and issues. The Indian authors could be divided into two groups. One group has the same general viewpoint as the British and gives it qualified support. The second group, comprising mainly of hardcore Hindus, reject the British view of Hindu history. We give below a popular and more accepted view of Hindu history, as it prevails at present.

The oldest civilization in India is the Indus Valley Civilization; it flourished in North West India in the area of Sindh and united pre-partition Punjab, from say 3000 to 1800 BC. The majority view is that that civilization may not have been Hindu; though, there are dissenting voices. Figurines of Rudra (Shiva) like god and mother goddess have been found, as also remains of altars where worship to gods might have been offered. This civilization collapsed around 1800 BC for unknown reasons.

A German working as a professor at Oxford University, Max Muller postulated the Aryan migration theory during mid 19th century AD. As per this theory, there were a set of migrating people; they spread out of Central Asia or Southern Steppes of Russia to Europe, Greece, and Iran. A sub-group from Iran, further traveled to India during 2000–1500 BC; these people have been called Aryans. The Aryans lingered in the pre-partition Punjab area for a few centuries. In all probability, it was during that period that Hinduism and Vedas evolved, starting with the *Rig Veda*. Gradually, the Aryans advanced to river Jamuna and then the Ganga (Ganges) basin. The Aryan migration theory is discussed in detail in Chapter 15.

However, many Hindu historians do not agree with the above thesis. They rubbish the Aryan Migration theory, which is being increasingly discredited these days. They place the origin of Hinduism in the 4000–3000 BC bracket; some even quote earlier figures. The Indus Valley Civilization is considered to be part of the ongoing Hindu civilization. This theory would place the Hindu civilization at par with the Sumerian/Akkadian civilizations in Mesopotamia, which flourished around 3500 BC. Keeping the totality of circumstances in view, there is no reason to believe that Hindu civilization is, in any way, less ancient than the Sumerian civilization. Thus, for the purpose of this book, we would place the start of Hinduism at around 4,000–3,000 BC.

Then, there is the orthodox Hindu view contained in the *Puranas*. As per that view, Hinduism is an ongoing religion from the very beginning of humankind. There is talk of four Yugas, each having its duration in hundreds of thousands of years. In this theory, we are talking of an entirely different level of mythology, perhaps with very little history. We will discuss this theory in detail in Chapter 7.1.

Ancient Languages

Spoken: At this stage, we may record a few words about the languages in ancient Bharat. When we talk of Hindu literature, Sanskrit comes to our mind. Almost the whole of ancient Hindu religious literature is in Sanskrit; it has a lot of commonality with Latin, Greek and some other Indo–European languages like German. As per Max Muller, Aryans brought the earliest form of this language with them when they came to India. Actually, Sanskrit was a very exclusive language, largely monopolized by Brahmins and the elite. The common people had very little knowledge of Sanskrit (which means refined and perfected); they spoke a language called Prakit (which means natural and unrefined). There is a lot of debate on the relative age and historicity of Sanskrit and Prakrit; the position is not very clear. The Prakrit spoken in North Bihar was Pali; Buddhist literature, including the Jataka tales, is in Pali. Prakrit spoken in South Bihar was Magdhi; the Jain literature is in Magdhi.

Script: The oldest script in India is the Indus Valley script, which, however, is not deciphered till date; it pre-dates 2,000 BC. Then, there was the Brahmi script which has been traced to around the 5th century BC. It would appear that both Sanskrit and Prakrit were written in this script. Ashokan edicts (3rd century BC) in the East and South of India are in the Brahmi script. Another script was Kharoshti (also around the 5th century BC), which was written from right to left; it was in use in the North West of India. This script appears to be derivative of some West Asian script like Aramaic. Some of the Ashokan edicts in North West India are in the Kharoshti script. It would appear that at some stage, Hindus completely forgot the Brahmi script; we have recorded earlier that a Britisher deciphered the Brahmi script in 1837 AD. Devnagri (literally, city of gods) is another ancient script which emerged out of the Brahmi script at some stage, and gradually replaced it. Presently, almost all Hindu literature is in the devnagri script. Of course, there are many regional scripts.

General Comments

The reader may keep the following types of aspects in view whilst going through the text:

- There is no easy way that a layman can get a grip over intricacies of the Hindu religion. As he starts understanding the issues, he comes across passages, which sends one hurtling back to the very start.
- There is hardly a statement or version about Hinduism, of which you do not find a different version, sometimes a contrary one.
- In Hinduism, there is not much difference between religion and mythology, i.e. Hinduism is basically a mythology based religion.

- As the reader goes through the text, he may feel that some of the stories therein are, prima-facie, not in the realm of practical possibility. By including these stories, we are not saying that we necessarily believe in those stories, or the events described therein actually happened. We are telling the tales as they appear in our ancient texts. Tales are there to be told; the reader may draw his/her own conclusions, depending upon his/her personal approach.
- Readers would come across passages where we talk of gods, demons, *dev-rishis* and other exalted persons undertaking austerities and severe penance, for hundreds, even thousands of years. There could also be hints of life spans being in hundreds of years. The reader should take such statements in his stride, keeping in view the strong mythological base of Hinduism.
- Most of the stories told herein have many other versions. We have selected the version(s) which we think is/are the most acceptable. That does not mean that only those versions are correct, and other versions are wrong.

4.2 Some Common Hindu Terms

In Hindu religious literature, there are repeated references to gods, demons and otherworldly beings. Right from the *Rig Veda*, innumerable tales are wound around them. *Devas* is the word most commonly used for gods, and *asuras* for the demons.

Before we proceed further, we may give some clarifications about the word *asura*. In the beginning of *Rig Veda* the word, 'asura' was used for gods. In Zoarastrian Avesta of Iran, gods are *ahuras* (Iranian equivalent of Sanskrit *asura*). Varuna, Indra, Agni, *Surya* have been repeatedly listed as *asuras/ahuras*. It is understood the word 'deva' in Iranian stands for demons. However, towards the end of the *Rig Veda* (Book 10), gods came to be called *suras* in India, though they are still called *ahuras* in the Iranian Avesta. *Devas, asuras* and humans were all progeny of a common father *dev-rishi* Kashyap (Chapter 7.3), though from different mothers like Aditi, Diti and Danu.

In the following paragraphs, we define and clarify some of the Hindi terms commonly used in Hindu texts.

Devas (gods)

Devas are the celestial beings who are benevolently inclined towards humankind. The three Sanskrit names used for 'gods' are —

Devas — or *devtas*; the most commonly used names for gods
Suras
Adityas — sons of mother Aditi (The Vast Expanse)

These gods are different from the Supreme Spirit God (with capital G) Brahman. Though gods (*devas*) reside in the heavens, they can visit the earth as per their convenience; gods freely inter-mingle with humans and may have human progeny. Both gods and humans are descended from *dev-rishi* Kashyap and mother Aditi. Thus, the two are made of the same material, and have the same general characteristics. The gods and humans may not be as distinct beings, as is commonly believed. However, gods are considered immortals, whilst humans have finite life.

Devas/Devtas: This is the most common generic name used for all types of gods, including *suras* and *adityas*. Later on, more powerful gods like Brahma, Vishnu and Shiva emerged; thereafter, *deva*s could be classed as 'lesser gods'. *Devi* is the feminine of *deva*. However, the word *'devi'* is more commonly used for the spouses of Shiva; they are also called mother goddesses or *Shakti* (chapter 6.11). Tvstr (Vishvakarma) is the architect of the gods.

Suras: One of the items to emerge out of the 'Ocean Churning' (Chapter 9.1) was *sura* — wine or spirits. After the *deva*s drank that *sura* (wine), they began to be called *suras*. There could be other explanations also.

Adityas: They are a sub-group amongst the gods. As per one version, *adityas* were sons of mother Aditi and father *dev-rishi* Kashyap. By various accounts, Aditi gave birth to 8 or 12 *adityas*. As per another version, Tvastr/Vishvakarma (architect of the gods) took up Vivaswat (Surya, the sun; a son of Aditi) for grinding on his lathe, to reduce his brilliance. In that process, some fragments fell on the ground; 12 *adityas* emerged out of those fragments. *Adityas* are the names of the sun in its various forms, i.e. Martanda, Vivaswat, Savitre, Pusan, Aryaman, etc (see 'Aditi' in Chapter 7.3). Vivaswat's son was Manu, the progenitor of humankind; hence, human beings are descended from Vivaswat, one of the *adityas*. *Aditya*, in singular, is a name of the sun.

Asuras (Antigods)

The position in respect of demons is a bit confusing. In Sanskrit/ Hindi, the following names are in use for the demons:

Asuras	— Most commonly used name for demons
Daityas	— Sons of mother Diti, father *dev-rishi* Kashyap
Danavas	— Sons of mother Danu, father *dev-rishi* Kashyap
Rakshasas	— Descended from *dev-rishi* Pulastya

Though all the above types were 'evil' beings, many of them got boons from the great gods Brahma and Shiva. These boons were obtained through severe *tapasya* (austerities/penance), which might have lasted hundreds, even thousands of years. A few of these boons were for 'near or almost' immortality. Perhaps *asuras/rakshasas* were not all that evil as are made out to be. Maya was the architect of the demons — counterpart of Tvastr/Vishvakarma of the gods.

Asuras

Generally, the word *asura* is used loosely to indicate all types of demons. *Asuras* came to be so called as they did not drink the *'sura'* (wine, spirits) which came out during the 'Ocean Churning'. *A-sura* is the opposite of *sura*. Sometimes, *asuras* have been called the fallen gods. *Asuras* are made of the same material as *devas/suras*, even sharing their overall cultural values. However, *asuras* are the enemies of *devas/suras*. There is constant struggle between *devas* and *asuras* for supremacy, and for control of the universe. Many a time, *asuras* are able to defeat *devas*; but ultimately the *devas* prevail, mostly with the help of Brahma, Vishnu or Shiva. *Asuras* may be dwelling in close proximity of *devas*, i.e. on the fringes of heaven. It is believed that the River Rasa separates the world of *devas* (gods) from that of the *asuras* (demons).

There are various theories on the lineage of *asuras*. As per one version, *asuras* emerged out of the *'asu'* (breath) of Prajapati (Supreme God). Another version says that Prajapati made gods, humans, gandharvas and others, from water. In that process, some drops fell on earth; these were made into *asuras*. Sometimes, *asuras* have been identified as belonging to the land of Assyrians. It is also expressed that in the distant past, there might have been an *Asura* tribe.

Daityas/Danavas

Daityas were sons of mother Diti and *danavas* of mother Danu, with *dev-rishi* Kashyap being the common father. *Daityas* and *danavas* are generally grouped together. They are *asura* like beings, but with much more strength and very long life; they are the Titans. Unlike the *asuras*, they do not share the cultural values of gods; they have a specialization for wrecking sacrifices made to gods. Human heroes can take on the *daityas/danavas*, but not the *asuras*. It is believed that *daityas/danavas* were expelled from heaven, at the end of the Treta Yuga. They perhaps live in *Patala* (the nether world).

The most famous and the first pair of *daityas* were the two brothers, Hiranyaksha and Hiranyakasipu, both sons of mother Diti. Hiranyakasipu was father of Bhagat Prahlad, who was a good *daitya*. A grandson of Prahlad was the famous *daitya* king Bali (more about him under the Vishnu *avatar* 'Vamana' in Chapter 6.9).

Sometimes, *daityas* and *danavas* are seen as a sort of pre-Aryan wild tribes, who were opposed to the Aryans. There are many accounts of marriages between *daityas/danavas* and humans. Aniruddha, grandson of Lord Krishna married the granddaughter of the great *daitya* king Bali.

Rakshasas (Ogres)

Tradition says that *rakshasas* are descended from *dev-rishi* Pulastya. However, sometimes their lineage is traced to *dev-rishi* Kashyap, the common progenitor of everyone else. Some texts list Khasa, daughter of Daksha, as mother of *rakshsas* (and *yakshas*).

Rakshasas are believed to dwell on earth or in *patala* (under-world), and are the enemies of humankind (as *asuras* are the enemies of the *deva*s). In some texts, *rakshasas* are clubbed with *asuras*. The most famous *rakshas* was Ravan of the *Ramayana*; he was the son of Visravas and grandson of *dev-rishi* Pulastya. However, there were *rakshasas* more senior to Ravan, e.g. Sumali, maternal grandfather of Ravan.

Rakshasas have huge bodies and hideous shapes. They drink human blood and are fond of their flesh. They are considered killers and drunkards; in short, they are evil characters. *Rakshasas* revel in disturbing and destroying *yajnas* (fire-sacrifices) and other sacrifices made by the humans, for the gods. They are anti-social elements that may live in places like cemeteries; some of them do not appear during the day. Parvati, wife of Shiva, gave *rakshasas* the boon that they would attain majority (adulthood), immediately on birth. *Rakshasas* can inter-marry with humans, e.g. the *Mahabharta* hero Bhima married a *rakshi* Hidamba; they had a son who played a significant role in the *Mahabharta* war. In the *Ramayana*, Ravan wanted to marry Sita.

Apsaras (Nymphs)

They are the celestial nymphs who are considered to have emerged at the time of the 'Ocean Churning' and later adorned Inder Lok, the heavenly court of Indra, the Ruler of gods in the heaven. They are meant for entertainment and pleasure of gods. Unlike the humans, *apsaras* have no scruples regarding sexual morality. Indra uses them to disturb the austerities of sages. The most famous cases were of *apsaras* Menaka and Rambha, sent by Indra to disturb the *tapasaya* (asceticism) of *dev-rishi* Vishvamitra. Whilst Menaka was successful, Rambha failed in her mission. Out of the union of Vishvamitra and Menaka, was born Shakuntala, whose son became the famous king Bharata, after whom our country Bharat is named. Some texts count the number of *apsaras* in millions.

Ghandharvas (Centaurs)

They are the male spirits of the air, forests and mountains. Ghandharvas are accomplished musicians and dancers; they are considered mates of the apsaras; they play musical instruments to accompany their dances. Ghandharvas, whilst being males may also have female qualities. They have the power to cast illusions. Sometimes, ghandarvas are projected as half man and half animal. In some texts, the word ghandharva is used only for musicians; dancers have been called *Kinnaras*.

Yakshas

They are a class of supernatural beings in the form of male guardian spirits, appointed by Indra to guard the treasures hidden in roots of trees on earth. Yakshas may also be the symbol of fertility and prosperity. They are given the status of semi-gods, as guardian of cities, lakes and wells. Yakshas are also called the 'tree spirits'; they are the good people, called *punya-janas*. Lord Kubera, god of wealth, is the chief or king of the yakshas; he was half-brother of Ravan; both being sons of Rishi Visravas, through different mothers. It seems that images of *yakshas* were the first images to appear when the idol worship started in India, around the 6th–5th century BC.

Yama

Yama is the god and judge of the dead, with whom the departed souls dwell. He is the ruler of the next world, i.e. of the dead. As per one version, Yama was the first man to die. His record-keeper is Chitragupta and his messengers are *yama-dhutas,* who take away the dead. Yama's mount is the buffalo.

Yama is the son of Vivaswat (the Sun) who is also the father of Manu; thus, Yama is brother of Manu. Yama had a twin sister called Yami. Sometimes, Yama and Yami are looked upon as the first human pair, from whom all others descended. Another female counterpart of Yama is the dark skinned Nirrti, who later emerged as Kali the spouse of Shiva. In the *Mahabharta*, Yama is depicted as Dharma or Dharamraj, the god of Justice, considered to be the father of Yudhistra. Yama controls the following domains:

- World of Fathers (*Pitra-Lok*); Souls with good *karma* dwell here.
- House of Clay: Dwelling of souls with bad *karma*

The above dwellings may be of temporary nature till the soul is allotted a permanent abode, which could be any of the following:

- *Swarg* (Heaven): For the good souls, to live in permanent bliss
- *Naraka* (Hell): For the condemned souls to suffer in perpetuity
- Merge with the Brahman: For the few exalted souls only
- To be reborn on earth as per their *karma* (Actions)

In Hinduism, Fathers (Pitras) or Ancestors are highly respected. They have been equated with the following:

- Fathers: Vasus — Eight in number, attendants of Indra
- Grandfathers: Rudras — 10 or 11 in number, derived from Rudra/Shiva; sometime called Maruts
- Great Grandfathers: *Adityas* — 12 sun gods, sons of Aditi/ Kashyap

This world is called *'mrityu-lok'*, wherein *'mrityu'* (death) is a certainty.

4.3 *Devas* and *Asuras* (gods and demons)

In the preceding sub-chapter 4.2, we have given definitions for the words *deva*s (gods) and *asuras* (demons). It is a complex issue and we may have been only partially successful. Early Hindu mythology revolves around *deva*s and *asuras*, and the interplay between them. It will be in order to spend some more time on these entities. We will also add humans to that inter-play. It will help us in understanding Hinduism a bit better. In our use of the word *asuras*, we will include all types of other-worldly demons including *daityas*, *danavas*, but generally not the *rakshasas*, who are the worldly demons. The word *'deva'* used in this chapter shall generally exclude the great gods Brahma, Vishnu and Shiva.

There are various versions of creation/birth of *deva*s, *asuras* and humans. We give below the two more popular versions.

Creation *(Devas & Asuras)*
Version I

At the very beginning of the Universe, the great Lord Brahma, through his mind-power, created seven *dev-rishis* (for details see Chapter 7.3). One of them was *dev-rishi* Kashyap; he married 13 daughters of Prajapati Daksha. Out of three of his wives were produced the following:

- Wife Aditi: She was mother of 27 *deva*s — including 12 *adityas* (sun gods), 8 *vasus* (Indra's assistants) and 11 rudras (maruts-wind gods). She is also considered the mother of gods Indra and

Vishnu (both sun gods). One of the *adityas* was Vivaswat (Sun), whose son was Manu; he became the progenitor of the human race. Thus, both *devas* and humans emerged out of Aditi.

- Wife Diti: She became mother to *daityas* (a type of demons). The first two *daityas* and sons of Diti were Hiranyaksha and Hiranyakasipu.
- Wife Danu: She became mother to *danavas* (demons).

Version II

At the beginning of time, there was the supreme god Prajapati. At some stage, he desired to have progeny; he acquired the power to do so. So from his mouth, he produced the gods. They entered *divam* (sky); hence they came to be called *devas* (from *divam*). When *devas* were created, there was daylight; that was called day. Thereafter, Prajapati, by his *asu* (downward breath) created the demons; they came to be called *asuras* (from *asu*). The *asuras* entered the sub-terrainian waters. There was darkness when *asuras* were created; thus, their evil nature. Prajapati made that into night.

Additional Versions

The Taittriya Aranyaka says that Prajapati made gods, humans from water. In that process, some drops fell on earth; these were turned into *asuras*, *rakshasas* and *pisachas* (one of the worst types of demons). Another version says that Prajapati became pregnant and *asuras* emerged out of his abdomen. Vishnu Purana says that *asuras* were produced from the groin of Brahma.

General Comments

We make the following further comments on *devas*, *asuras* and humans:

- *Devas*, *asuras* and humans are made of the same material; they have emerged from the same common stock. They may not be entirely different beings, which generally they are believed to be.
- While having some differences, the above three entities have many common characteristics. *Devas* and *asuras*, being sons of a common father, have common cultural values; so do humans.

However, *daityas* and *danavas* do not have those cultural values.

- It may come as a surprise to most readers that *asuras* are also capable of extreme asceticism. There are references to *asuras* doing harsh austerities for hundreds, sometimes thousands of years. As a result, they were recipients of extra-ordinary boons from the three great gods Brahma, Vishnu and Shiva. Some of these boons were for 'almost or near' immortality. The great Lord Brahma once said that there was no 'total immortality'.
- *Devas* were immortal, perhaps without 'total immortality'. The *asuras* and humans had finite lives.
- *Devas* and *asuras* differ from the humans in some ways. *Devas* and *asuras*:
  — Do not blink; they also do not perspire.
  — Their feet do not touch the ground.
  — Dust does not stick to them, and their garlands do not wither.
- *Devas* and *asuras* have the following common characteristics:
  — Both are driven by greed.
  — They have a great desire to dominate the triple worlds.
  — Both want to grab/share the sacrifice offered by the humans.
  — They wanted to get hold of *amrit* (ambrosia) to attain immortality. Whilst *devas* got it, the *asuras* could not.

To attain the above ends, *deva*s and *asuras* were mostly at each other's throat, and did not hesitate to adopt unfair means to gain dominance over the other.

Important *Asuras*

We will be covering *deva*s (gods) in detail later in Chapter 6. Here, we list some of the more important daityas and *asuras* (we do not come across many names of *danavas* as such).

Daityas: They started with the twin brothers Hiranyaksha (golden-eyed) and Hiranyakasipu (golden-robed), both sons of mother Diti. Vishnu, in the form of 'Boar' (3rd *avatar*) retrieved the earth from Hiranyaksha, and in the form of 'man-lion' (4ᵗʰ *avatar*) killed Hiranyakasipu.

Hiranyakasipu — Devotee of Shiva who helped him as 'Sarabha'

son

Prahlad — Devotee of Vishnu; a good *daitya*

son

Virochna

son

Bali*

son

Bana (Vairochi) — Friend of Shiva & enemy of Vishnu

daughter

Usha**

Asuras: Some of the more important *asuras* were:

Mahishasura/*Durga:* Generally these two names are considered interchangeable. Tradition says that following the Ocean Churning, there was a mass slaughter of *daityas* by the *devas*. Diti, mother of daityas was most unhappy and upset. She requested for the boon of a powerful son who could avenge the death of her sons. Her boon was granted and she gave birth to a son with body of a man, and head of a buffalo, called Mahishasura. He triumphed over all the gods. Thereupon, the gods were most worried; they surrendered their individual energies to a powerful goddess, later called Durga. She battled Mahishasura for nine days and nine nights, and killed him. Sometimes, Mahishasura has been called the demon Durga; as the goddess killed him, the goddess came to be called Durga.

Sumbha and Nisumbha: They were two powerful *asura* brothers, who were devotees of Shiva. Through severe penance over 1000 years, they obtained the boon of great strength as well as riches from Shiva.

---

*   Bali was a good daitya; he defeated Indra, and ruled over the triple worlds; Vishnu as 'Dwarf' (5ᵗʰ *avatar*) won the the world back from him.

**   Usha married Aniruddha, grandson of Krishna; that caused a war between Shiva and Krishna (Vishnu).

The demons greatly harassed the gods. Finally, they were killed by *Devi/Shakti*, in the form of Durga (or Kali).

Chanda and Munda: They were commanders of Sumbha and Nisumbha. The two demons were killed by *Devi/Shakti* in her form which came to be called Chamunda (combination of Chanda and Munda).

Rakta-Bija: He had the boon that when a drop of his blood fell on the ground, a new *asura* would emerge. Thus, he was quite invincible. *Devi/Skakti*, in her form of Sinha-Vahini (riding a lion) killed him.

Taraka: One of the most powerful daityas – who was a terror to the gods. Brahma expressed that Taraka could be killed only by a son born of Shiva. That led to the birth of Karttikeya (Skanda), who slew Taraka (see chapter 6.13)

Bhasmasura: Shiva gave him the boon that if Bhasmasura placed his hand on any person's head, that person will die. Having obtained this boon, Bhasmasura wanted to destroy Shiva himself. It was through some clever planning by the *dev-rishi* Narad that Bhasmasura could be destroyed by his own hand over his head.

Namuchi: A very powerful demon who had made Indra his prisoner. He released Indra with some conditions. Indra violated those conditions, and killed Namuchi. Tradition says that his head follows Indra uttering, 'You, wicked slayer of your friend'.

Devas and the Truth

When created, the *devas* and the *asuras* spoke truth as well as falsehood. However later, *devas* retained only truth, giving up falsehood. On the other hand, *asuras* gave up truth and retained falsehood. As *devas* spoke only the truth, they became very weak. As a result, *asuras* were able to over-power *devas*. In such a situation, the great gods Brahma, Vishnu and Shiva, would invariably help the *devas* to get out of the tight spot. To cope with the situation, having given up falsehood, the *devas* adopted 'deceit' as their weapon. That helped them deal with the *asuras*. The philosophy of 'deceit' is also strongly recommended in Chanakya's *Arthshastra* of the 4th century BC (see chapter 17.2). Hindu mythology records many cases in which the *devas* tricked the *asuras* of their rightful share. One such case was during the 'Ocean Churning', where the *devas* tricked the *asuras* of their agreed half share of *amrit* (ambrosia). None other than the great gods Vishnu and Shiva helped the *devas* do that (see chapter 9.1).

Devas/Asuras' Priests

Priest is a central figure of Hindu religion; everything revolves around him. Priests conduct the sacrifice, which will fulfill desires of the devotee, and get him boons. Without the priest, the ancient Hindu was completely at a loss. *Devas* and *asuras* had their respective priests.

Brhaspati

He was the son of *dev-rishi* Angiras, and priest to the *devas*. He is also called lord of the 'sacred speech'. His wife was Tara who was abducted and raped by Chandra, the Moon. A son Budha (different from Gautam Budha) was born; Budha married Ila, daughter of Manu, and became a Chandravanshi king. Later Tara was recovered by Brhaspati.

Brhaspati's sons were Bhardwaj and Raca. Raca played a major role in helping the *devas* overcome the *asuras* (see the Chapter 7.3).

Usanas Shukra

Usanas Shukra, son of *dev-rishi* Bhrigu was the priest of the *asuras*; he was a Bhargava (*gotra*). He was also known as Kavya or Shukracharya.

As the *asuras* were finding the going tough against the *devas*, Shukra went to the great lord Shiva to get boons from him. Shiva asked Shukra to undertake austerities for 1000 years, hanging downwards on the sacrificial fire and inhaling thick smoke. Shukra did that and came out successful. Shiva was highly pleased. He taught the *'mrityu-sanjivini' mantra* to Shukra; this *mantra* can revive the dead. As per a second version, Shiva swallowed Shukra. He stayed inside for 100 years, and finally may have come out of Shiva's *lingam* (penis); that is why he is called Shukra (which means seed or semen in Hindi). Tradition also says that Shukra may have learnt the revival *mantra* during his stay in Shiva's stomach.

Shukra, with his knowledge of the revival *mantra*, would revive the *asuras* whenever they were killed by the *devas*. Thus, the *asuras* started defeating the *devas*. Worried by this, the gods approached Raca (Racha), son of their priest Brhaspati. They asked him to pick up the '*mrityu-sanjivini'mantra* from Shukra; Raca agreed to do that. He went to Shukra and became his pupil for 1000 years. Shukra's daughter Devyani fell in love with Raca. When the demons came to know of this arrangement, they killed Raca; but Shukra brought him back to life twice. When the demons killed Raca the third time, they burned him up and mixed his ashes in wine. That wine was given to Shukra; he drank it, unaware of its contents. Thus, (the dead) Raca got lodged inside Shukra. If Raca was to be revived now, he would have to burst out of Shukra; that would mean Shukra's death. Therefore, Shukra was forced to teach the *mrityu-sanjivini mantra* to Raca, so that he could revive Shukra. And Raca did revive Shukra. Shukra had helped Raca for the sake of his daughter.

Having learnt the *mantra*, Raca decided to depart. He turned down the request of Devyani for marriage on the plea that being his guru's daughter, she was a sister to him. The knowledge of the *mantra* by Raca helped the gods to finally defeat the demons. Later, Shukra's daughter Devyani was married to the great Lunar king Yayati.

Devas/Asuras' Conflicts

Early Hindu mythology revolves around conflicts between *devas* and *asuras*; it is the central theme of Hindu mythology. The *devas/asuras* battles represent the conflict between righteousness and evil, and between light and darkness.

There may have been as many as 12 conflicts between the *devas* and *asuras*. Duration of conflicts has been mentioned to vary between one to 100 'celestial' years; one celestial year equals 360 human years (the reader need not take these figures too seriously). Described below are two of those conflicts.

Ocean Churning

This appears to be one of the very early conflicts; it was during the second *avatar* (incarnation) of Vishnu. The conflict is described in detail in Chapter 9.1. Before the start of the churning, it was agreed that *amrit* (ambrosia) would be shared equally between the *devas* and *asuras*. However, in keeping with their ideology of 'deceit', the *devas* tricked the *asuras* of their share; for that, the *devas* were helped by the great lords Vishnu and Shiva.

Having drunk the ambrosia, the *devas* acquired immortality; they indulged in all-round wholesale massacre of the *asuras*.

Vishnu's Budha *Avatar*

This is an ingeniously devised fascinating story, purporting to explain the reasons behind the emergence of Buddhism and Jainism in India. It displays a high degree of anti-Buddhist bias, by trying to project Buddhism as a sort of evil doctrine.

The story starts with the *devas*, as usual, finding the going tough against the *asuras*. They went en-bloc to the great Lord Vishnu, and expressed their inability to defeat the *asuras* as there was hardly any difference between them, viz. —

- *Devas* and *asuras* were made of the same material, both having emerged out of the great Lord Vishnu himself.
- Like *devas*, the *asuras* also believed in the Vedas, and the rite of *yajna* (fire-sacrifice).
- The *asuras* like *devas*, could also undertake any amount and degree of asceticism to please the three high gods.

The *devas* expressed that they were thus at a loss to understand as to how they could over-power the *asuras*. The great Lord stated that in order to defeat the *asuras*, some way may have to be found so that the *asuras* lose their faith in the Vedas and sacrificial rites.

At that time, Shukra, the priest of the *asuras* was with the great Lord Shiva. The *devas'* priest Brhaspati took the form of Shukra and went to the *asuras*. There, he started weaning the *asuras* away from the Vedas and the sacrificial rites. In addition, lord Vishnu manifested himself in his ninth *avatar* as the 'Great Deluder', i.e. Gautam Budha. He also started preaching a path away from the Vedas and the sacrifices. Under the urging of Brhaspati (in the form of Shukra), the *asuras* took to the new path advocated by Budha. That is how Buddhism came to be established in India (as per this story). This story would appear to imply that it were largely the *asuras* who converted to Buddhism.

Once the *asuras* gave up the Vedas and the sacrifices, they were weakened. Therefore, the *deva*s were able to over-power the *asuras*. That would be equivalence of restoration of Hinduism, after driving out Buddhism. It might have taken the Hindus some 800 years to do that.

Devas' Behavior

During the Epic and Puranic period, there appears to be some deterioration in the sexual behavior of the gods, viz.:

- We come across many sexual indiscretions by Indra, the ruler of the gods. For example, he seduced Ahelya, the wife of *dev-rishi* Gautama, and had to suffer for that act.
- Chandra, the moon abducted and raped Tara, wife of Brhaspati, the priest of gods; the result was the birth of Budha (different from Gautam Budha). Later, Brhaspati accepted Tara back.
- The five Pandavas were born to Kunti with help from the gods.

It may be a bit surprising to note that we do not come across many tales of sexual indiscretions by the *asuras*, who are supposed to be all evil. Though Ravan abducted Sita, he treated her with respect and did not force himself on her.

Asuras' Residence

In Hindu mythology, there is constant conflict between the *deva*s and *asuras*. That would appear to suggest that they could have been living in close proximity of each other. Some texts refer to the River Rasa as the dividing line between the residences of these two entities.

However, on a more conventional note, demons are supposed to be living in the nether-world or infernal regions, called *Patala* in Hindi. *Daiytas, danavas, nagas,* etc are supposed to be dwelling in the *Patala,* There are seven layers of *Patala,* viz:

Atala
Vitala
Sutala — Supposedly ruled by the great *daitya* king Bali

Talatala — Ruled by Maya

Mahatala — Residence of *nagas,* the serpents

Rasatala — Residence of *daityas* and *danavas*

Patala — The seventh layer: the *naga* king Vasuki rules over it; it is the residence of the snake-gods.

It is possible that some types of *rakshasas* may also be living in one of the above places.

Tradition has it that the *dev-rishi* Narad once visited *Patala*. He found everything in perfect order and was very impressed with the place.

5
THE HINDU RELIGIOUS LITERATURE

5.1 Background

Hindus have, possibly, the most voluminous religious literature of humankind. A few general points need to be made before we go on to study this literature:

- In the initial stages, the Hindu religious thought was carried forward in the oral tradition from generation to generation. That process may have been going on for thousands of years. That could be due to two reasons:
  — Shortage and difficulty of writing material those days
  — To keep up purity of the religious thought and its secrecy within a clan, or a group
- There is scope for lot of uncertainty as to when certain events actually took place; how long were their stories carried forward in the oral tradition, and when were these written. Once written, the works might have been subject to revisions, updating and interpolation. Sometimes, that might have been undertaken by interested parties to give a different spin to events.
- In many cases, authors of texts are either unknown or uncertain. In some cases, the presumed authors lived centuries before the estimated date of writing the particular text. For example, it has been expressed that *Manusmriti* was written (in its present form) in the closing centuries of BCE, or early AD. However, its presumed author Manu lived thousands of years earlier, as he was the first man in Hindu mythology. Such anomalies are not easy to resolve.
- It is generally believed that the written script in Bharat started

around 5th/4th century BC; the actual historical evidence may be available only up to 3rd century BC, in the form of the Ashokan edicts.

Hindu Religious Literature

Ancient Hindu literature is divided into two broad categories, *Sruti* and *Smriti*, viz:

- *Sruti:* That which God revealed and *rishis* (sages) heard, i.e. the literature is of divine origin. Tradition has it that this literature emanated from the breath of the Supreme Spirit Brahman, which was taught by Him to Brahma the Creator god, who in turn transmitted it to the *rishis* (sages) for the benefit of humankind.
- *Smriti:* That which is remembered, i.e. of human authorship. The *Smriti* takes its main authority from *Sruti*; it is written to explain and elucidate *Sruti*, for ease of understanding for the common people.

Sruti

There are four series of texts under *Sruti*. There is wide divergence of opinion as to when these texts may have evolved. We give below the generally accepted (very) approximate dates against each category.

- *Four Vedas (Samhita):* These four are the proper original Vedas, all in *mantra* (hymn) form. These hymns of incantation are addressed to the gods and deities. The Vedas are believed to have evolved between 1500 to 1200 BC, starting with the *Rig Veda*; however, orthodox Hindus consider these to be much older.
- *Brahmanas:* These are prose texts explaining *mantras* of the *Samhita* of the four Vedas. Brahmanas prescribe rules for performance of rites and sacrifices. (Likely period 1200 to 1000 BC).
- *Aranyakas:* These texts were evolved by and for the 'forest dwelling' people, i.e. those who have withdrawn from the world. Aranyakas are sort of a link between the Brahmanas and the Upanishads. (Likely period 1000 to 800 BC).
- *Vedanta Upanishads:* These evolved at the end of the Vedas. Upanishads give the most detailed accounts of the religious and spiritual Hinduism. (Likely period 800 to 600 BC).

The written script in Bharat started around the 5th/4th century BC. Thus, the above literature must have been reduced to writing much after the periods we have mentioned above.

The term 'Vedas' has been used in two ways:

- In the 'restricted' sense, it means only the *Samhita* portion of the four Vedas, i.e. *mantras* (hymns).
- In the 'extended' sense, the term 'Vedas' includes the four Vedas (*Samhita*), the Brahmanas, the Aranyakas and the Vedanta Upanishads. In this sense, each of the four Vedas has four parts, i.e. each Veda has its own set of Brahmanas, Aranyakas and Upanishads. Another term for the 'extended' Vedas could be 'Vedic literature'.

The more common use of the word 'Vedas' is in the restricted sense, meaning the *Samhita* (mantras/hymns) portion. The combination of the four Vedas (*Samhita*) and Brahmanas is called *Karma Kanda*; *Karma* means action or deeds. Upanishads are *Jnana Kanda* — the knowledge portion of the Vedas.

Smriti

The following texts fall in the category of *Smriti*; the estimated dates these may have been evolved/written are given alongside:

1. *Ithas, Epics — Ramayana* and *Mahabharta:* The actual events of the two wars may have taken place in 2nd millennium BC, or even earlier. However, the two texts may have been written anytime between the 4th–3rd century BC to the 3rd/4th century AD. *Mahabharta* has been called *Ithas*; sometimes, it is given the status of the 5th Veda. *Ramayana* is called *Kavya* (poetry).
2. *Puranas: Puranas* are believed to have been written from the 2nd century BC to 12th century AD. However, there is an alternate view that the writing of the *Puranas* may have started in the Gupta Period in the 4th century AD.
3. *Kalpa Sutras:* 5th to 2nd century BC
4. *Dharam Shastras:* 5th to 4th century BC
5. *Tantras:* 7th to 14th century AD

6. *Darshanas:* 6th to 4th century BC (Oldest *Smriti*?)
7. *Miscellaneous Literature*

In considering the above dates, we have to keep in mind that writing in India is believed to have started around the 5th/4th century BC. The actual evidence of writing is available only from the 3rd century BC, in the form of Ashokan edicts; these are in the Brahmi script. That would suggest that the entire Vedic literature, including the Brahmanas and Upanishads, were carried forward for long periods in the oral tradition. That must have been a formidable (almost impossible) task, for which there is no parallel.

The above classification between *Sruti* and *Smriti* should not be taken too rigidly. Sometimes, the Upanishads are classed as *Smriti*. In the opinion of the author, only the Vedas (*Samhita*) are the true *Sruti*. Other Vedic literature evolved over 1,000 years following the Vedas (*Samhita*), may be more like *Smriti*.

5.2 The Vedas (*Samhita*)

Vedas form the base and origin of Hindu civilization and religion. These constitute the highest authority on religious and social issues of the Hindus. The Vedas are of divine origin, the true *sruti*. The Supreme Spirit Brahman conceived and transmitted the Vedic knowledge to Brahma, the Creator god. In turn, Brahma gave this knowledge to the *dev-rishis* for the benefit of humankind. The word Veda springs from the Sanskrit *'vid'* meaning knowledge. These most holy Hindu religious books consist of mantras (hymns), addressed to a galaxy of *deva*s (gods); these were mostly in the form of forces of nature. In the Vedas, the term *'rita'* is used for the word *'dharma'*; the latter term came into common use in later times and literature.

There is a lot of debate as to when the Vedas were evolved. In the initial stages, the Vedas were transmitted in the oral tradition from the guru to the pupil; that process must have lasted over centuries. Most Western historians place the age of the Vedas in the region of 1500 to 1200 BC, with *Rig Veda* being the oldest. However, many scholars the especially Indians, consider these much older, going up to 3000 BC, or even earlier. It has been expressed that at some stage, Ved Vyas (author of the *Mahabharta*) re-arranged the Vedas; he might have split the original one Veda into four.

Each Veda has two parts, i.e. *Samhita* and Brahmanas, jointly called *Karma Kanda*:

- *Samhita:* It consists of the *mantras* or hymns and forms the core of the Vedas.
- *Brahmanas:* These are attached to the *Samhita* of each Veda. They are the expository prose, explaining the mantras or hymns.

In this chapter, we would study the *Samhita* portion of the Vedas, taking up the Brahmanas in Chapter 5.3.

There is a view that initially there may have been only three Vedas; Manu talks of three Vedas only. The fourth, Atharva Veda was added later on. It is possible that the Atharva may have been accepted as a Veda with some reluctance. The four Vedas are:

- *Rig Veda:* This is the real Veda and was the first to evolve. The Yajur and Sama are mainly derivatives of the Rig, and consist largely of the *Rig* hymns re-arranged for different types of sacrificial rituals. The hymns are incantations to various gods to shower prosperity on the devotee, and bring him good luck. The *Rig* is discussed a few paragraphs later in detail.
- *Sama Veda:* It is a metrical rearrangement of the hymns of the Rig. It has the *Rig Veda* hymns set to music for singing during sacrificial processes. There are some 1,500 hymns in this Veda; out of that, only about 75 hymns are not found in the Rig.
- *Yajur Veda:* This again is the Veda of sacrificial rituals; it has some 4,000 hymns. Yajur contains sacrificial formulae in verse and in prose, for chanting at the time of sacrifice. Generally, two priests were involved in the sacrifice; one carried out the actual sacrifice, whilst the other chanted mantras. The *Samhita* of the Yajur is split into two parts:

> Taittiriya, Black or Krishna Yajur
> Vajasaney, White or Shukla Yajur

The word 'Taittiriya' is derived from '*titar*' — the word in

Hindi/Sanskrit for partridge. Tradition has it that in a class teaching the Yajur Veda, the teacher became angry with one of the pupils; (this pupil might have been Yajnavalkeya who is the author of the White Yajur). The teacher asked him to disgorge the Yajur that he had been taught. As the pupil was disgorging the Veda, other students turned themselves into partridges, and picked up the disgorged Yajur; that came to be called the Taittiriya Yajur. The Vajasaney Yajur is named after its author, the great sage Yajnavalkeya (Chapter 7.3); his first name was Vajasaney.

- *Atharva Veda:* This Veda consists of the practical arts and sciences, and tries to explain the nature and intricacies of life. It has medicinal formulae. A substantial part of this Veda is full of charms and spells for destruction of the enemies, and help to friends. The Aharva guides man on material aspects and issues of daily living. It has some 6,000 hymns.

We summarize the four Vedas as follows:

Rig: The 'Veda of the Vedas'; a source for other Vedas
Sama: Mostly *Rig Veda* hymns set to music
Yajur: Hymns chanted at the time of fire sacrifice
Atharva: Veda on material issues, and of charm and spells

Rig Veda

The *Rig Veda* is of primary importance. It has 1028 *shuktas* (hymns) consisting of 10,580 *shloka*s (verses) and about 153,800 words. It is divided into ten mandalas or books. 2nd to 7th books of the *Rig* are the older books; 8th and 9th books fall in the intermediate range: 1st and 10th books were the last to be evolved. Most recognized historians consider that the *Rig* was evolved around 1500–1200 BC. However, many quote earlier dates, i.e. 4000–3000 BC; a respected Hindu leader, Lokmanya Tilak has suggested 6000 BC. The oldest manuscript of the *Rig* presently available to us is from 1,500 AD only. A German scholar, Fredrich Max Muller published the *Rig Veda* for the first time in the mid 19th century AD. To each hymn of the *Rig* is prefixed the name of the *rishi* (seer), to whom it was revealed, e.g. Bhardwaja, Vasishta, Vishvamitra, etc.

Let us try to fix a historical date of the *Rig Veda* in relation to the *Ramayana* and the *Mahabharta*. The *Rig* refers to a few kings of the Solar Line (the *Ramayana*), but has hardly any specific references to kings of the famous Lunar Line (the *Mahabharta*). However, some kings of an off-shoot of the Lunar Line, the North Panchala are referred. North Panchala was a kingdom somewhere in UP, north of the river Ganga. Some of the N. Panchala kings listed in the *Rig* are — Mugdala, Divodass, with special prominence being given to King Sudas. A battle of Sudas with 'ten kings' is highlighted. *Rig* also states that Vishvamitra and Vasishta were, in turn, priests of Sudas.

We will see later in Chapter 8.3 that King Sudas has synchronism with the 68th king (from Manu) of the Solar Line, i.e. three generations after Lord Rama who was the 65th generation. That would place Sudas just after the *Ramayana*, but some 26 generations before the *Mahabharta*, which is placed corresponding to the 94th Solar generation. We may not be wrong in assuming that the *Rig* may have evolved not much after the king Sudas. That may give some rough indication of the *Rig Veda*'s location in history, not in absolute, but in relative terms. (This reasoning need not be taken too seriously; there could be some flaws in this line of thinking).

The *Rig* lists 33 gods. Of these, the three central gods were:

Agni: Fire and light on earth
Vayu: Air in the atmosphere
Surya: Sun in the sky

In the very beginning, Varuna may have been a major god. Intially, Varuna was god of the sky; later, Varuna came to be associated with the ocean. At some stage, Indra emerged as a very powerful god; he is the god of war, of rain and thunder, and ruler of the gods in heaven.

Other popular gods were Usha (Dawn), Prithvi (Earth), Maruts (Storm gods), two Aswins (sons of Surya) and an intoxicating drink called Soma. Then, there was Rudra, the howling, furious god, who ruled the tempest and storms. Surprisingly, Soma was an important god; both humans and gods were fond of it. Indra is often shown inebriated with

Soma; he used to gain strength after drinking Soma. However, Max Muller says that one *Rig* poet has mentioned a figure of 33,339 gods.

Generally, the last two books of the Rig, the 1st and 10th, are considered to be the beginning of monotheism in the Vedic *dharma*. There is a lot of debate on polytheism vs monotheism of the *Rig Veda*; no finality has been reached and the jury is out. The major part of the *Rig* is devoted to the gods referred above; there are hymns after hymns addressed to these gods, seeking boons, blessings and riches for the devotee. Of the 1028 *Rig* hymns, some 250 are addressed to Indra, and approximately 200 to Agni. Priests praised these powers of nature in the struggle between good and evil; they sought from the gods prosperity for themselves and protection of their flock.

The references to one single God in the *Rig* are somewhat muted and limited. By the side of a few passages referring to the one single God, there are hundreds of passages in which other (smaller) gods are praised. There are one or two hymns, which project the various gods as manifestation of the single God, e.g.

'They call it Indra, Mitra, Varuna, Agni; may be the heavenly bird *Garumat* that flies. The wise speak of that one in numerous ways; they call it Agni, Yama, Matarisvan'. RV, 1, 164, 46.

Note: Matarisvan is the assistant or messenger of Agni. Sometimes, it is identified with lightening in the sky, which is also a form of fire.

A view has been expressed that the concept of one single God (later called Brahman in the Upanishads) can only be inferentially drawn from the Rig. The concept does not get as much emphasis as it came to acquire in the Upanishads. However, Swamy Dayanand of the Arya Samaj is of the view that the Vedas are mainly about one single God. We are in no position to disagree with such a great *rishi*.

In the *Rig Veda*, there is the god Vishnu, but as a minor deity; nowhere near the exalted position he came to acquire later in the epics and the *Puranas*. There is reference to Vishnu covering the universe in three steps (RV, 1, 154). We will see a similar story later in Vishnu's *avatar* of Vamana (dwarf). There are no specific references to Shiva as such. However, Shiva is identified with the *Rig Veda* god Rudra, and sometimes perhaps with Varuna. The word *'brahman'* appears at some places in the Rig; however, it does not appear to mean the one single universal God, as we understand it today. There is no specific talk of *Trimurti* (the Hindu Triad), i.e. Brahma, Vishnu and Shiva — it was to emerge much later. There is no reference to transmigration of soul. In the Vedic times, there were no idols (of gods); hence no idol worship. However, we cannot be too sure of that; may be some lower classes were worshipping images. It would appear that those days there were no temples, as we understand the concept today. There is no reference to any segregation of women, or child marriage. *Rig* is a supremely religious and revered book of the Hindus. It throws a lot of light on the social and economic beliefs of the Hindus of the early times.

During the Vedic period, man was required to propitiate the gods, mainly through rituals; animal sacrifices were carried out at altars in open spaces. That would appear to imply a lot of bloodshed, perhaps to no particular purpose. Regular animal sacrifices are not a great advertisement for the concept of *ahimsa* (non-violence), which is (wrongly) projected these days as a hallmark of Hinduism.

There is some controversy regarding killing of cows and eating of beef in the Vedic times. It cannot be denied that there are some references in a few texts like the Brahmanas, listing bull/cow as sacrificial pashu (animal) along with horse, goat, ass, etc. There may also be an odd reference that bull/cow may be killed for a guest. On the other hand, at least one passage in the Brahmanas strictly prohibits eating of beef. Pradha, a son of Manu had to become a shudra as he killed the cow of his guru. The author is of the view that cow veneration is in the very DNA of the Hindus; it is most unlikely that Hindus could have been killing cows any time in the past. The passages permitting killing of bulls/cows may be later interpolations, of which there is no dearth in the Hindu religious literature.

Ashwamedha (horse-sacrifice) was of special and extra-ordinary significance. There are detailed and repeated references to the procedure for conducting the horse-sacrifice. Sometimes it was preceded by the sacrifice of a goat to announce the following horse-sacrifice to the gods. The horse meat was cooked and eaten; Indra and the other gods appeared and took their share. That would imply that in those days, both humans and gods were non-vegetarians; everybody ate horse meat. The horse-sacrifice was considered to remove all sin.

There is a *Rig* Vedic hymn relating to sacrifice of the Cosmic or Primeval Man, called Purusha (in Hindi, *purush(a)* means man). As per that hymn, all creation may have been the result of the sacrifice of the Purusha by the gods; all things were created by dismemberment of Purusha. It was out of the sacrifice of Purusha (sometimes also called Prajapati) that everything emerged, i.e. the gods, humans, animals, serpents, birds, and everything else. (See 'Purusha' under Chapter 6.12).

The supreme sublimity of the thought process of the *rishis* of the *Rig Veda* can be judged from the following verses of the 'Creation Hymn' of the *Rig* — Book X, Hymn 129 (sourced from *The Vedas* by Fredrich Max Muller):

Verse 1 —

Nor Aught nor Naught existed; you bright sky
Was not, nor heaven's broad woof outstretched above;
What covered all? What sheltered? What concealed?
Was it the waters' fathomless abyss?

Verse 7 —

He from whom all this creation came,
Whether his will created or was mute,
The Most High Seer that is in the highest heaven,
He knows it — or perchance even He knows not.

Another translation of the 'Creation Hymn' reads as follows (source *Wonder that was India* by AL Basham, Sidgwick and Jackson, London):

Verse 1 —

Then even nothingness was naught, nor existence.
There was no air then nor the heavens beyond it,
Who covered it? Where was it? In whose keeping?
Was there then cosmic water, in depths unfathomed?

Verse 6 —

But, after all, who knows, and who can say
Whence it all came, and how creation happened?
The gods themselves are later than creation,
So who knows truly whence it has arisen?

Verse 7 —

Whence all creation had its origin,
He, whether he fashioned it or whether he did not,
He, who surveys it all from highest heaven,
He knows — or maybe even he does not know.

5.3 The Brahmanas

Brahmanas are the prose (non-*Samhita*) portion of the Vedas. The word 'Brahmana' means that which belongs to the Brahmans (the Hindu priests whom we spell as Brahmins). These are the works meant for the guidance of Brahmans (Brahmins) in the day to day use of the hymns of the *Samhita* of the Vedas. Brahmanas are also classed as *Sruti*. There are one or more Brahmanas for each Veda (total 13):

- *Rig* has two Brahmanas — Aitareya and Kaushitaki   (2)
- Black (Taittiriya) Yajur has the Taittiriya   (1)
- White Yajur has Satapatha Brahmana   (1)
- Sama has eight Brahmanas   (8)
- Atharva has Gopatha   (1)

                               Total   13

There is a viewpoint that the Brahmanas in their totality embrace the treatise called the Aranyakas and Upanishads, which we cover in the following sub-chapters. The Brahmana of the White Yajur, the Satapatha is the oldest; next come the two Brahmanas attached to the Rig.

The Brahmanas contain details of the Vedic ceremonies, with full explanations of their meaning; they are the ritualistic religious writings in prose, for the purpose of worship. These give instructions as to how to use a particular Vedic verse. In them are found details of mythology and old legends and their explanations. The Brahmanas maintain the central character of the Veda to which they belong. Brahmanas throw a lot of light on the process of development of Hindu culture of those days.

5.4 The Aranyakas

Aranyakas were compiled by and for the forest dwelling people, who had retired from the world. These are the 'forest books' which in a way form the concluding portion of the Brahmanas. Presently, there are four surviving Aranyakas, i.e. Brihad, Taittriya, Aitreya and Kaushitaki. The Aranyakas are closely associated with the Upanishads (described in sub-chapter 5.5), thus —

- The Brihad is sometimes called Aranyaka and sometimes as Ayanyaka-Upanishad. It is attached to the Satapatha Brahamana.
- The Kaushitaki Aranyaka consists of three chapters; of that, the third chapter is the Kaushitaki Upanishad.

Aranyakas form a sort of link between the 'ritualistic' Brahmanas and the 'speculative' Upanishad, i.e. a stepping-stone from the former to the latter. The Aranyakas talk about the soul in all its manifestation; these also explain the origin and elements of the universe. Whilst laying a lot of stress on rites and rituals, Aranyakas also mark the transition from the ritualism of the Vedas to the philosophical thought processes, which climax in the Upanishads. Some Upanishads are incorporated in the Aranyakas; others form their conclusive separate portions. Sometimes, Upanishads are considered as the expository appendices of the Aranyakas. On their own, Upanishads are the greatest source of Hindu thought. Most Hindu laymen would have heard of the Upanishads; not many may have heard of the Aranyakas and the Brahmanas.

5.5 Vedanta Upanishads

The word Vedanta consists of two parts — *Ved* and *Anta*; in Hindi '*anta*' means 'end'. As the Upanishads evolved towards the end of the Vedic period, these were called Ved-anta, i.e. Vedanta. 'Anta' has a second meaning — objective or goal. As such, Upanishads could be considered as the goal of the Vedas. The word the Upanishad comprises of three syllables, i.e. *Upa+Ni+Shad*, which means 'Come, sit, near me'.

Western historians put the age of the Upanishads at about 800 to 600 BC. Others have suggested 1200 to 1000 BC. As usual, Hindu scholars consider the Upanishads of even older vintage. There might have been 150 to 200 Upanishads, out of which some 108 have survived; of these, 14 Upanishads are rather important.

As stated earlier, Upanishads are attached to the Aranyakas, which are linked to the Brahmanas, which from a part of the Vedas (in the 'extended' sense):

- *Rig Veda* has the Aitareya Upanishad attached to the Aitareya Brahmana.
- Taittiriya (Black) *Samhita* of the Yajur has the Taittiriya Upanishad.
- Vajasaney (White) *Samhita* of the Yajur has the Isa or Isha Upanishad.
- Satpatha Brahmana has the Brihad Aranyaka-Upanishad.
- Sama Veda has Kena and Chhandyoga Upanishads.
- Atharva has the Katha, Mundaka, Prasna, Mandukya and innumerable others.

Upanishads excel in expounding Hindu philosophy. Max Muller has called the Upanishads the most wonderful composition of the human mind. The Upanishads give the vivid account of the religious and spiritual thoughts of the Hindus. They give detailed explanations of the relation between God, soul and matter. Doctrines of *Karma* (Deeds/ Actions) and *Moksha* (Salvation) are explained; the ways to attain these have been elaborated. Without the Upanishads, Vedic literature and Hinduism itself would have been barren or fallow. The Upanishads contain the very kernel of the Hindu thought process.

The main emphasis of the Upanishad philosophy is on Monism;

this doctrine states that there is only one Supreme Being and the Rest (His creation) is all *Maya* (Illusion). That would imply that the duality between the Creator and his creation is unreal. In other words, mind is a manifestation of matter, i.e. the brain cannot exist apart from matter (Dualists believe that there is a real distinction between mind and matter).

The One Absolute Supreme Spirit called Brahman pervades the entire cosmos; that is the Outside Reality. That Brahman lives in the individual consciousness as *atma* (the soul); that is the Inside Reality. Both Brahman and *atma* are everlasting and indestructible; man should strive for unity between the two. That path will ultimately lead to *moksha* (salvation), i.e. freedom from *samasara*, the cycle of birth and death.

Tat Tuam Asi (that thou art), i.e. *Atma* is Brahman, i.e. the unity of all things in the One Absolute Supreme Being. Truth is within us, it may not be sought outside.

The Upanishads lay heavy emphasis on finding out the truth; these also concentrate on mental adventures, and a spirit of inquiry, i.e. to know more and more. In the Upanishadic philosophy, gods are secondary, perhaps even unnecessary; man is at the centre of all activity. The *Rig* sacrifices promised mainly earthly bliss; perhaps even heaven. The Upanishads emphasized the search for *moksha* (release from earthly existence), which is not stressed in the *Rig Veda*. The Upanishads lay a lot of emphasis on meditation and knowledge. Vedic rituals and sacrifices yield place to human thought and ethics. Instead of rituals, man was advised to have knowledge of the new doctrines. Of the new doctrines that emerged, one of the most important ones was that of 'Transmigration of Soul'. The Upanishads developed the concept of Monotheism to its finality; the concept, if at all, gets only a passing reference in the *Rig Veda*.

The Yoga Systems
The route to the union of *Atma* with Brahman called 'self-realization' is through Yoga, which implies spiritual discipline. Broadly speaking, there are four major paths of Yoga:

- *Jnana* (Knowledge): This system involves the route of meditation, self-enquiry and self-realization; that leads to the knowledge of God and to wisdom. The Upanishads cover this yoga system comprehensively.
- *Bhakti* (Devotion): The route of *Bhakti* is to concentrate on search and praise of the Supreme Spirit Brahman; in the process one may lose one's self-identity. It is a way of emotional bonding with God. Srimad-Bhagavatam covers this system.
- *Karma* (Action/Deeds): It emphasizes the need for *karma*, i.e. righteous action, without any fear of the result, i.e. *nishkam karma* (selfless action). The *Karam-kanda* covers this branch of yoga.
- *Raja* (Spiritual Practices): *Raja* Yoga means conduct of mental and physical exercises to achieve a degree of oneness with the Supreme Being. It is a way of transcendental awareness and psychic discipline. For this branch of yoga, we have the incomparable Yoga-Sutras of sage Patanjali.

All the above four systems of Yogas are covered brilliantly in the *Bhagwad Gita;* it combines the four Yogas into a single, unified path. There are also some other forms of Yogas, out of which Hath Yoga (a Tantric practice) is perhaps the most important.

5.6 Charvaka

Before we go any further, we may make a very brief reference to an atheist philosophy, which may or may not fall under the overall umbrella of Hinduism. It is called Charvaka, after the name of its founder; it emerged around the 6th century BC.

Based on purely materialistic tenets, the Charvaka philosophy holds knowledge can come only out of direct perception. The philosophy considers God as a myth, and rather unnecessary. The philosophy regards reality only in terms of matter's interactions, without any spiritual component. The Charvaka philosophy is considered to encourage selfish behavior.

We may note in passing that two other religions, i.e. Buddhism and Jainism also emerged in the 6th century BC. These two religions, like the

Charvaka, also have no need of God. Thus, the three religious thought processes without any need of God, emerged just after the Upanishads. That may be indicative of a deeper problem. It is possible that the common Hindu folks of those days found it a bit difficult to relate to the lofty concept of a Nirguna and Nirupa Brahman of the Upanishads; they went in for not one or two, but three systems without the need of God. Both Buddhism and Jainism flourished in India; Jainism, though in diluted form, still does.

5.7 *Advaita* and *Dvaita*

Some 1000 years after the advent of the Upanishads, three different relationships came to be established between the Outside Reality Brahman and the Inside Reality *Atma*, viz. —

Advaita (Non-Dualistic) by Adi Shankaracharya (788 to 820 AD)

Shankaracharya expressed that Brahman, the Universal Soul and *Atma*, the individual soul are one and the same, i.e. they are identical; that would imply God is within man. *Moksha* (liberation from cycle of birth and death) can come only by merging the *Atma* with the Brahman. The main obstacle to that is *maya*; that creates the illusion of plurality and hence is the cause of all suffering. The route to *moksha* is through *jnana* (knowledge) which comes through meditation and *tapasya* (asceticism).

Shankaracharya was born in the small town of Kalady in Kerala. He traveled on foot all over Bharat several times in his short life span of 32 years. He was responsible for reviving, almost single-handedly, the teachings of the Upanishads, which had been over-shadowed under the onslaught of Buddhism. He established the first Hindu monastic order by setting up *ashrams* (hermitages) in the four corners of the country at Sringeri, Badrinath, Dwarka and Puri.

Shankaracharya wrote detailed commentaries on the *Bhagwad Gita* and the Upanishads. He worked tirelessly towards the restoration of Hinduism, which was gasping for breath under Buddhism. He was highly successful in his efforts, and Hinduism could finally emerge out of the clutches of Buddhism. He codified doctrines of Hinduism and grouped the then prevailing innumerable methods of worship into six major groups; these involve worship of the deity stated against each:

Saiva	—	Shiva
Vaishnava	—	Vishnu
Shakta	—	*Shakti*, the mother goddess
Ganpatya	—	Ganpathi, Ganesha
Saura	—	Surya, the Sun
Kaumara	—	Kumara, Skanda, Murugan, Karttikeya

Vishisht Advaita (Qualified Non-Dualism) by Ramanujacharya (1017 AD)

The philosophy states that Brahman and *atma* are both unitary and dual. Howsoever high the *atma* may rise it always remains, to some extent or the other, apart from Brahman.

Brahman, the Single Universal Reality has three qualities: the individual soul, the insensate world and the Supreme Soul. The first two are controlled by the third. Release for the individual soul can come only out of its realization as a part of the Supreme Soul. The path to this release is through a combination of knowledge and religious practices, resulting in *Bhakti* (devotion to God).

Dvaita (Dualism) by Madhavacharya (1199 AD)

Brahman and *atma* are similar, but always separate. Salvation lies in the *atma* dwelling close to the Brahman, but it does not merge in the Brahman. Madhava disputed Sankara's concept of oneness of all souls.

We may note that all the three Acharyas were from South India.

5.8 The Epics

5.8.1 The Historical Context

Like the *Iliad* and the *Odyssey* of the Greeks, Hindus have their great epics, the *Mahabharta* and *Ramayana*. The *Mahabharta* is the longest story in history, some seven times the combined length of the *Iliad* and *Odyssey*. The *Ramayana* is about ¼ the size of the *Mahabharta*. Both are stories of royal ruling dynasties of the pre-historic (mythological) times.

The origin of the *Ramayana* and the *Mahabharta* dynasties are traced to Manu. Out of the many dynastic lines emerging out of Manu, the two most important were the solar and lunar lines (see Chapter 8.3 for details).

Solar Line

Manu's eldest son was Ikshvaku. He founded a dynasty which ruled in the general area of Central UP; the dynasty came to be called Ikshvakus, Suryavanshi or the Solar dynasty. Vikukshi, the eldest son of Ikshvaku inherited the part of the kingdom, called Kosala, with its capital at Ayodhya. The 64th and 65th generations of this dynasty ruled in Kosala during the *Ramayana* age. The *Ramayana* story revolves around Lord Rama, the seventh *avatar* (incarnation) of Vishnu.

Another son of Ikshvaku, Nimi, inherited the kingdom of Videha (literally meaning bodiless), with its capital at Mithila. In the *Ramayana* age, King Janak ruled in Videha. His daughter was Sita, who became wife of Rama. Ikshvaku, possibly, had 98 other sons, who might have ruled over various other parts of Bharat *varsha* of those days.

Lunar Line

Manu's daughter was Ila who founded a dynasty which has been called the Chandravanshi, or the Lunar dynasty, or Pauravas/Purus (named after two early kings of this name). The Chandravanshis were split in many lines. The main line ruled in Central/Western UP; initially, its capital was at Allahabad, which was later shifted to Hastinapur, near Delhi. It is a bit difficult to give serial numbers to Chandervanshi kings, as there were major breaks in this dynasty. One such break lasted 22 generations (about 400 years), after the 20th lunar generation. It is, therefore, customary to give serial numbers to lunar kings in synchronism with kings of the solar dynasty, which ran without a break. We will adopt this procedure in our study.

The *Mahabharta* war was fought between two groups of cousins of a generation, which corresponded to the 94th generation of kings of the solar line. In the lunar line that would be about the 50th kingly generation, i.e. some 44 (94–50) generations of lunar kings were missing in different batches. Lord Krishna, a Yadava (an off-shoot of the lunar line) played a stellar role in the *Mahabharta*. He was the eighth *avatar* of Vishnu.

Age of the Epics

There is lot of debate about the age and historicity of the two epics. First, when did the actual events take place? Second, when were these written in the form of the epics? Was there a major gap between the occurrence and the writing? The texts might have been updated (or even re-written) many times. Some interpolations could also be expected to have taken place. The various dates/periods given below are more informed estimates, and less authentic history.

As per the orthodox Hindu view, the *Ramayana* is placed towards the end of the Treta Yuga, and the *Mahabharta* at the end of the Dvapar Yuga; that is pure mythology. After the *Mahabharta* war, Lord Krishna was killed by a hunter's arrow piercing his heel. The death of Lord Krishna is associated with the end of the *Dvapar* Yuga, and the beginning of the (present) *Kali* Yuga; sometimes, the date of 3102 BC is associated with that event (see chapter 9.5). This could be on the border line of mythology and history.

On a more realistic timeframe, but still in the context of pre-history, the following possible periods (just informed estimates) have been suggested for actual occurrence of the two wars, but without proof of any sorts:

- *Ramayana:* 2300 to 2000 BC; say between 700 to 1000 years before the *Mahabharta.*
- *Mahabharta:* 1400 BC, or 900 BC; we may also keep 3102 BC of the preceding paragraph at the back of our mind. A bit arbitrarily, we will assume the middle date of 1400 BC, for the purpose of our analysis.

It has been expressed that the two epics might have been written anywhere between the 4th/3rd century BC to the 3rd/4th century AD. A Greek writer Megasthenes visited India around 320 BC; he does not refer to the two epics. That has led some historians to opine that the epics may have been written around 2nd/1st centuries BC. It is possible that the epics were written over a period of few centuries; the stories may have continued to evolve as the writing progressed.

At some stage, may be during 1st/2nd century AD, the two texts appear to have undergone major updates, when;

- *Ramayana* got increased from the earlier 12,000 *shloka*s (verses) to the present 24,000 *shloka*s; it is believed that the first *Valmiki Ramayana* had only about 6,000 *shloka*s.
- *Mahabharta* was increased from the earlier 24,000 *shloka*s to the present 100,000 *shloka*s; the *Bhagwad Gita* may have been embedded in it at that time.

5.8.2 Ramayana

As per tradition, the seventh Manu Vaivaswat is ruling the present universe; he was the son of Vivaswat (an aditya-Surya, the sun). Manu's son was Ikshvaku who founded a ruling dynasty, which came to be called Suryavanshi, after his grandfather Vivaswat (Surya). One line of this dynasty ruled at Kosala with its capital at Ayodhya; Vikukshi, the eldest son of Ikshvaku inherited it. The second line ruled the kingdom of Videh with its capital at Mithila; Nimi, one of the younger sons of Ikshvaku, was its head. Both kingdoms appear to have been located in present day UP. Genealogy of the Suryavanshi dynasty is as follows:

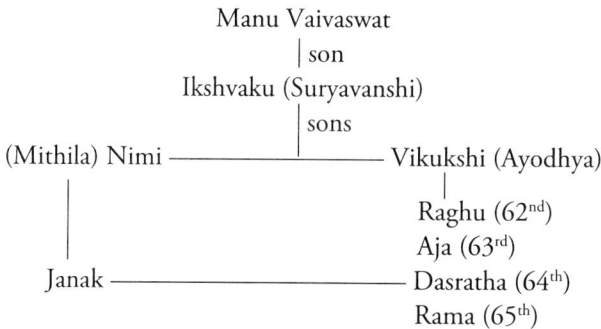

```
                    Manu Vaivaswat
                       | son
                  Ikshvaku (Suryavanshi)
                       | sons
(Mithila) Nimi ────────┴──────── Vikukshi (Ayodhya)
         |                            |
         |                       Raghu (62nd)
         |                       Aja (63rd)
     Janak ───────────────────── Dasratha (64th)
                                 Rama (65th)
```

The following comments are offered on the above table:

- The figures in bracket indicate the generation from Manu, i.e. Lord Rama was the 65th generation from Manu.
- Raja Janaka ruled at Mithila, as a contemporary of Dasratha. He was father of Sita who became Rama's wife.

Ramayana is the story of the Suryavanshi clan, more specifically of Lord Rama, the seventh *avatar* of Vishnu born on earth to show the path of righteousness. The *Ramayana* has been called *Adi kavya*, the first poetic composition of the world. It is also called *Ramkatha*. In its extant form, it has some 24,000 verses, divided into seven books called *kandas*.

Tradition has it that initially Rishi Narada told the story of the *Ramayana* to Valmiki, who wrote it down on the urging of Brahma. The original *Valmiki Ramayana* was in Sanskrit. A Tamil version was written in the 9th or 12th century AD, by Kambhan; it is a called the *Kambhan Ramayana* or Iramavatram. Krittivasa wrote the *Ramayana* in Bangla during the medieval times. There is a *Ramayana* in Marathi by Sridhar. Tulsidas authored a Hindi version of the *Ramayana* around 1575 AD; it is titled Ramcharitmanas and is very popular in North India. Then there are the *Adhyamta* and *Ananda Ramayana*s, both estimated to have been written in the 15th century AD. In addition, there are *Ramayana*s in many other regional languages. The Valmiki and Tulsi *Ramayana*s could perhaps be considered more authoritative, though everyone may not agree. There are numerous variations in the stories told in the various *Ramayana*s. Many events are mentioned which were not recorded by Valmiki; some events are given in a different version.

The *Ramayana* is very popular in South East Asian countries. The *Ramayana* is also told in a Buddhist Jataka tale, called Jataka-Dasratha. The story in this text is substantially different from the mainline *Ramayana* story; there is no role for Ravan and Hanuman in that tale. A view has been expressed that Valmiki treated Lord Rama more as a king, and less as a god. It was Tulsidas who fully deified Rama.

The dynasty ruling at Ayodhya was called Raghu-Kul or Raghu-Vansh, after King Raghu (62nd generation from Manu). The grandson of Raghu was King Dasratha, whose son was Lord Rama. Rama had three other brothers, of whom Lakshmana was his constant companion and ardent admirer. In another dynasty, King Janaka ruled in Videha (capital Mithila) at that time. His adopted daughter was Sita; she was found as a baby girl in a furrow in the field. Sita was the incarnation of Lakshmi, the wife of Vishnu. Janaka announced that anyone who could string (Shiva's) *Haradhanu* bow held by his clan would get the hand of his daughter in marriage.

Rama could not only string the *Haradhanu*, but broke it into two. He married Sita. Due to a court intrigue, Rama was banished to the forests for 14 years. His brother Lakshmana and wife Sita, accompanied Rama to the forests, though they were not required to do so. In the last year of their stay in the forest, the demon king Ravan of Lanka abducted Sita. Rama got help from the monkey king Sugriva and his commander Hanuman. Rama attacked Lanka; a fierce battle raged for many days. Finally, Ravan was defeated and killed; Rama had to use Brahma's weapon, the *Brahmastra* to kill Ravan.

Sita was rescued; but then fate intervened. When Sita came in Rama's presence, he refused to accept her. At that time, the following words have been put in Rama's mouth in the *Ramayana* by Griffith, as reproduced in *Gods of India* by Osborn Martin.

> Lady, 't was not for love of thee
> I led my army o'er the sea.
> I battled to avenge the cause
> Of honor and insulted laws
> My love is fled, for on thy fame
> lies the dark blot of sin and shame;
> And thou are hateful as the light
> That flashes on the injured sight.
> The world is all before thee; flee;
> Go where thou will, but not with me,
> For Ravan bore thee through the sky,
> And fixed on thine his evil eye;
> About thy waist his arms he threw,
> Close to his breast his captive drew;
> And kept thee vassal of his power,
> An inmate of his ladies' bower.

That was not the welcome Sita had expected. All present were embarrassed, including the gods who had descended in large numbers to facilitate Rama for his great victory. Tradition has it that at that stage, Brahma told Rama of his divine origin, of which he was, perhaps, unaware until that time. By intervention of gods, it was agreed that Sita would go through a 'Fire Test'; Lakshmana prepared the pyre. As Sita walked through fire, the god of fire *Agni* himself appeared and declared Sita to be pure and spotless. Reconciliation took place and all returned to Ayodhya, where Rama was anointed the king.

The *Kambhan Ramayana* gives a bit different version. As per that version, Rama used very harsh language for Sita. The explanations and entreaties of Sita had no effect on Rama. Thereupon, Sita decided to burn herself and asked Lakshmana to prepare a pyre. Sita got into the fire. Due to the force of Sita's chastity, it was the fire god *Agni* who was singed; Sita emerged harmless. *Agni* went protesting to Rama, for his (wrong) action. *Agni* expressed that so exalted was the status of Sita (Lakshmi) that one furrow on her head could stop *Surya* (sun) from shining and the wind from blowing. At that stage, both Brahma and Shiva informed Rama of the divine origins of himself and of Sita, as being incarnations of Vishnu and Lakshmi. Shiva then called King Dasratha from the heavens to talk to Rama. On the counseling of Dasratha, Rama agreed to take back Sita.

Happiness was short-lived. Rama overheard one *dhobi* (washer man) talk disparagingly of Sita's character (there are different versions). Rama banished Sita to the forests, where she got refuge in the hermitage of Valmiki, the author of the *Ramayana*. There, her two sons Luv and Kusa were born. When the boys grew up, they were united with their father Rama. At that happy moment, Sita decided to descend into the earth, from which she had risen at her birth.

An introduction to the *Tulsidas Ramayana* says, "He who reads and repeats this holy, life-giving *Ramayana*, is liberated from all sins and exalted, with all his posterity, to the highest heaven". Though Rama was of divine origin, he lived the life of an ordinary mortal, going through all the trials and tribulations. Unlike Lord Krishna, not many miracles are attributed to Rama.

Rama is 'god of gods', who is entitled for the highest worship. Tulsidas portrays Rama as the complete incarnation of the Absolute and the Supreme Spirit. Tulsidas expresses that the Supreme Being became incarnate to relieve the world of sin. Rama has been called the '*Maryada Purshottam*', i.e. the 'Ideal Man' whose footsteps need to be followed by all.

At this stage, we would like to record a few words about Ravan, who had an important role to play in the *Ramayana*. He was the demon king of Lanka, who was considered an evil incarnate. Ravan was a Brahmin, being the grandson of *dev-rishi* Pulastya (Chapter 7.3). The Rishi had a son called Visravas who had two wives; Ravan was born from his second wife Nikasha/Kaikasi (daughter of *Rakshas* Sumali). Visravas had a son named Kubera (god of wealth), from his first wife. Ravan was the king of *rakshasas*, and his stepbrother Kubera, the king of *yakshas*.

Sumali (maternal grandfather of Ravan) was the original ruler of Lanka. Kubera took over rule of Lanka, after Sumali was forced out. On the urging of Sumali, Ravan snatched the kingdom of Lanka from Kubera and expelled him from there. Ravan's two brothers were Vibhishan and Kumbhkaran; one of his many sons was Meghnad, also called Indrajit, as he had recorded '*jit*' (victory) over Indra.

Though a *rakshas*, Ravan was a learned man who had studied all the scriptures, including the four Vedas. He is called Dashanan, i.e. with ten heads, to indicate his high level of knowledge. Ravan was an ardent devotee of Brahma. By severe austerities, Ravan obtained boons both from Brahma and Shiva; these made him invulnerable against gods and demons. Ravan was so powerful that he had no fear of humans. Therefore, he had sought no protection from Brahma against humans; that proved to be his undoing.

We record below some random events of interest, relating to the *Ramayana*:

Kshatriyas in the Ramayana
The general belief is that the *Ramayana* war was a conflict of the Kshatriyas. However, it appears from the totality of events that in that great war there were only two Kshatriyas, i.e. Rama and Lakshmana. All the rest were non-Kshatriyas, whose numbers could be in hundreds

of thousands. One version of the *Ramayana* talks of Ravan's troops in millions.

We know Ravan, his brothers and his many sons were all Brahmins of the blue-blooded type. His army was of the *rakshasas* who were non-Kshatriyas, whatever their actual caste, if any. On the side of Rama were monkeys and bears; they are unlikely to have been Kshatriyas. So, where does all that leave the Kshatriyas? We have no answer. It is most likely that after the general massacre by Parshurama (the *avatar* preceding Rama), Kshatriyas had been highly weakened.

Bali Killing

Once Brahma's teardrop fell on the ground; the result was the birth of a male monkey Riksharaja. By a dip in a lake, Riksharaja was changed into a beautiful woman. Indra and *Surya* (Sun) could not resist her charms. Indra's semen fell on the hair (*Bal* in Hindi) of the 'woman'; result was birth of Bali. Surya's semen fell on her shoulder (*griva*); the result was the birth of Sugriva. After the two births, the 'woman' again became the male monkey Riksharaja. *Dev-rishi* Atri and his wife Ansuya, brought up Bali and Sugriva. A different version says that it was Aruna, the charioteer of Surya, who turned himself into the beautiful woman. The rest of the story is largely the same.

Bali became the powerful king of Kishikindhya. Once Bali went to fight a demon; Sugriva, rather hurriedly assumed Bali to be dead. He took over Bali's kingdom as well his wife Tara. Later, Bali came back; he defeated Sugriva and took back his kingdom and wife Tara. He also took Sugriva's wife Ruma, to punish Sugriva for his earlier actions. Later, Sugriva allied with Lord Rama, and asked him to kill Bali; Rama agreed.

Bali had the boon that half the strength of any adversary coming face to face with him, would get transferred to Bali. To evade that, Rama decided to attack Bali from behind a tree. Rama aimed an arrow at Bali, whilst Bali was engaged in combat with his younger brother Sugriva. Both the actions were against the rules of combat. As he lay dying, Bali enquired from Rama why he had broken the 'warrior code', especially when Bali had done no harm to Rama. Rama replied that Bali had taken Sugriva's wife, and that the 'warrior code' was for the human race; it was not applicable to the *Vanar* (monkey) race. Bali expressed that it was not a crime in the *Vanar* race to take another's wife.

When Bali was killed, Tara, highly desolate, cursed that for the (unjustifiable) act of killing Bali, Rama would have to suffer long separation from his wife Sita; we know that actually happened. However, Tara started living with Sugriva. Sugriva was so happy to get Tara that he forgot that he was to help Rama recover Sita. Lakshmana had to give him a terse reminder.

Shambuk Killing

During the days of Ram Rajya, a Brahmin lost his 5 year old son. The Brahmin was very upset and coming to Rama's Court expressed that they could not be in the Treta Yuga, if the son died before the father. After checking, the courtiers informed Rama that the cosmos' balance had been disturbed, as one Shudra named Shambuk was practicing austerities, which were not permitted to his caste. On hearing of that, Rama went and cut off the head of Shambuk. (This incident is described in the *Valmiki Ramayana*, Uttar Kanda, Book 7, sargas 64–67).

There are different versions of this event in the various *Ramayana*s. In one version, *dev-rishi* Narada told Rama that events like the 'son dying before the father' could take place only if some Shudra was to practice austerities. Thereafter, Rama set out on a journey, to trace such a Shudra. In another version, untimely deaths of seven people of all castes, male and female, were reported, instead of only one Brahmin's son. Some versions refer to Rama having explained to Shambuk as to why he had to kill him. Shambuk even asked for two boons, which Rama granted to him; these included Shambuk's entry into heaven.

The incident has also been given the shape of struggle between the Brahmin North and Dravidian South. In 1920, a Telugu playwright, Ramaswami Chaudri published a play 'Shambuka Vadha', based on this incident; that resulted in considerable turmoil and agitations.

Brahm-hatya

Brahmn-hatya (Brahmin-cide) is one of the most heinous crimes in Hinduism, for which there is hardly any repentance or forgiveness.

Now, Ravan was a Brahmin, and of the highest class at that, being the grandson of a *dev-rishi* (celestial seer). In addition, he had knowledge of all scriptures, including the four Vedas. Was Ravan's killing a case of *Brahm-hatya*, for which Lord Rama may have had to suffer? We are not qualified to answer that question; we just record two facts:

- First, Rama had to spend the rest of his life without his beloved wife Sita.
- Second, his dynasty appears to have come to a rather premature end. Even the *Mahabharta* people do not particularly talk of the Suryavanshi dynasty of Lord Rama.

Was there a divine design in getting *Brahm-hatya* executed at the hands of Rama? Were Kshatriyas to suffer for the same? We may note that the sixth Vishnu *avatar* Parshurama (preceding Rama), had wiped out the entire Kshatriya clans 21 times; that must have been as per the divine will. Could it be that the divine plan was still in operation? In Chapter 9.9, we will see that the power of the Kshatriyas gradually ebbed away after *Ramayana*. Even in the *Mahabharta*, the two main warring factions were only half-baked Kshatriyas (see under *Mahabharta*).

Dussehra Festival

In North India (mostly West and Center), it is an annual Hindu custom to symbolically burn paper effigies of Ravan, his brother Kumbhkaran, and his son Meghnad, on the occasion of Dussehra. Does that amount to *Brahm-hatya*? We do not have an answer. However, we cannot help noticing the ground reality that Hindus are the only civilization in the world, to have suffered 750 years of continuous slavery. Is there some linkage between the annual burning of Ravan and Hindu slavery? The simple answer is, 'no, no way'. However, there could be a more complex answer.

Even if there is a reason to burn Ravan, there does not appear to be enough justification for meting that treatment to Kumbhkaran and Meghnad. They were just doing their duty to their king. We agree that killing of Kumbhkaran and Meghnad was justified during the actual war. Do we need to keep on burning them every year, for thousands of years? Was their crime so grave? We may note that some parts of South

India, especially Tamil Nadu, do not take too kindly to this North Indian obsession of burning Ravan annually. Even East India is not too enthusiastic.

The festival of Dussehra is projected as a victory of 'Good over Evil'. Hindus have been celebrating Dussehra now for thousands of years. However, we do not find any reduction in evil. Actually 'evil' is the single most predominant factor in present day Bharat. This annual ritual does not appear to be having any beneficial effect.

By burning those 'paper effigies' of Ravan and others, Hindu youth tend to get a false macho feeling of their military prowess. Facing a real (not paper) enemy in an actual war situation calls for an entirely different type of mindset.

Shabri's *Bers* (berries)

During the days of exile of Lord Rama, he came to the hut of a low caste (Shudra) woman Shabri. She had heard of his coming and was collecting the fruit '*ber*' (green/red berries) to offer to the Lord. She would bite each berry to test its sweetness. On arrival, Rama had no hesitation in eating the berries bitten by a low caste Shudra woman.

Shiva to meet Rama

Due to his Rama *avatar*, Vishnu had been absent from heaven for a long time. Shiva decided to pay a visit to Vishnu (Rama), to pay his obeisance; Parvati accompanied him. When Shiva and Parvati arrived, Sita had been just abducted, and Rama was in a highly desolate state. Shiva thought it best not to disturb him in that state and wanted to go back. But Parvati would have none of that. She wanted to know who was this man in such a bad way, to whom her husband, the master of the universe, wanted to pay obeisance. She took the form of Sita and appeared before Rama. The great Lord recognized Parvati instantly, and addressed her as mother.

5.8.3 Mahabharta

Manu had a daughter named Ila. Her brother Ikshvaku had founded the dynasty of Ishvakus or Suryavanshis. Ila also founded a dynasty, which has the following lineage:

```
                Rishi Kashyap           Rishi Atri
                     |                       |
                   Manu          Chandra/Soma (Moon)
                     |                       |
                   Ila ——— husband ——— Budha*
                     |son

                Pururavas
                     |
            Yayati (6th generation)
                     |
        ┌─────────── sons ───────────┐
      Yadu              Puru (7th generation)
        |                       |
     Vasudeva           Dushyant (43rd)
        |son                    |
   Lord Krishna         Bharata (44th)
   (94th generation)            |
                          Hastin (51st)
                                |
                        Kuru (70th generation)
                                |
      Satyavati —— wife —— Shantanu —— wife —— Ganga
                         (91st
                        generation)
        ┌──────────grandsons──────────┐        |son
   Dhritrashtra            Pandu        Bhisham Pitama
        |sons                   |sons
   Kaurvas—100 brothers    Pandavas—5 brothers (94th generation)
   Eldest—Duryodhna        Eldest—Yudhistra
```

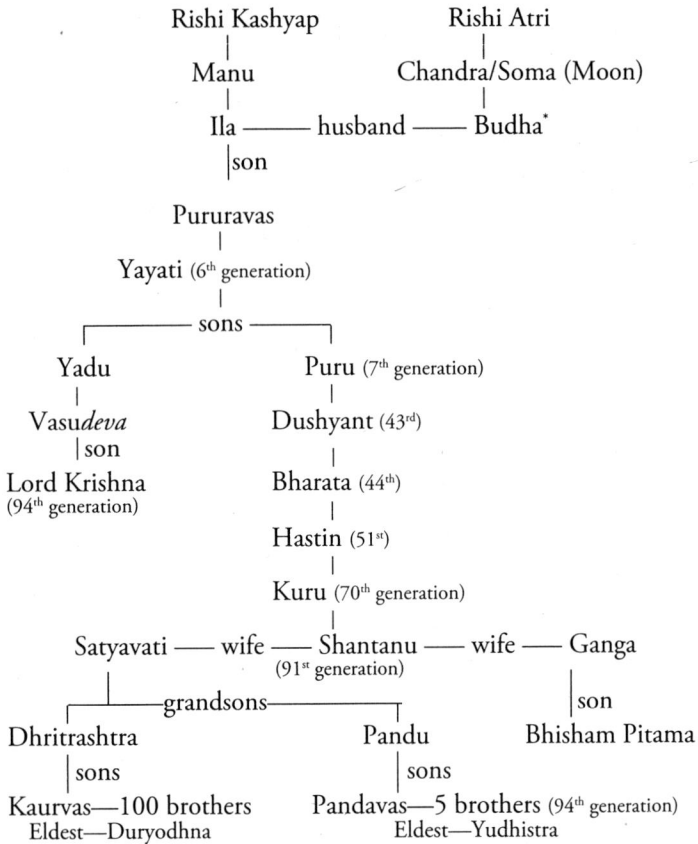

Ila, having founded the dynasty, married Budha, son of Chandra/
Soma, the Moon. The dynasty is called the following:

- Chandravanshi or Lunar Line after *Chandra*, the father of
  Budha.
- Ailas, as derived from Ila
- Paurvas, as derived from Pururavas son of Ila
- Purus, named after King Puru

The following comments are offered on the above table:

---

*   Nothing to do with Gautam Budha

- There were two major breaks in the Paurva line — one of that was for 22 generations before Dushyant who re-established the dynasty; the second break was after Ajamidha, son of Hastin.
- Due to the reason of these breaks, it is difficult to give serial numbers to the Paurava (lunar) kings. In order to establish synchronism between the solar and lunar lines, serial numbers of kingly generations given above relate to the corresponding serial numbers of kings of the solar dynasty, which ran without a break.
- *Chandra/Soma* (Moon) was the son of Atri. Chandra forcibly took away, and seduced Tara, the wife of Brhaspati, the priest of the *deva*s (gods); Budha was born. That was a major transgression by *Chandra*. Later, Indra helped Brhaspati recover Tara.
- Ila established her reign before she married Budha.
- Yadu and Puru (7th generation) were brothers. Puru continued the main Paurva line. Yadu founded the Yadava clan in which was born Lord Krishna as the 94th generation from Manu. At some stage, the clan of Haihayas emerged out of the Yadavas.
- The people in the main Chandravanshi line were called Paurvas, Purus or Kurus, after the names of three kings in the line. The Kauravas and Pandavas, the adversaries in the *Mahabharta* war, were the 94th generation in this line.

The extant *Mahabharta* has some 100,000 *shloka*s (verses), divided into 18 *parvas* (books). The *Mahabharta* story is much more colorful and complex than that of the *Ramayana*, with a number of sub-plots.

At the 20th generation of the lunar line, there was king Tamsu. Thereafter, there was a break of 22 generations. In the 43rd generation, there was Dushyant, an heir in exile. At that time, King Marutta of the Turvasu (a Paurva off-shoot) dynasty was ruling; he was childless (This Marutta was different from another Marutta Aviksit of Vaisali). Marutta Turvasu adopted Dushyant as his son. Dushyant went on to recover his kingdom, and re-established the Paurava line; he made major conquests. Dushyant romanced and married Shakuntla, daughter of *dev-rishi* Vishvamitra through *apsara* Menaka, sent by Indra to disturb austerities of the great *rishi*. Their son was Bharata, one of the greatest monarchs to rule India; our country Bharat is named after him. His descendents are called the Bharatas — a name, sometimes, used for all residents of Bharat.

King Bharata had three wives; but he killed all his sons, as they were not worthy, and not up to his standard. Thus, Bharata was left without an heir; he undertook *yajnas* (fire-sacrifices) to get a son. Finally, rishi Bhardwaj, named Vidathan Bhardwaj, was offered to the king, and he adopted him as his son. Now Vidathan Bhardwaj may not have been the original *dev-rishi* Bhardwaj (son of Bhraspati), but a descendent of the *dev-rishi*. On the death of King Bharata, a son of Vidathan Bhardwaj, called Vitatha, ascended the Paurva throne. If this tale is true, it indicates induction of Brahmin blood in the great Paurva/Puru or Chandervanshi line.

As the 91st generation (solar synchronism) from Manu, was the great Paurva King Shantanu. He was married to Ganga (otherwise a river), who, in her human form, was daughter of Himmavat, king of the Himalayas (Ganga's sister was Parvati, wife of Lord Shiva). Shantanu and Ganga had a son who became famous as Bhisham Pitama. After the birth of Bhisham Pitama, Ganga left Shantanu.

One day, king Shantanu happened to see a nubile young fisherwoman; her name was Matsyaghangi (also called Kali). She was born of a fish with the seed of King Vasu (other names are also in use) of Cedi (a Paurva off-shoot), when the seed accidentally fell into the river. The fish actually was the *apsara* Adriki. Matsyaghangi was extremely beautiful. Shantanu could not resist her charm and proposed marriage. Her father put the condition that after Shantanu, her children should ascend the throne. When Bhisham Pitama came to know of the condition, he surrendered his right to the throne. Additionally, he took a vow of celibacy, so that no progeny of his could lay a claim to the throne in the future. After marriage, Matsyaghangi became famous as Queen Satyavati.

As it happened, the two sons of Satyavati died childless, leaving two widows, Ambika and Ambalika. Satyavati asked Bhisham Pitama to impregnate the two widowed queens, so that the dynasty may continue.

Bhisham Pitama refused, citing his vow of celibacy. Now Satyavati, before her marriage to Shantanu (i.e. as Matsyaghangi), had a son through Rishi Parasara. That son's original name was Krishna Dvaipayna, but who became famous as Ved Vyas (author of the *Mahabharta*). He was called to perform the rite of *niyog* on the two widowed queens. As a result two sons were born. The elder was Dhritrashtra who was born blind; the younger was Pandu who suffered from physical ailment(s). Dhristrashtra and Pandu were the 94th (solar synchronism) generation from Manu. A third son Vidur was born through a low-caste maid.

Dhritrashtra had 100 sons, the eldest of whom was Duryodhana. Pandu was incapable of fathering children, though he had two wives; the senior one being Kunti or Pritha. Help of the gods was sought and five sons were born; three to the elder queen Kunti, namely Yudhistra, Bhima and Arjuna, respectively with the help of gods — *Dharamraj* (god of Justice), *Vayu* (Wind god) and *Indra* (Ruler of gods).

As the elder brother was blind, the younger Pandu sat on the throne; but he died rather early. So, the blind Dhritrashtra ascended the throne. When it came to the next generation, a dispute naturally arose as to who should inherit the throne, i.e. Duryodhana, the eldest of Dhritrashtra, or Yudhishtra, the eldest of Pandu. Among other things, Duryodhna argued that the Pandavas were not the biological sons of Pandu; hence, they could lay no claim to the Chandravanshi throne. That is what led to the great *Mahabharta* war between the 100 sons of Dhritrashtra, called Kauravas on one side, and the five sons of Pandu, called Pandavas on the other, along with their armies and allies.

Now Kunti had not been inactive before her marriage to Pandu. She had produced a son before marriage with the help of Surya, the Sun god. His name was Karan who was abandoned at birth. He grew up to be the greatest fighter of his time, perhaps, next only to Bhisham Pitama. Tradition says that the great Parshurama himself was his guru. Karan became a friend and a close ally of Duryodhana, the eldest Kaurava. Karan was a donor without parallel. Though Karan was a half-brother of the Pandavas, he fought in the *Mahabharta* war on the side of the Kauravas. Arjun was an arch enemy of Karan.

There were some of the greatest fighters on the side of the Kauravas. These included Bhisham Pitama, guru Dronacharya (a Brahmin), Karan, Duryodhana himself and many others. Bhisham Pitama was the greatest fighter of his times, perhaps of all times. Though sympathies of Bhisham Pitama lay with the Pandavas, he had to fight on the side of the Kauravas. On the Pandava side were mainly the two brothers Bhima and Arjuna. But, the Pandavas had the peerless Lord Krishna, who however, did not carry arms. He planned every move in the war, to ensure victory for the Pandavas.

On the very first day of war, Arjuna saw his kith, elders, guru and others lined on the opposite side of the battle line; he lost the will to fight. It was at that moment that Lord Krishna, the eight *avatar* of Vishnu, gave his discourse of the *Bhagwad Gita*, which we shall cover later. In short, the great Lord reminded Arjuna of his duty as a Kshatriya to engage in fight, without worrying about the result.

The *Mahabharta* war lasted 18 days. By the clever plans of Lord Krishna, all the 'undefeatable' warriors on the side of the Kauravas could be defeated and killed. The great Bhisham Pitama suggested his own way of destruction. The Pandavas emerged victorious; Yudhishtra sat on the throne.

Dhritrashtra and his wife Gandhari deeply mourned the death of their 100 sons; Dhritrashtra exchanged harsh words with Bhima. Ghandhari held Krishna responsible for the carnage and cursed him. Dhritrashtra, Gandhari, Kunti and some ministers left for the hermitage in the forests, where they all perished in a fire, after about two years. Deep sorrow and remorse seized the Pandavas. Yudhistra abdicated soon thereafter, and along with his four brothers and Draupadi, set upon a journey in the Himalayas, towards Indra's heaven. A dog followed them. The four Pandava brothers and Draupadi fell one by one en-route, due to their problems of ego and vanity. Yudhistra along with the dog was the only one to reach the gates of Heaven. The dog was Dharamraj, the biological father of Yudhistra. In heaven, Yudhistra was surprised to find his four brothers and Draupadi missing. Most surprising was the absence of Arjuna, a great warrior and a son of Indra, the ruler of heaven.

We give below some events of interest from the *Mahabharta*:

Kshatriyas in the *Mahabharta*

The Kauravas and Pandavas are generally known as Kshatriyas of the highest order. However, in this connection, we need to note the following points:

- We have brought out in our above narrative, that King Bharata did not have any heir left. He adopted rishi Bhardwaj as his son. After the death of Bharata, Vitatha, a son of Bhardwaj, ascended the throne. This tale suggests that at that stage, Brahmin blood was inducted in the Paurva line, and replaced the Kshatriya blood. Descendents of Bharata have been called Kshatriya-Brahmins.

- Bharata himself was grandson of *dev-rishi* Vishvamitra, being the son of Vishvamitra's daughter Shakuntla. Thus, he was half Brahmin.

- Dhritrashtra and Pandu were biological sons of Ved Vyas who was born out of the union of Rishi Parsara and Matsyaghangi, a low caste fisherwoman. Parsara was a blue-blooded Brahmin, being the grandson of *dev-rishi* Vasishta. Thus, Ved Vyas had no Kshatriya blood in his veins, but a mixture of Brahmin and low caste blood. His progeny could not be Kshatriyas.

- Matsyaghangi, a low caste fisherwoman, was the grandmother of Dhritrashtra and Pandu in two ways:
  — Biologically, through her son Ved Vyas.
  — Nominally as Queen Satyavati through her two royal sons, who died childless.

- All the five Pandava brothers were biological sons of gods who may not have any specific caste.

From the above, we conclude as follows:

— Pandavas had almost nil Kshatriya blood.
— Kauravas may have had some limited Kshatriya blood.

In patriarchal society, it is the father's bloodline that counts. If that yardstick is adopted, the amount of Kshatriya blood in the Pandavas and Kauravas veins would reduce further.

Now, as a result of the *Mahabharta* war:

— The Kaurva clan was completely destroyed.
— Some time later, all male Dwarka Yadavas were killed in an internecine drunken brawl.

The surviving victorious Pandavas relinquished their rule within a few years and started their journey towards heaven. Arjuna's grandson Parikshit ascended the throne; Parikshit's son was Janmejaya. One version says the dynasty was finished after Janmejaya. However, other versions talk of continuance of the dynasty for many generations.

A view has been expressed that after the *Mahabharta* war, the Kshatriyas may have been severely weakened. There is a (conspiracy) theory that says that the *Mahabharta* war was fought to finish off the Kshatriyas. From the story of Parshurama, we know that Lords Vishnu and Shiva were not favorably inclined towards the Kshatriyas. In the *Mahabharta*, Ved Vyas tells Yudhistra as follows — "You will become throughout your living days, the sole cause of destruction of the entire species of the Kshatriyas." (*Mahabharta* II, 46,12). However, we may note at this stage that apart from the Paurvas, there were many other dynasties like Ishvakus, Kasi, Magadh, etc, who might have continued to rule, in some form or the other. We have very little information.

We give below some incidents/anecdotes of interest from the *Mahabharta*:

Arjuna in Heaven

During the exile days of the Pandavas, Lord Krishna asked Arjuna to go to heaven to get divya-astras (heavenly weapons) from Indra (Arjuna's father), for their forthcoming fight against the Kauravas. In heaven, Arjuna met the heavenly nymph Urvashi. She was besotted with Arjuna and asked him to bed her. Arjuna refused, saying that she was like a mother to him, as she was bedding his father Indra. Urvashi replied that that type of human logic was not applicable to the heavenly apsaras (nymphs) who were created for the amusement of the gods. Arjuna was not impressed. Urvashi then cursed him and made Arjuna an eunuch. On the intervention of Indra, the period of eunuch-hood was reduced to one year. That came to the help of Arjuna, when the Pandavas had to spend the last year of their exile incognito.

Karan, the Donor

Karan, being the son of the Sun god Surya, was born with integral protective body-armor and earrings, called '*kavach and kundals*'; that made Karan impregnable. Lord Krishna sent Indra in the guise of a Brahmin to ask Karan for the donation of his 'protective elements'. Now, Karan was the greatest donor of his time; he readily parted with those elements, which were a part of his body. When Indra was recognized, he gave Karan his weapon Vajra or Indra-*Shakti*, whose deadly attack was always fatal. The weapon was only for a one time use by Karan; so, Karan kept that weapon reserved for Arjuna, his arch opponent. However, Lord Krishna drew such plans that Karan was forced to use Indra's weapon against Gatochkatch, son of Bhima through a female *rakshasin*. Thus, Arjuna was not harmed.

Duryodhana — Body Toughening

Before his final battle with Bhima, the eldest Kaurva Duryodhana went to his mother to get his body toughened by a glance from her, who had tied a band over her eyes throughout her married life. Duryodhana was to go 'in the buff'. But Lord Krishna asked him as to how he could appear in that condition in front of his own mother. Therefore, Duryodhana wore underwear. The whole body of Duryodhana, with the exception of the part covered by the underwear, was toughened. During the fight, Lord Krishna asked Bhima to hit at the very part that was not toughened, though that was against the rules of war; Bhima did that and Duryodhana was fatally wounded.

Yadavas' Fight to Finish

After the war, Lord Krishna, his elder brother Balarama and the Yadava army returned to Dwarka. At some stage, there was a drunken brawl among the Yadava clan and all the male Yadavas, with the exception of Krishna were killed. Those killed included Balarama and Krishna's son Pardyumna. Only an infant son of Pardyumna, called Aniruddha, survived. Soon thereafter, the great Lord himself died by means of a hunter's arrow piercing his heel. That was a sorry end to a great period. The moment of death of Lord Krishna is considered the end of the Dvapar Yuga and start of the present *Kali Yuga*. Ghandhari, wife of Dhritrashtra had held Krishna constructively responsible for the death of her 100 sons and total destruction of the Kaurava clan. She had cursed Krishna that he would suffer a similar fate, i.e. wiping out of his clan and death of his son(s). The Yadava tragedy described in this paragraph may be the result of that curse.

Arjuna's Ego

Following the destruction of the Yadava clan (described in the preceding paragraph), only women-folk were left in Dwarka. Lord Krishna asked Arjuna to escort the women and take them to the Pandava kingdom. Having won the *Mahabharta* war, Arjuna was on ego trip. Instead of coming with an army, he came alone. On their return journey, tribal people (perhaps Bheels) waylaid the caravan of the Yadava women, in spite of being escorted by the great Arjuna himself.

Shudras in the *Mahabharta* Age

Once four murderers, one each a Brahmin, a Kshatriya, a Vaishya and a Shudra were brought up for trial in the court of Dhritrashtra. At that time, Duryodhana and Yudhistra were young princes and aspirants to the throne. They were asked to give their opinion on the sentences that could be awarded. Duryodhana opined that as each had committed a murder, all four were to be awarded death sentences. In his turn, Yudhistra expressed as follows (not an exact quote):

'The Brahmin cannot be given death sentence as that would amount to *brahm-hathya* (killing of a Brahmin), a crime more heinous than the original one. The Shudra cannot be sentenced, as he is so stupid that he does not understand the result of his actions. The Kshatriya and Vaishya could be given death sentences; the Kshatriya was to be treated more harshly, as it was his duty to defend.'

Lord Krishna

Vasudev Krishna, son of Vasu*deva*, belonged to the Yadava clan founded by King Yadu, in the Chandravanshi clan of Ila, daughter of Manu. He was the eighth *avatar* of Vishnu. He is the author of the celestial song *Bhagwad Gita,* which he delivered on the battlefield of Kurukshetra. The *Gita* contains the most original and sublime thoughts that humankind may have ever known.

Krishna had a central role to play in the *Mahabharta*. In that war, Krishna was on the side of the Pandavas, though he did not bear any arms; he chose to drive Arjuna's chariot. His army fought on the side of the Kauravas.

Krishna is the most celebrated Hindu hero. He has been called the complete *avatar* of Vishnu. He is sometimes equated with the Supreme Spirit Himself, as he declares in the *Gita* that all things emanate from Him, and ultimately rest in Him. As a child he showed the whole cosmos in his open mouth, to his mother. The Hari-Vansha and the Bhagvata Purana expand further the divine concept of Lord Krishna. He is famous for carrying his *sudarshan-chakra* (divine revolving disc), with which he punishes the wicked and the unworthy. The *Mahabharta* and Hari-Vansha are full of exploits of Krishna, which indeed were innumerable.

Lord Rama being in the Treta Yuga was full of modesty, virtue and truth. Lord Krishna appeared towards the end of the Dvapar Yuga, when evil forces were prevalent. Whilst full of all virtue, Lord Krishna did deviate here and there for a good cause, and for the higher aim of ensuring victory of 'good' over 'evil'; in fact, Krishna lists five such 'good' causes.

Krishna also had a playful side. He would have open dalliance with gopies (milkmaids); he would pick-up their clothes when they were taking their bath in the stream and then dare them to come out to reclaim the clothes. Tradition has it that Krishna had 16,108 wives, divided into two groups of 16100 + 8 normal wives. There are two versions of the 16,100 wives, viz —

- Krishna defeated and killed demon king Naraka (aka Bhunasura). The king had 16,100 maidens in captivity with a view to forced marriage. Krishna married these maidens so that they were not stranded.
- Once Brahma, the Creator happened to create 16,100 daughters in the age of the *Ramayana*. The daughters insisted on marrying Rama, who, however, expressed that in that life, he was committed only to Sita. Rama added that he would return later

in another incarnation, and could then marry them. That is how Krishna came to marry 16,100 wives.

- Tradition has it that once Rishi Narada came to check how Krishna could be staying with 16,108 wives. He found Krishna staying with each of the wives separately.

Draupadi's Vastra-Haran (Disrobing of Draupadi)

The disrobing of Draupadi is possibly the most shameful episode in Hindu history cum mythology. Every tenet of 'dharma' was made to stand on its head in this episode — false or a delusionary sense of dharma prevailed.

Once in a game of dice, Yudhistra lost everything, including his four brothers and their common wife Draupadi, to Duryodhana, the Kaurava. Yudhistra did not consider it necessary to take the consent of the five (4+1), before putting them on the table, as if they were mere property or slaves. Duryodhana asked his brother to bring and disrobe Draupadi in full view of the whole Court, which included the patriarch Bhisham Pitama.

Whilst Draupadi was being disrobed and otherwise insulted (Karan called her a prostitute), the great Pandava warrior brothers Bhima and Arjuna stood as mute spectators. This has been justified on the specious ground that they had been traded off into slavery by their elder brother, and it was their dharma to act as slaves. Nothing in Hindu scriptures gives any such right even to a father, leave alone a brother. Neither can any husband barter off his wife, especially when he has only an informal partnership of 1/5th. Draupadi was legally married only to Arjuna; the rest was an informal arrangement. In any case, the wife is not property; neither then not now. Among the marriage vows is a vow to defend the wife under all circumstances.

By not acting at that crucial moment, Arjuna and Bhim sullied their image for all times to come; Draupadi tells as such to Arjuna. Lord Krishna informs Arjuna that he and Bhima had failed grievously in their dharma in not stopping Duryodhana in his nefarious action. Without any shadow of doubt, 'Saving the honor of a woman' (especially of one's wife) is the greatest of all dharmas, over-riding all other dharmas. A society that does not follow this basic tenet cannot call itself civilized or honorable. Nothing can justify the non-action by Bhima and Arjuna; traded or not, it was incumbent on them to protect the honor of their wife.

The story gets even worse. The great and peerless Bhisham Pitama was present in the court. Now, as per Hindu mythology cum history, Bhisham Pitama is:

- The greatest warrior ever of all times. He even fought Parshurama.
- The greatest defender of *dharma*.

However, the great Pitama sat tongue-tied and did not exercise his authority, both moral and nominal, to stop that despicable act of infamy and calumny. Draupadi, most tearfully, appealed to the great Pitama again and again for his intervention, to prevent that brazen act of injustice and indecency, which would shame the whole of humanity, and not only the great Chandravanshi clan. She got no response from the Pitama. The scene has been enacted most poignantly in the TV serial 'Mahabharta'. The great Lord Krishna had to come to the rescue of Draupadi; he provided her with an unlimited length of her sari.

The episode was undertaken under the orders of Duryodhana, who was just a prince in waiting, and had no legal or formal power; at best, Duryodhana was an extra-constitutional authority. Bhisham Pitama was not bound by any loyalty whatsoever to him. His loyalty, if any, was to Dhritrashatra who was the king. Actually, the Paurva throne legally and rightfully belonged to the Pitama, who had renounced it. Dhritrashtra and Pandu, and their progeny did not even have Chandravanshi blood in their veins, and thus their claim to the throne was rather tenuous. The Pitama could have exercised that right of his for all of the five minutes, to save humanity from the naked dance of *adharma* (unrighteousness).

That was one occasion when, regardless of any other factor, the great Pitama should have unsheathed his sword, and told Duryodhana to back off, or else; that would have been true Hinduism, in all its glory and splendor. Bhisham Pitama failed to unsheathe his sword at a most crucial point of Hindu civilization, and in defense of the greatest *dharma* of mankind, i.e. defending the honor of women.

That one act of *a-karma* (non-action) of the Pitama sent all the wrong messages to future generations of Hindus, which reverberated for thousands of years. From 1000 AD onwards, with the start of Muslim invasions, it became a norm to dishonor Hindu women, by the thousands. Not one Hindu ruler unsheathed his sword and marched to Ghazni, Ghur or Iran, to teach a lesson or two to those molesters. Were these rulers drawing their inspiration from the non-action of Bhisham Pitama, in similar circumstances? The Hindu rulers perhaps argued that the Hindu masses would find some (non-existing) justification for their inactions, as they had done in the case of the Pitama.

If the Pitama had unsheathed his sword at that crucial moment, that would have been real *dharma*; the Hindu nation would have been electrified for eons to come. It would have taught the Hindus how to deal with violators and tormentors, with speed, resolve and promptitude; and not to look to the scriptures for support for their *a-karmas* (non-actions).

The great Pitama did not even walk away to show his disgust and disapproval; he kept sitting glued to his seat watching the entire scene. Guru Dronacharya, *kul-guru* (family head priest) Kripacharya and the Prime Minister, the great Vidur gave company to the Pitama. What were these worthy leaders afraid of? Their perks, or their life? Every one chose to hide under the pretext of some false or delusionary *dharma*; true Hinduism was let down. With the sole exception of the great Lord Krishna, everybody who was anybody failed in that (easy) test of *dharma*. If the Hindus care to look carefully, they may find, at least partially, reasons of their subsequent downfall, in this episode of Draupadi's *vastra-haran*.

The '*Mahabharta*' TV serial shows Draupadi, at one stage, on the point of pronouncing a 'curse' on the entire Chandravanshi clan. Gandhari, mother of Duryodhana and a woman of substance, is shown intervening to prevent Draupadi from pronouncing the curse, calling her '*putri*' (dear daughter), in the process. But she does nothing to stop her son Duryodhana in his dance of *adharma* (unrighteousness), and save her '*putri*' from utter disgrace. She could have achieved her aim in innumerable ways, e.g. a threat to disrobe (or kill) herself if Duryodhana went ahead with his nefarious plan; it would have stopped Duryodhana in his tracks. That would have been real *dharma* in all its glory. Even without being pronounced, Draupadi's curse appears to have worked; the Chandravanshi clan appears to have been wiped out not much after the *Mahabharta* war.

Two other popular works connected with the *Mahabharta* are:

Shanti Parva: It forms a part of the *Mahabharta*. After being wounded in war, the great Bhisham Pitama lay on a bed of arrows. At the end of the war, Bisham Pitama gave a lengthy discourse to Yudhistra on *Dharma* or righteous conduct, statecraft and the duties of the king and related issues. This text covers that discourse in full.

Hari Vansha: It is a supplement to the *Mahabharta* attributed to a much later period. It gives the genealogy of Hari or Vishnu in a long poem of 16,374 verses. It covers in detail the life and adventures of Krishna. The text talks of the future of the world and the oncoming dark deeds of the *Kali Yuga*.

General Comments on the Epics

It may not be an exaggeration to say that the two epics have entered the very souls of Hindus. Presently, after thousands of years, Hindus get in a trance when they listen to recitations of the *shloka*s (verses) of the epics. At any time, hundreds of sermons on the epics go on, which are attended by hundreds of thousands of Hindus. When a Hindu finds himself in a moral dilemma, his first instinct is to go to the epics for guidance and to look for a precedent; invariably, he finds something to guide him. Without the epics, Hinduism would have been* rather insipid.

It is quite apparent that the solar and lunar dynasties were throughout ruling simultaneously, except for some breaks in the lunar line. The two dynasties must have been ruling in different, but adjoining areas of Bharat varsha. As such, one would expect copious references to the *Ramayana* in the *Mahabharta*. However, no major character in the *Mahabharta* makes any reference to events and actions of the great stalwarts of the *Ramayana*. No attempt is made in *Mahabharta*, to draw any lesson, or take any guidance from the events of the earlier epic. It would appear that the *Mahabharta* people may have had only rudimentary knowledge of the *Ramayana* episode; or perhaps none at all. That sounds incredulous, keeping in view the fact that the *Ramayana* might have preceded the *Mahabharta*, just by 700 to 1000 years. Presently, after thousands of years, we are aware of every detail of the two epics.

There is a reference in Book 3 of the *Mahabharta* to a concise but complete story of the *Ramayana*; this could be a later interpolation (of which there is no dearth). Tradition says that during the *Mahabharta* days, saint Markandeya told the story of the *Ramayana* to *Yudhistra*. That would imply that until the time Yudhistra was told of the story, he was unaware of the same. Some texts record that Vibhishan (brother of Ravan) was present as an honored guest at the 'Rajasuya' *yajna* conducted by Yudhistra. This may not be taken too seriously. People who had lived thousands of years earlier have been made to be present at the Rajasuya of Yudhistra.

There is another interesting story. Once, the Pandava Bhima went on a long journey in search of the flowers of paradise for his beloved. Towards the far end of the journey, Bhima came across an old monkey who was in no position to move. He turned out to be none other than the great monkey god Hanuman of *Ramayana*; but Bhima was in no position to recognize him. Hanuman informed Bhima that he was the son of the Wind god *Vayu*, as was Bhima, and hence they were brothers. Hanuman went on to tell Bhima of the four Yugas (see chapter 7.1) and that in the Treta Yuga, Hanuman was all powerful; but, near the end of the Dvapar Yuga, Hanuman had lost all his powers.

Normally, one would have expected the dynamic dynasties of *Ramayana* and *Mahabharta* with divine backgrounds, to continue for thousands of years. That does not appear to be the case. Though names of later kings of both the dynasties are mentioned in some texts, nothing much is known about their rule and exploits.

5.8.4 The *Bhagwad Gita*

The *Bhagwad Gita* is embedded in the *Mahabharta* in the form of a dialogue between Lord Krishna and his friend and pupil Arjuna. In sublimity of thought and originality of concepts, nothing comes anywhere near the *Gita*. Hindus consider it be their holy book, though it is not the final book of adjudication, like the Bible or the Koran.

The *Gita* is a comprehensive synthesis of Upanishadic and Vedic concepts. *Gita* preaches the doctrine of yoga, full resistance to evil and the gospel of the divine. It requires man to engage in selfless action (nishkam *karma*). A Kshatriya must engage in his duty of righteous war — to kill, or be killed. If victorious, the Kshatriya would rule over this world. If killed, he would enter heaven, without let or hindrance. The *Gita* does not provide any option of surrender or retreat in the battlefield.

A few *shloka*s from Chapter 2 of the *Bhagwad Gita*, addressed by Lord Krishna to Arjuna, are given below:

1. The soul is neither born, nor does it die. Having been in existence for ever, it never ceases to be. It is eternal and ever-lasting. It is not killed when body is killed. BG.2.20

2. If you do not perform your religious duty of fighting, then you will certainly incur sin for neglecting your duty and then lose your reputation as a fighter. BG.2 33

3. *Hato va prapsyasi svargam jitva bhoksyase mahim*
   Tasmad uttistha kaunteya yuddhaya krta-niscayah BG.2.37

If you are slain in battle, you will attain heaven. If you gain victory, you will enjoy fruits of this earth. Therefore, arise, O Arjuna and get on with the fight.

4. *Karmanye Evadhikaraste, ma phalesu kadachan* BG.2 47

Your right is only to perform action, and not to its fruit.

5. Every time *dharma* is weakened, and non-*dharma* prevails
   I manifest myself O descendent of Bharata BG.4.7

To protect the righteous and punish the unrighteous
To re-establish *dharma*, I am born again and again   BG.4.8

Though the *Gita* was delivered right in the midst of the *Mahabharta* war, it has been expressed that it was written in its present form in the closing centuries of BCE.

The three most Hindu holy books in order of appearance are:

The *Rig Veda*
The Upanishads (A set of books)
The *Bhagwad Gita*.

Each of the above perhaps delivers a slightly different message:

- The *Rig Veda* (the first to appear): Its stress is on happiness in this life through ceremonial (animal) sacrifices to a galaxy of gods.
- The Upanishads: Their main emphasis is on meditation, and not on action (i.e. sacrifices). Upanishads spell out and define with clarity the concept of one universal, omnipotent and pre-existing Supreme Being.
- The *Bhagwad Gita* (the last to appear): It achieves synthesis by reconciling the need for action with the necessity of meditation. One, who merely thinks without doing anything, may not succeed. On the contrary, a thoughtless action could result in bad results.

5.9 The *Puranas*

In all, there are 18 *Maha-Puranas* and 18 *Upa-Puranas*. The figure 18 appears to have some special significance in Hinduism. There are 18 *Dharma Shastras*, 18 *parvas* (books) of the *Mahabharta* and 18 chapters of the *Bhagwad Gita*. The 18 *Maha-Puranas* contain some four lakhs *shlokas* (hymns). *Puranas* are the great storehouse of Hindu mythology and legends, and of ancient Hindu history. Without the *Puranas*, Hinduism would have served a very insipid fare.

Of the 18 Maha-Puranas, six are nominally dedicated to each of the three principal deities — Brahma, Vishnu and Shiva, viz:

- Brahma Puranas:

  Brahma, Brahmanda, Brahmavaivarta, Vamana, Markandeya and Bhavishya — The quality of *Rajas* or Passion prevails in these texts.
- Vishnu Puranas:

  Vishnu, Bhagvata, Naradiya, Garuda, Padma and Varaha — These have the quality of *Sattwa* or Purity.
- Shiva Puranas:

  Vayu (or Shiva), Linga, Skanda, Matsya, *Agni* and Kurma — Quality of *Tamas* or Gloom/Ignorance predominates in these.

  Vayu Purana is considered connected with the Shiva Purana, and generally only one of these is mentioned in any list of the *Puranas*. Of all the 18 *Puranas*, Skanda is the longest, having some 80,000 verses.

We cover below the more important *Puranas*:

Markandeya: It tells tales of lord Rama and lord Krishna, sings praises of *Surya* (Sun) and Agni. It has the legend of Raja Harishchandra. It has 1000 stanzas and was composed in around 3rd century AD.

Vishnu: It is the oldest Purana composed in 2nd century BC, having some 6,000 stanzas. It details the stories of Lord Krishna and of other royal kings who might have ruled from time to time.

Bhagvata: Perhaps the most important Purana; it has 18,000 stanzas. The Purana tells the story of Krishna, including the origin of the *Raslila*. It covers in detail the philosophy of Bhagvata *Dharma*.

Padma: It was composed in 4th century AD and has 48,000 stanzas. It stresses the importance of Vishnu worship.

Vayu (Shiva): It has some 24,000 stanzas covering various aspects of the great lord Shiva. The text is divided into four sections; the first deals with creation, and the last talks of the future ages to come.

Though a Purana may be listed under one particular deity, none of the *Puranas* is devoted exclusively to that one deity alone; most of them cover more than one deity. Vishnu and his *avatars* Ram and Krishna occupy the maximum space in the *Puranas*. Tradition says that Brahma narrated the Shiva Purana to *dev-rishi* Pulastya, who passed it on to rishi Parasara, grandson of *dev-rishi* Vashista. In turn, Parasara gave this Purana to his disciple Maitreya, for the benefit of humankind.

The word 'purana' in Hindi means old/ancient; however, the texts known as the *Puranas* were the last to be written, even after the epics. The *Puranas* cover a whole range of ancient stories, myths and legends; hence perhaps the use of the name 'purana'. An approximate version as to when the *Puranas* may have been written runs as follows:

'The bulk of the *Puranas* were written during 200–650 AD, the majority being during the Gupta period (320–520 AD). Some early *Puranas* were written during 200 BC to 200 AD, starting with the reign of the Brahmin Shunga and Kanva dynasties, which replaced the Mauryan Empire, around 180 BC. The later *Puranas* were penned during 650 to 1100 AD; may be some as late as the 13th/ 14th century AD. However, there is an alternate view that writing of the *Puranas* may have started only during the Gupta period, i.e. around 4th century AD.'

Tradition has it that in the first instance, Brahma gave the knowledge of the *Puranas* to his mind-born sons, the four Sanat Kumars (Chapter 7.3). In turn, they passed on the knowledge to sage Narada. He transmitted it to Ved Vyas, who then composed the *Puranas* (we know Ved Vyas as the author of the *Mahabharta*). The *Puranas* were composed over some 1000 years and there must have been innumerable authors. As such, to attribute authorship of the *Puranas* to one author (Ved Vyas) is a bit difficult to accept. May be there were many Ved Vyasas, as that might have been a common name those days. Alternately, the various authors gave the name of Ved Vyas to their writings, to give these ancient touch, and therefore higher respectability and acceptability.

The *Puranas* are the storehouse of ancient knowledge, consisting of a vast range of mythology and legends. The parables and fascinating stories belonged to very ancient times — mostly pre-history; however, these came to be recorded much later. *Puranas* are an artful mix of history and mythology. Even the most prescient minds cannot tell where history ends and mythology begins, and vice versa. We, the (pseudo) modernists can only be stuck with awe at the wide range over which the thought processes of those *rishis* and sages spread. We can only feel humbled by the enormity of knowledge, some of which may be a bit farcical and unacceptable to the modern mind.

The *Puranas* are in the form of dialogues between the exponent and an enquirer. The *Puranas* generally cover the following five subjects:

- *Sarga* — Creation of the Universe
- *Pratisarga* — Re-creation of the Universe, after its destruction
- *Vamsa* — Genealogies of gods and sages, both real and mythological
- *Manvantra*—The Manu period; reign of 14 Manus, mythological progenitors of humanity
- *Vamsanucanta* — History of the Solar and Lunar dynasties

Later, the following types of subjects were also covered in the *Puranas*:

- Stages of life and duties of various classes
- Activities of gods and Vishnu *avatar*s
- Laws of *Karma* and practice of Yoga
- Realization of the Supreme Being
- Pilgrimages to holy places
- Knowledge relating to all aspects of the soul
- Constructions of temples and worship of idols

The *Trimurti* (the Hindu Triad) concept of the three gods Brahma, Vishnu and Shiva came in focus and reached its finalization in the *Puranas*. These also give the first clear description of the cult of idols, including their methods of manufacture, installation and worship.

The function of the *Puranas* is to personalize and make more understandable to the ordinary folks, the rather abstract truths and higher teachings of the Vedas and the Upanishads. To achieve that aim, the *Puranas* have deviated substantially from the Vedas and the Upanishads. The *Puranas* stress God's power and grace; these emphasise on God's merciful and compassionate approach. The Puranic teachings form the main base of present-day Hinduism in Bharat; hence, we can call it the Puranic Hinduism. It is substantially different from the Vedic *dharma*. Salient differences are:

- The Vedic *dharma* was largely without idols and temples. The Puranic religion revolves around temples and idols.
- Vedic *dharma* revolved around *yajnas* (fire-sacrifice) and animal sacrifice, conducted at altars in open spaces. The Puranic religion offers worship to deities with flowers and fruits, in the confines of temples; there is no longer any room for animal sacrifice.
- The Puranic gods are almost entirely different from the Vedic gods. Indra, *Agni* and *Surya* were the major gods in the Vedic times. Presently, Shiva and Vishnu, with the two incarnations of Vishnu, Rama and Krishna, occupy the central space.
- Ganesha and Hanuman were nowhere on the horizon during the Vedic times; they emerged later and are dominant gods now.
- There was no concept of *avatars* (incarnation) in the Vedic times. In the Puranic Hinduism, Vishnu's *avatars* receive major worship.
- The Vedic people did not have the concept of transmigration of soul; Puranic people have a rigid belief in this.
- The message of the Hindu holy book, the *Bhagwad Gita* was delivered much after the Vedas; it dominates the present Puranic age.

The *Puranas* may not be for the high-end scholars and those intellectually evolved. These try to convey wisdom of the ancient Hindu scriptures, including the Vedas and the Upanishads, in simple language for the benefit of the layman. It is difficult to explain a complex religion like Hinduism in simple language; and that may have involved many compromises with the original sublime theories of Hinduism. For this reason, the *Puranas* have come in for criticism by the intellectuals, for presenting a somewhat distorted view of Hinduism. The great Rishi Swamy Dayanand of the Arya Samaj refused to associate himself with the *Puranas*.

5.10 The Kalpa Sutras

Literally translated, 'Sutra' means 'threads'. The Kalpa Sutras are strings of Rules of Righteousness. These are a part of the human tradition which is possibly sourced from *Sruti*. Sutras consist of instructions in prose on human behavior, morals and ethics. The Sutras comprise of the following:

1. *Srautra Sutras:* The public ceremonies described in these and are considered to be based on *Sruti*. The rituals described present a rationalized system of symbols for the relations governing the cosmos.

2. *Griha Sutras:* These detail the domestic day-to-day religious ceremonies of the householder. The texts also include a large number of rules for religious life, i.e. birth, marriage, funeral rites, conception, thread ceremony, etc. There are 16 Griha Sutras, which are believed to have been composed during the 6th century BC to 3rd century AD.

3. *Dharam Sutras:* The aim of Dharam Sutras is to teach *dharma* (righteousness). These also lay down the rules of conduct in spiritual matters, and give rules of marriage, regulations for the life of a student, and other connected issues.

5.11 Dharam Shastras

Dharam Shastras are the 'Law Codes' which include almost the entire body of Hindu law; these lay down rules applicable for all of society. Dharam Shastras are the texts on righteousness containing instructions in verse on Hindu religion and morals; these are the books on religious laws which amplify the Vedas. Dharam Shastras are generally divided into three parts — Rules of Conduct and Practice, Judicial Matters and Penance.

In all, there are 18 Dharam Shastras, the more important of which are as follows:

- Manusmriti or Mannav Dharam Shastra (Chapter 8.4)
- Yajnavalkeya and Parasara's Shastras on Social and Legal matters

- Vatsayana's Kamasutra: For human pleasure and enjoyment, including the sexual
- Kautilya's *Arthshastra* (Chapter 17.2)

The Shastras authored by Manu and Yajnavalkeya are considered most important. Manusmriti is a very popular text (chapter 8.4). Tradition has it that it contains the *dharma* declared by Brahma to Manu, the first man; Manu passed it on to Rishi Bhrigu of the ten great sages. Though Manusmriti is better known amongst the common people, it is Yajnavalkeya's Dharam Shastra (Law Code) which forms the primary basis of the present Hindu law in Bharat.

5.12 The Tantras

Tantras are the treatise relating to a variety of mystifying and occult subjects. Tantras are post-Puranas (bulk of those), written most probably during 7th to 14th centuries AD. Tantras are considered an amalgam of Hindu and Buddhist thoughts. These may not be an integral part of the Hindu/Vedic scriptures; these perhaps run alongside. One view is that the tantras are for the *Kali Yuga*, when Vedic rules may not apply. Tantras may also have some links with the Atharva Veda.

Tantras are the texts, which are based on the female energy of the god, called *Shakti*, *Devi* or mother goddess (Chapter 6.11). In the concept of *Shakti*, god is passive; his wife *Devi* is active. Tantras are generally in the form of dialogue between Lord Shiva and his wife Parvati. The texts are called Agamas, where Shiva is the teacher. When Parvati becomes the teacher, the texts are called Nagamas. There are also Vaishnava Agamas.

There are 64 Tantras relating to the three deities Shiva, Vishnu and *Shakti* or *Devi*. It is believed that Shiva revealed the 28 Saiva Tantras to his devotees. The texts relating to *Shakti* or *Devi* are called Shakta Tantras; these represent the fertility and creation aspects of Brahma and are considered the most important of the Tantras. However, the Shakta Tantras have been highly abused. Many of the undesirable tantric practices presently reported from time to time, could be attributed to the Shakta Tantras. Of the 64 Tantras, a few are Mahanirvana, Kulsara and Kularnava.

There is the concept of 64 *yoginis*, mostly prevalent in Orissa; it was the expression of an extreme form of Atticism, starting around the 8th century AD. The *yoginis* are the mother goddesses; their origin is shrouded in mystery The Markandeya Purana states that gods created the *yoginis* to assist Durga in killing the demon Raktabija. Some of the more important *yoginis* are Brahmni, Vaishnavi, Varabi, Indrani, Chamunda and Maheshwari.

Believers in tantric philosophy say that it is a religious system, which can help mankind in spiritual growth. Tantras deal with the spiritual knowledge required for worship of the deity. These have rules for temple construction and sculpting images, and for conduct of festivals. The Tantras give special prominence to the female energy, the *Shakti* of Shiva; she is the energy concerned with magical powers, and sexual relationships. It has been expressed that there are five requisites for tantric worship, the five makaras, or the five Ms: *Mansa* (flesh), *Matsya* (fish), *Madhya* (wine), *Mudra* (parched grain or mystical gesticulations) and *Maithuna* (sexual intercourse).

The Saktas (Shaktas) or worshippers of *Shakti*, are divided into two categories:

- *Dakshinacharya* — the right handed
- *Vamacharya* — the left handed

The worship of Vamacharis is for the fierce (or black) form of *Shakti* (Durga, Kali), and is considered licentious. The female principle is worshipped both symbolically and in actual woman; promiscuous intercourse may form part of the orgies. The five Ms listed above relate to the Vamacharis. The Dakshinacharis worship the gentle (or white) form of *Shakti* (Uma, Parvati).

5.13 The Darshanas

Hinduism has six schools of philosophy; these cover some of the most complex and fascinating subjects that humankind might have seen. The philosophies are in search of linkages between God, Truth, Soul, Nature and Matter. Such complex issues are generally well outside the concern and understanding of the *aam admi* (common man). Therefore, we will just list the philosophies, without trying to go into any details.

There is one Darshana for each school of philosophy. The word 'Darshana' means 'vision' or 'to see the self'. Darshanas try to answer mystical questions relating to mysteries of the Universe, e.g. who created the Universe? What is the nature of the Soul? And innumerable such questions. The Hindu sages and *rishis* (seers) tried to answer these questions during 6th to 4th centuries BC. The result was the six Darshanas, which could be divided in three groups, viz.:

First Group: These elaborate on the system of logical reasoning and structure of matter.

1. *Nyaya (Sage Gautama):* It is called the Logical school, which uses logic and analysis to arrive at a conclusion. Gautama expressed that the world was constituted of five elements — earth, sky, fire, air and ether.
2. *Vaishesika (Sage Kanada):* This philosophy is classed along with the Nyaya School and supplements it. It has been called the Atomic school, i.e. the world is composed of eternal atoms. The philosophy says that the world had originated from Actions/Deeds.

Second Group: This group covers various aspects of discrimination between Spirit and Matter.

3. *Yoga (Sage Patanjail):* The philosophy lays down the means by which the soul may finally merge with the Supreme Spirit Brahman,. That can be achieved through meditation. The philosophy talks both of the individual soul and the Supreme Soul.
4. *Samkhya (Sage Kapila):* This atheist philosophy has no need of a god. Kapila also expressed the dualistic theory of Purusha (Soul) and Prakriti (Matter). Its text book is called Sankhya-Karika.

Third Group: This deals with the essence of the Vedas, including the Upanishads. The two could be grouped under the heading 'Vedanta'.

5. *Mimansa or Purva (prior) Mimansa (Sage Jamini):* It covers the philosophy contained in *Karma Kand* i.e, Veda (*Samhita*) plus Brahmanas. The text concerns with the correct interpretation of the speculative and practical aspects of the Vedas.

6. *Vedanta or Uttara (later) Mimansa (Sage Vyasa):* It covers the philosophy of Jnana Kanda, i.e. Upanishads. It expresses that the omniscient and omnipotent Brahman, the Supreme Spirit is the cause of existence, continuance and finally the destruction of the universe. The all pervading, unmanifest Supreme Spirit exists in manifest creation.

The following clarifications are offered:

- The texts at 1 to 4 are not related to the Vedas, though that does not mean these are anti-Vedas.
- Sankhya is an atheist philosophy. Another atheist (Hindu) philosophy was Charvaka (Chapter 5.6).
- Darshanas are perhaps the oldest texts under *Smriti*, having been evolved during the 6th to 4th century BC.

All the six schools of philosophy have the central aim of emancipation of the soul from the wheel of *samsara* — the cycle of birth and death. These also recommend the merger of the individual soul with the Supreme Soul.

5.14 Miscellaneous Literature

In addition to the texts listed above, there is a vast range of other Hindu religious literature. We cannot possibly cover all of that in this book. Of those, we list (without any details) below some of the more important ones:

Six Vedangas

Vedangas means the limbs of the Vedas. These tell us how to read the Vedas correctly, understand their texts, and cover their proper application to sacrificial rituals. The six Vedangas are:

- Siksha (by Panini): Science of phonetics and pronunciation
- Vyakarna (by Panini): Grammar
- Nirkuta: Linguistics and Etymology
- Chandas Shastras: Prosody — rhythm and sound in poetry
- Kalpa Shastra: Ceremonies and Rituals
- Jyotisha: Astronomy and Astrology

Four Up-Vedas:

- Ayur Up-Veda: Dealing with medicine, health and human well-being
- Dhanur: Science of weapons and archery
- Ghandharva: Music and dance
- *Arthshastra*: Polity, economics and state administration

Panchatantras

These are collection of fables and stories mainly having animals and birds as their characters. These texts have been translated into innumerable languages all over the world.

Jatakas

These are Buddhist folk-tales relating to the former incarnations of Budha, when he lived as a Bodhisattva. It is believed that as a Bodhisattva, Budha had innumerable earlier lives. He was born as a gardener, archer, Brahmin, sweeper, etc. He also took the form of animals like lions, lizard.

6
THE HINDU GODHOOD

6.1 Background

The Hindu godhood is an extremely complex phenomenon; there is almost no limit to the variety and range of Hindu gods. At one end is the concept of the Supreme Spirit Brahman. On the other extreme, are gods and goddesses of various hues, perhaps in hundreds, perhaps in thousands; no one is willing to commit to exact numbers. It requires a prescient mind of a sage to understand the underlying concept of the Hindu godhood; an average person can just about skim the surface.

The Hindu layman rushes from one temple to another, worshipping one god here and another there. Sometimes, in hallowed precincts (which presently include TV studios), he is informed that the multifarious gods are but a manifestation of the One Supreme God. He surely hears this message, though we are not sure that he understands it, or is willing to adopt it in his day-to-day life. For an average Hindu, Rama or Krishna, Hanuman or Ganesha, are gods in their own right, who would listen to and answer his prayers. It is seldom that an average Hindu appeals directly to the One Supreme God.

In the ensuing pages, an attempt has been made to present a simplistic view of Hindu godhood, to which hopefully, an average layman can relate. We would study the Hindu gods under the following headings:

- Pre-Vedic gods — Pre 2000 BC
- *Rig Veda* gods — 1500–1200 BC
- The Upanishadic God — 800–600 BC
- Buddhist 'no-God' — 500 BC

- New Polytheistic Gods — Indeterminate Date (Gods of Epics & Puranas)

6.2 Pre-Vedic Gods

As per the orthodox Hindu theory of the four Yugas, the universe and the Hindu religion started millions of years ago, with the Kritya Yuga (Chapter 7.1). Without taking a specific position on that, we start our study with recorded history. As per this version, there was initially the Indus Valley Civilization in the North-West of Bharat (Harrapa and Mohenjodaro). That civilization possibly flourished from 3000 to 1800 BC (approximately), when it collapsed. That was near about the (presumed) arrival of the Aryans in Bharat around 2000–1500 BC. Nothing much is known about the religion and gods of the people of the Indus Valley. We are not very sure whether they were Hindus, or followed some pagan religion and practices, as was the wont in other parts of the world at that time.

Figurines of Rudra or Pashupati type gods have been found on the seals found in the ruins of Mohenjodaro. Some seals show a figure seated in a yogic posture, sometimes with feet crossed; a buffalo, a tiger, and an elephant are shown nearby, perhaps in attendance. Seals with a god like figure holding a trident accompanied by a bull, have also been found; that suggests a Rudra type god of the *Rig Veda*, who was later identified as Shiva. Figurines depicting the mother goddess were also found in the ruins. There are also depicted some altar type spaces; it is presumed that some sort of worship in the form of *yajna* (fire-sacrifice) might have been carried out at these places. All this may suggest manifestation of an earlier type of Hinduism, though we cannot be sure.

The orthodox Hindu view is that the Indus Valley Civilization was a part of the ongoing Hindu civilization which may have started around 4000–3000 BC.

6.3 *Rig Veda* Gods

As per the generally accepted view of history, the *Rig Veda* evolved during 1500 to 1200 BC. The *Rig Veda* was the first to introduce the concept of (Hindu) god.

The *Rig Veda* divided the universe into three broad spheres, each with its own god as its presiding deity, viz:

1.	The Celestial Sphere	Surya or Savitre (Sun)
2.	The Atmosphere	*Vayu* (Wind)
3.	The Terrestrial Sphere	*Agni* (Fire)

The *Rig Veda* refers to 33 gods; hymn 8.30.2 says as follows:

"Therefore, you are worthy of praise and of sacrifice, you thirty-three gods of Manu, arrogant and powerful."

Sometimes, the 33 gods are expressed as 3x11 gods, i.e. 11 gods in each of the three spheres. We give below names of the more important of the *Rig* gods:

1.	Celestial gods	Surya (the Sun) Varuna (Sky/Ocean) Usha (Dawn) Ashwins
2.	Atmospheric gods	*Vayu* (Wind god) Indra (god of Rain/Thunder) Rudra (Storm god) Maruts (Wind Spirits)
3.	Terrestrial gods	*Agni* (Fire) Prithvi (Earth) Soma (Intoxicating Drink)

The above classification may not be taken too rigidly. The functions, role and even names of some of the gods changed from time to time, and from text to text, viz:

- Varuna, the god of sky later became the god of ocean; sometimes, he represents the mysterious creation.
- Soma was an intoxicating drink. Later, the name came to be identified with Chandra, the moon, which is also called Soma. The drink soma may have been stored at Chandra; hence the

name association. There are references to Indra in the form of a falcon (or riding a falcon), getting soma from Chandra.

- Indra is the ruler of gods in the heavens. He has been described as the god of rain, thunder and storm; sometimes even cloud. He is the god of war and the commander of heaven's army. Sometimes, *Vayu* is clubbed with Indra.
- The god *Agni* (Fire) is sometimes identified with Shiva.
- Surya, *Agni* and *Vayu*, as the three important first gods, have been called the 'first Triad' of Hindu gods of the Vedic period. Sometimes, Indra replaces *Vayu*.
- It appears that at the very beginning, Varuna may have been the chief deity; perhaps, he received the maximum worship in the very early times.

Surya, the Sun god

Surya the Sun, was one of the more important gods, if not the most important (at one time). Initially, *Surya* had 12 names: Svitre, Vivaswat, Martanda (literally dead-egg), Pusan, Mitra, Aryaman, Dhatr, Daksha (otherwise a Prajapati), Vajur, Asvina, Bhaga, and Amsa. The 12 names listed are the 12 *adityas*, sons of mother Aditi and father *dev-rishi* Kashyap (see chapter 7.3). Vivaswat was perhaps the most important *aditya*; he was the father of Manu, the progenitor of the human race. The figure 12 may have something to do with the 12 months of the year, indicating different phases of the sun. Aditya (in singular) is a name of the sun. Some other names of *Surya* are: Prabhakara, Divakar and Bhanu.

Sometimes, Varuna is identified with Surya. Varuna is generally grouped with Mitra, and the two are often mentioned together as Varuna-Mitra. In some texts, the Vedic Vishnu is identified as an aditya. Svitre is a very important aditya; it is associated with the invisible or nighttime sun, perhaps the setting sun. A view has been expressed that the famous Gayatri Mantra is addressed to Svitre. Indra also gets mentioned as an aditya.

Agni (Fire)

It would appear that at the very beginning Varuna was the main deity; he received maximum worship. Surya, the sun was another important god. However, later, the god *Agni* (Fire) may have emerged at the top; that appeared natural for the following reasons:

- The Vedic people offered worship through *agni* (fire) in a havan or *yajna* (fire sacrifice or burnt offering). It was the god Agni, who accepted the offerings of *ghee* (clarified butter) and other ingredients, on behalf of the gods and conveyed these to the gods. Therefore, *Agni* is often called the 'oblation-eater'.
- It was Agni, who conveyed the dead bodies to their intermediate or final resting place.
- *Agni* helped clear forests, as other tools were in their rudimentary stage in the Vedic period.
- *Agni* cooked food, which was a necessity for life.
- *Surya* (Sun) and lightening in the skies were also a form of Agni.

Indra

Indra is also known as Sakra and Vasava. The later name is derived from Vasus — attendants of Indra; they were eight in number. Vasus were cursed by a rishi to be born on earth. They were born to Ganga (otherwise a river), who in her human form was the wife of the Paurava king Shantanu. In order to shorten the stay of the Vasus on earth, Ganga drowned the first seven in the river Ganga. The eighth Vasu was the great Bhisham Pitama of the *Mahabharta*. Indra has also been called the 'lord of a hundred sacrifices', 'having a thousand eyes' and 'destroyer/shatterrer of cities'. He has many other names.

Initially, the Aryans were wandering tribes, who lived primarily by hunting animals. At that early stage, their main deities must have been Varuna, *Surya*, *Vayu*, and *Agni*. At some stage, the Aryans settled down and became agriculturists; rain and water became very important to them. Hence, the role of the main deity got switched from *Agni* to Indra, who was the god of rain, thunder and storm. Indra is also the god of war and commander of the army of the gods (heaven's army); later on, that job was given to Skanda/Karttikeya. Indra is a sun god. Sometimes, he has been called a god of creation and fertility.

Indra's Parentage

There are various versions of the parentage of Indra:

- Indra is projected as son of Dyaus Pitra (Sky god) and Prithvi, the mother earth. Zeus (Greek) and Jupiter (Latin) are the equivalents of Dyaus Pitra (Aryan), which also means the Father god, or Father Heaven.
- Indra was born as an aditya (a sun god) to mother Aditi and father *dev-rishi* Kayshap (see chapter 7.3). Aditi could be another name for mother earth.
- Some texts refer to Prajapati (chapter 6.12), sometimes called Ka (Who?), as the father of Indra.
- The most commonly accepted father of Indra is Tvastr (Vishvakarma), who is sometimes referred as a god, other times as a prajapati. Tvastr is the architect of the gods. We will take this as standard version for our text.
- In addition to Aditi and Prithvi, names of Indra's mother are variously given as Nistgiri, Savasi, or Ekastaka, daughter of Prajapati; in this case, Prajapati would be the grandfather of Indra.

It has been expressed that Indra may have been born as a fully-grown man from the side of his mother. Tradition says that the mother of Indra kept him hidden in her womb from the wrath of his father who wanted to kill him; a figure of one thousand months has been mentioned; Verse 4 of RV hymn 4.18 states:

'Why has she pushed him out, whom she carried for a thousand months and many many autumns? For there is no other one who is his equal among those who are born, and those who will be born in future.'

Indra is the king of gods in heaven; he is the ruler of the kingdom of heaven. Indra's court is full of *apsaras* (heavenly nymphs), who are primarily for entertainment of the gods. Sometimes, Indra uses *apsaras* to disturb *tapasya* (asceticism) of *rishis* (sages). Indra fears that if the *rishis* are successful in their *tapasya*, they may displace him from his throne. Indra entertains gods with ambrosia and heavenly music. Indra's wife is

Indrani, Sachi or Prasana. Indra had some 5–6 sons, including Arjuna of *Mahabharta*.

Indra & Vrtra

It would appear that Tvastr the father, for some unknown reason, wanted to kill his own son Indra, even before his birth. As Tvastr could not kill Indra before his birth or in infancy, he created a three-headed son called Tri-sira, a very powerful person. However, Indra succeeded in killing Tri-sira. That further angered Tvastr. Therefore, he created, through a demoness, a powerful and awesome demon called Vrtra. Sometimes, Vrtra has been called a dragon. Being the son of Tvastr, Vrtra was half Brahmin. Tvastr told Vrtra, "Kill Indra". The story of the demon Vrtra is very famous, and appears in many texts, including the *Rig Veda*.

Vrtra has been considered to be the demon of draught. Tradition says that he took away rain, water, and may be even flowing rivers, with him. As such, if civilization was to survive and prosper, it became essential to kill Vrtra. Indra made many attempts to kill Vrtra, but was not successful. During one of those conflicts, Vrtra swallowed Indra; he managed to come out when Vrtra yawned.

Thereafter, Indra sought the help of (Vedic) Vishnu, who agreed to help. Vishnu entered Indra's thunderbolt in order to kill Vrtra. Thus, it was with great effort, and only with the help of Vishnu that Indra could kill Vrtra. Tradition says that Indra struck his thunderbolt at the back of Vrtra; that was against the rules of engagement. The killing of Vrtra is considered a major event of mythology and a great victory of Indra; after that, Indra came to be called Mahendra — the great Indra. After the death of Vrtra, huge showers of water and rain were released; rivers started flowing again, and (Aryan) civilization prospered. There is a version that says that humankind may have started only after water had been released from the clutches of Vrtra. As Vrtra was half-Brahmin, Indra had to suffer for long for the sin of Brahmincide. Indra is also credited with the killings of many other demons.

Indra & Soma

Soma is a sort of intoxicating drink in Hindu mythology. Both gods and humans were very fond of it; Indra had a special fascination for it. Sometimes, soma has been classed as an ambrosial offering to gods, which imparted them immortality. Tradition says that soma might have been stored at Chandra, the moon; hence Chandra is also called Soma. There are references in the *Rig Veda* to Indra, in the form of a falcon, getting soma from the moon; alternately, Indra could have ridden a falcon to the moon.

It is believed that soma was produced from a plant which grew abundantly in those days. However, tradition also says that Tvastr (father of Indra) had a cow which gave unlimited quantities of soma. At some stage, there was conflict between Indra and Tvastr over soma, and Indra killed Tvastr, his father. Verse 12 of the *Rig Veda* Hymn 4.18, addressed to Indra, says as follows:

'Who has made your mother a widow? Who wanted to kill you when you were lying still, or may be moving? Which god helped when you caught hold of your father by the foot and crushed him?'

It would appear from the above that Indra killed his father Tvastr, and his two half-brothers, Tri-siras and Vrtra (both sons of Tvastr).

Indra's indiscretions

Indra had the facility to assume any shape or form that he wanted. At the *yajna* (fire sacrifice) of King Marutta of Vaisali, Indra assumed the shape of a peacock to escape the attention of Ravan whom he feared greatly. By the time of the Epics-Puranas period, many stories came to be woven around Indra, which showed him in bad light; we give below some of those stories.

The great Indra had a licentious side; he seduced Ahelya, wife of *dev-rishi* Gautama, by deceit as per one version, by her consent as per another. Rishi Gautama cursed his wife and turned her into a stone. Later, Lord Rama removed the curse and Ahelya again became a woman. Indra also had to suffer for that act of adultery; his scrotum was detached. Later, through the intervention of other gods, a ram's testicles were grafted on Indra. Indra was father of Arjuna through Kunti who was otherwise the wife of Pandu. Indra was also the father of monkey king Bali, born out of union with a monkey named Riksharaja, who had turned himself into an attractive woman, through a dip in a magical lake.

Indra was also not above a theft, or two. There was a great Suryavanshi king Sagara (41st generation from Manu), who had 60,000 sons (the figure may not be taken too seriously). The king had already performed 99 *ashvamedha yajnas* (horse-sacrifices) and was preparing for the 100th. As per tradition, after 100 *ashvamedhas*, Sagara could have dethroned Indra. Therefore, Indra stole the sacrificial horse of Sagara and tied it in the hermitage of sage Kapila. The sons of Sagara went in search of the horse and saw it in the rishi's hermitage. They prepared to ransack it. Coming to know of that, Kapila burnt the princes to ashes. A great great grandson of Sagara, King Bhagirtha was told that his 60,000 ancestors could get deliverance only if river Ganga (which flowed in heaven at that time) was brought down to earth. Through severe penance to goddess Ganga, Lord Vishnu and later Lord Shiva, Bhagirtha succeeded in bringing Ganga on earth. The river flows out of the toe of Vishnu in the heavens; it came to earth in seven streams, one of which is the Bhagirthi. Shiva softened the impact of the fall of the Ganga on earth, by taking the fall in his *jatas* (matted hair). Thus, the souls of the 60,000 ancestors of Bhagirtha were delivered and exalted to heaven. Indra also stole the sacrificial animal(s) at the *yajna* (fire-sacrifice) of King Ambarisha of Ayodhya at Pushkar (For more details, see 'Sunassepa' in chapter 7.3).

We may mention here that (the river) Ganga in her human form was the sister of Parvati, wife of Shiva. The two sisters were daughters of Himmavat, king of the Himalayas. Ganga was married to the Suryavanshi king Shantanu; their son was the great Bhisham Pitama of the *Mahabharta*.

Indra and Maruts

Aditi and *dev-rishi* Kashyap have been listed above as among the possible parents of Indra. Aditi's sister was Diti, who was also married to Kashyap; Diti was, thus, a step-mother and aunt of Indra. Diti was the mother of daityas (demons).

After the Ocean Churning, the gods became immortal. They undertook large scale slaughter of daityas. Diti was highly displeased and agitated. She asked her husband Kashyap to give her a most powerful son, who would avenge the killings of her *daitya* sons. Kashyap agreed and Diti got pregnant; that got Indra worried. One day, seeing Diti in a vulnerable position, Indra entered her womb. He cut her embryo into seven parts. So, instead of one powerful son, seven sons were born to Diti. They were the seven Maruts, sometimes also called Rudras. The seven Maruts are the seven types of wind spirits, which blow on this earth, in heaven, and the spaces in-between, and many other places. Diti did not hold that against Indra, and was eventually reconciled with him. (In chapter 6.11 *Shakti*, we have given another version of birth of a son Mahishisura to Diti).

Ashwins

Vivaswat (Surya, the sun), son of *dev-rishi* Kashyap and Aditi, was married to Samjna, daughter of Tvstr (Vishvakarma). As Samjna could not stand the brilliance of Vivaswat, she left him, and went to her father's place. After some time, Vivaswat followed her there. Samjna took the form of a mare and went to the forests. Vivaswat turned himself into a horse, and went after her. When they met, (mare) Samjna became pregnant by the smell of the (horse) Vivaswat. Another version says that Vivaswat's seed fell on the ground and Samjna became pregnant by smelling it. The result was the birth of the two Ashwins, named Nastya and Dasra. They grew up to be excellent physicians.

Worship of the Gods

Along with Indra and Agni, *Surya* and Varuna were the recipients of major worship. Surprisingly, an intoxicating drink Soma has the whole of the ninth book of the *Rig Veda* dedicated to it. Out of the 1028 hymns of the *Rig Veda*, the following approximate numbers are dedicated to the following gods:

Indra	250 hymns
Agni	200 hymns
Soma	113 of 114 hymns of the ninth book

Vishnu appears as a minor god in the *Rig Veda*; he helps Indra in defeating the demon Vrtra. Sometimes, the word Indra-Vishnu has been used. There is also the deity Rudra (Howler), who was later identified with Shiva; the term Rudra-Shiva appears in later texts. There is some usage of the word 'brahman' (not Brahmin, the priest) in the Rig; however, that usage does not appear to mean the one supreme universal God 'Brahman', as defined in the Upanishads.

The *Rig Veda* refers to 33 gods. Later in the *Puranas*, the number of gods got increased, from the original 33 to 33 thousand. There is a reference even to a figure of 33 crores (330 million) gods. It appears that whilst the figure '33' is from the *Rig Veda*, the zeros came to be added later on. The numbers of gods mentioned in this paragraph are just numbers, which need not to be taken seriously. On a practical note, the total numbers of Hindu gods may be somewhere between 50 to 100; 33 gods of the *Rig Veda*, plus a score or two of the Puranic gods.

One Supreme God

In later books (1st and 10th) of the *Rig Veda*, there may be a reference or two, to One Universal Supreme God. It has also been expressed that different deities or lesser gods are but names of the one and same Supreme God viz:

"They call him by many names like Indra, Mitra, Varuna, Agni; he may be the beautiful-winged heavenly Garutmat: that which is One, the wise call it by different manners: they call it Agni, Yama, Matarisvan." — *RV*, Book 1, Hymn 164, Verse 46.

Note: Garutmat is a celestial bird. Matarisvan is the assistant and messenger of *Agni* (fire), sometimes identified with wind; it may also mean *Agni* in the form of lightening in the sky.

The above verse lies immersed, somewhat obscurely, at serial 46 in a very long and complex hymn of 52 *shloka*s (verses). It is the modern (secularist) Hindus who have picked it up to project the concept of one Universal God in the Rig.

There is a lot of debate as to the role and status of One Single Universal God in the *Rig Veda*. Most (recognized) historians are of the view that references in the *Rig* to One Universal God are somewhat muted, and that too, in the later books. However, Swamy Dayanand of the Arya Samaj is of the view that the Vedas (*Samhita*) are mainly about one Universal God, and references to the other gods are but incidental. That could perhaps be explained by the argument that in Hinduism, polytheism and monotheism were not necessarily treated as incompatible. These may embody two different outlooks of the One Divine; one sees 'One in all', and the other sees 'all in One'. Polytheistic and monotheistic attitudes may be interwoven into Hinduism. Max Muller has tried to explain the issue by use of the words:

>   **Kathenotheism:** Worship of one god after another
>   **Henotheism:** Worship of single gods
>   However, the above concepts did not become popular.

It appears that there was no specific creator god in the *Rig Veda*. However, a creator god may have developed by the end of the *Rig Veda*; he was called Prajapati — The Lord of Beings. Sometimes, Prajapati is viewed as the primeval man Purusha, who existed before the universe. It was Purusha who was taken up for sacrifice, perhaps to himself. The universe was produced from the divine victim Purusha. Later, the god Prajapati was conflated with the creator god Brahma of the Trimurti (See Purusha and Prajapati in chapter 6.12).

6.4 The Upanishadic God

The concept of One Universal God, mentioned in the *Rig Veda*, may be in a passing manner, is developed further in the Brahmanas, which are a sort of expository appendices to the *Vedas* (*samhita*). Brahmanas recognize the Great Being as the Soul of the Universe.

A few centuries after the Brahmanas emerged the Vedanta Upanishads (800–600 BC). Upanishads perfected the concept of One Universal Supreme God, calling it BRAHMAN. He is Self-Existent, Absolute and Eternal. All things emanate from Him and finally rest in Him. As Brahman is not conceivable to material senses, He is *Nirguna* — without attributes; and Nirupa — without form. The Divine Essence Brahman is pre-existent, immaterial, *ananta* (without beginning) and *anadi* (without end). He was there before 'time and space', and would be there after 'time and space' cease to be.

As the common person may have had some difficulty in relating to the Nirguna and Nirupa form, the Brahman has also been called *Saguna* — with all attributes. It was perhaps felt that the common persons would be able to relate more easily to the *Saguna* Brahman.

We have over-simplified the concept of the Brahman by describing Him in a few paragraphs. In reality, hundreds of pages in the Upanishads are full of explanations of the concept of the Brahman.

It would appear that the Hindus did not offer any great direct worship to Brahman. At least, one does not find any specific references to that effect. Further, when temples were built for the gods, hardly any temples were built for Brahman. It would also appear that in spite of the loftiness of the concept of Brahman, the common Hindu folk could not relate to Him in any substantial way. We find that within a century or two of the Upanishadic concept of the Brahman, two new religions emerged on the Indian horizon. These were Buddhism and Jainism; both these religions had no need of a God. Surprisingly, Hindus took to these religions with gusto and they flourished. Almost simultaneously, the atheistic philosophy of Charvaka (chapter 5.6) developed within the folds of Hinduism. In the totality of circumstances, it would appear that the concept of the Brahman did not fire the popular imagination of the Hindus of those days. We admit that we may be under qualified to make such a profound statement.

6.5 Buddhist no-God

It would appear that in spite of the best efforts of the authors of the Upanishads (which are divine revelations), the common Hindu layman was unable to fully grasp the concept of the Abstract Brahman. The common person may have been somewhat confused between the contrary pulls of 33 gods of the *Rig Veda* on one side, and the one Universal Supreme God of the Upanishads on the other. That situation perhaps suited the founders of the two new religions.

In the 6th century BC, two great personalities, i.e. Gautama Budha and Vardman Mahavira, emerged on the Indian scene. They founded two new religions, i.e. Buddhism and Jainism, which have no specific need of God. Strange as it may seem, the two religions not only got a foothold in India, but also caught popular imagination and flourished. Especially Buddhism pushed Hinduism on the margins, not for a century or two, but perhaps for nearly a millennium. At this point of time, we may not be in a position to comprehend the strong hold that Buddhism got over the people of Bharat. It would appear that Hinduism might have been gasping for breath under the onslaught of Buddhism, for a prolonged period.

So strong was the sway of Buddhism that many of the kings of the barbarian tribes who invaded and ruled North-West Bharat from 200 BC to 300 AD, converted to Buddhism. One was the Indo–Greek (Bactrian Greek) King Milinda (or Menander). His book on Buddhism was published as *Milinda Panha — Questions of Milinda*. He ruled North-West India around the mid 2nd century BC. Another was the famous Kushana king Kaniksha, who in all probability, ruled from 78 AD to 114 AD. Kanishka became an ardent Buddhist; he called a council of Buddhists to sort out their differences.

The movement for revival of Hinduism was started in the 2nd/1st century BC, by the Brahmin Shunga and Kanva dynasties, which replaced the Mauryan Empire. The Gupta Kings (320 to 520 AD) also worked for the restoration of Hinduism. In spite of royal patronage from time to time, the process of the revival of Hinduism was slow. Adi Shankeracharya, in 9th century AD, gave a fillip to the revival of Hinduism in Bharat. Muslims delivered the final blow to Buddhism in the 13th century AD; they looted and burnt the Buddhist viharas (monasteries), both in the West and East of Bharat.

The sway of Buddhism in Bharat, varying in intensity from time to time and place to place, may have lasted some 800 years; that may have been a period of almost 'no-god' for a large part of the population.

6.6 New Polytheistic Gods
(Gods of Epics & Puranas)

It would appear that over the long term, the common Hindu folk could adjust:

- Neither to the concept of the Abstract Brahman brought into sharp focus by the Upanishads, around 800 to 600 BC

- Nor to the almost 'no-god' concept of Buddhism from 500 BC onwards

The result was that new types of polytheistic gods started emerging in Hinduism. These gods were different from the 33 gods of the Vedas, who were pushed to the background. The process may have started in the epics, which we have, somewhat arbitrarily, placed around the 4th/3rd century BC. That is the age we think the epics were written; the actual events of the epics might have been one or two thousand years earlier. The concept of the new polytheistic gods reached their finalization during the Puranic age; may be, in the first half of the first millennium AD, when the bulk of the *Puranas* were written. The periods mentioned by us may be taken as very tentative. We study these new gods in the following pages.

6.7 The Trimurti

Amongst the new gods to emerge were Brahma, Vishnu and Shiva, who came to occupy the highest position in the hierarchy of the new gods. The three gods have the following functions:

Brahma — The Creator, the god of Activity
Vishnu — The Preserver, the god of Goodness
Shiva — The Destroyer, the god of Darkness

At some stage, the three gods came to be identified as Trimurti (Hindu Triad or Trinity).

Brahma embodies *Rajo-guna*, the quality of passion by which the world procreates and was created. Vishnu represents *Sattva-guna*, the quality of goodness and of mercy, by which the world is preserved. Shiva has *Tamo-guna*, the attribute of darkness, which destroys the world. Sometimes it is said that the Supreme Spirit Brahman manifested Himself as Brahma, Vishnu and Shiva.

Tradition has it that the primeval god Brahma proceeded from the Supreme Spirit Brahman, when the latter was under influence of *maya* (illusion). Brahma, in turn created the three worlds, the living beings and all other things. However, the act of creation goes along with the acts of preservation and destruction; hence, Vishnu, the Preserver and

Shiva, the Destroyer came to be associated with Brahma. The three gods form the central group of deities around which the system of modern Hinduism revolves. The three letters composing the mystic symbol AUM, embody the three gods (see chapter 9.4).

At this stage, we may note that the three gods may not be confined only to the functions stated against each in the preceding paragraph. Each of the gods is quite capable, and indeed does perform the functions attributed to the other two. For example, Shiva besides his destructive side has creative functions, as demonstrated by his phallic worship. At the same time, the Padma Purana expresses that at the very beginning of creation, the great lord Vishnu, desirous of creating the world, became three fold — Creator, Preserver and Destroyer.

Tradition says that the Supreme Spirit Brahman produced:

— from the right side of his body himself as Brahma for creation
— from his left side Vishnu, for preservation
— from his middle Shiva, for destruction

The Padma Purana goes on to say that some persons worship Brahma, others Vishnu and Shiva. However it is Vishnu, one and three-fold who creates, preserves and destroys; therefore let the pious make no difference between the three.

The physical representation of the Trimuti is in the form of one body with three heads; Brahma in the middle, on the right Vishnu, on the left Shiva. This representation in the physical form is shown in a rock cutting at Elephanta Caves near Mumbai (Bombay).

To establish a hierarchy between these three gods is largely a fruitless exercise. Sometimes one is shown as the greater, sometimes another; more of it later as we study the three gods in detail. The great Indian poet Kalidasa talks of the Trimurti as follows:

'In these three persons, the one god has been shown each first in place, each last, not one alone; of Brahma, Vishnu, and Shiva, each may be first, second, and third amongst the three blessed ones.'

However, at some stage, the equality between the three gods was overthrown. Brahma having created came to be worshipped less and less. Vishnu and Shiva gained in honor and esteem. Vaishnavas consider their god Vishnu to be the greatest. On the other hand, Saivites express that Shiva is the highest. The two sects do not even refrain from taking pot shots at each other, in the name of their respective god.

6.8 Brahma

Brahma is the creator god of the Trimurti. He is represented as a four-headed god, though it is believed, he had five heads earlier. Brahma is often shown with four arms, in each of which he holds, a string of beads, a scepter, an alms bowl, and the bow called parivata. Sometimes, Brahma is represnted as holding a Veda in each of his hands. Brahma's abode is Mount Meru, which is considered the centre of the universe. His vehicle is the goose.

Brahma's spouse is Saraswati, goddess of wisdom and knowledge, having the qualities of good speech and high imagination. Sometimes, she is called the mother of the Vedas. She is also called Vagdevi — goddess of speech. Her other names are Vach, Brahmani and Savitri. Brahma had a second wife named Gayatri; he married her in an emergency, as will be explained later.

There are various versions as to how Brahma got his five heads and how he lost the fifth head —

- *Matsya Purana:* Once Brahma assumed a human form. He farmed out from one half of his body, without the body suffering any diminution, the most beautiful woman Shatarupa. Brahma was attracted towards her. However as Shatarupa was born from his body, she was like his daughter; Brahma was ashamed of his emotion. Shatarupa, understood the situation and was anxious to avoid his looks. She started going around Brahma, who caused a face to spring in the directions in which she moved. She moved in four directions and four faces were created. Shatarupa then moved upwards and Brahma developed the fifth face. These feeling of Brahma for his daughter annoyed Shiva; he burned up the fifth head of Brahma with his third eye, or he may have cut the head off. The story goes on to tell that thereafter, Shatarupa assumed the female form of various animals, birds, etc. Brahma would assume the corresponding male form and thus, various types of species were born.

- Another Puranic version says that once the *dev-rishis* (heavenly sages) asked Brahma to explain the true nature of godhood. At that time, Brahma was under the influence of the *asura* (demon) Mahishasura; Brahma declared that he was the greatest. That became a cause of a dispute between Brahma and Vishnu. When the matter was referred to the Vedas, they declared that Shiva was supreme. Brahma was highly agitated and he challenged the supremacy of Shiva. That annoyed Shiva, who, in his form of Bhairva, cut off the fifth head of Brahma with his left hand. As that action of Shiva was a sin, the head got stuck to the thumb of Shiva. Now, Varanasi is the holy of the holiest city. To expiate his sin, Shiva went to Varanasi; there, Brahma's fifth head was separated from Shiva's left hand.

- Still another story says that the behavior of Brahma towards Parvati, at the time of Shiva's marriage to Parvati, may not have been proper. As such, Shiva cut off Brahma's fifth head, or he may have burnt it off with his third eye.

At some time in the ancient past, Brahma received worship of the devotees. However, gradually his worship declined and worship of Vishnu and Shiva took over. Various reasons have been given for the decline of Brahma's worship. We give below two of these:

- Once Brahma was required to perform a *yajna* (fire-sacrifice) at Pushkar, Rajasthan; but his wife Saraswati was late in coming. As the auspicious time was approaching, Brahma asked Shiva to get a wife. Shiva got a passing beautiful milkmaid called Gayatri; she sat on the sacrifice. When Saraswati arrived, she was highly angry to see Gayatri in her place. She cursed Brahma that he would be either not worshipped, or may be worshipped only once a year. She also cursed Vishnu, Shiva and Gayatri. Later, on the intervention of the gods, Saraswati was reconciled with Gayatri; but, the curse on Brahma stayed.

- In the second version, Shiva once appeared in the form of a Pillar of Fire and asked Brahma and Vishnu to find its two ends. Brahma journeyed upwards to the head and Vishnu downwards to the feet. Tradition has it that Brahma could not find the top end, but lied that he found it. That infuriated Shiva and he expressed that thenceforth, Brahma would not be worshipped, or may be worshipped once a year. Vishnu frankly admitted that he could not find the bottom end. Gladdened by this, Shiva declared Vishnu as the greatest amongst the three gods. Another version of this story says that both Brahma and Vishnu could not find the top and bottom ends of Shiva's Pillar. Therefore, they conceded that Shiva was the greatest of the three.

In view of the above, Brahma is hardly worshipped these days. There are only two temples to Brahma in Bharat, one of which is at Pushkar, Rajasthan.

6.9 Vishnu

Vishnu is the second god of the Hindu Triad, the Trimurti; however, that does not imply any inferiority to Brahma. Rather many a time, a position of pre-eminence is claimed for Vishnu, especially by his followers called the Vaishnavas. Vishnu is the preserver god of the Trimurti. The Vaishnavas consider Vishnu the most influential of the three gods, who can perform all the three functions of creation, preservation and destruction.

Vishnu initially appears in the *Rig Veda*, but not as a major deity; he is a solar god, and the manifestation of solar energy. The Vedic Vishnu strode through the seven regions of the universe in three steps. (We will see another version of this episode later in the 'Vamana' incarnation of Vishnu). In the Rig, Vishnu is shown as an ally of Indra; he helps Indra defeat the demon Vrtra (drought). Sometimes, the two gods form one entity called the Indra-Vishnu.

After the Rig, Vishnu acquired new attributes in the Brahmanas, but was still well below the attributes and powers he was to acquire later in the epics and the *Puranas*. In the *Puranas*, he is even associated with the creation process itself. Vishnu is also associated with the primeval water, which spread everywhere before the creation of the world. He is shown floating on, or submerged in the primeval waters, reclining on the serpent Sesha. In that form, Vishnu has been called Narayana. 'Nara' means water, 'yana' means 'residence', or 'to move on'. Thus, Narayana means the entity, who has his home in water, or who moves on water. When Narayana woke up from that cosmic sleep, a lotus emerged from his navel; ensconced in the lotus was Brahma. Thus, as per this version, Vishnu has been given the first act in the process of creation.

For Vaishnavas, Vishnu is the Supreme Being from whom all things emanate and in whom all things may rest. In the *Mahabharta* and *Puranas*, he has been called Prajapati (Creator) and sometimes, even Supreme God. It has also been expressed that Brahma grew out of Vishnu's navel, and Shiva sprang from Vishnu's forehead. Though *Mahabharta* generally talks of the overall supremacy of Vishnu, it does make some exceptions, here and there. In the Upanishads, Vishnu in his form of Narayana has been equated with the Supreme Spirit Brahman Himself, and sometimes with Purusha (more about him later). Vishnu represents the Cosmic Pillar, considered to be the mystic centre of the Universe, which supports the heavens. Vishnu also owns the Yupa, the Sacrificial Pillar to which the victim is tied. Vishnu is a benevolent god, who is concerned with the good and welfare of humankind; he resists evil powers. He is the most humane of the three gods, with outstanding human qualities.

Vishnu's spouse is Lakshmi, the goddess of wealth. She is also called Sri. Lakshmi was born as Sita, wife of Vishnu's Rama *avatar* in the Treta Yuga. In the Dvapar Yuga, she was born as Radha; some say as Rukmani, wife of Krishna, the eighth *avatar* of Vishnu. Vishnu's vehicle is Garuda, a mystical bird. Vishnu is depicted as a tall, dark and extremely handsome man. Generally, he is shown with four arms, each holding a mace, a disc (sudarshan chakra), a conch shell and lotus.

A special feature of Vishnu is his *avatars* or incarnations. Though there are some references to Shiva's *avatars* (especially in the South), it is the Vishnu *avatars* who dominate Hindu mythology cum history. There is a story behind the Vishnu *avatars*. There was the perpetual conflict between *devas* (gods) and *asuras* (demons). Asuras' priest Shukra had gone to Shiva for the long term (see chapter 4.2). Once under high pressure from the *devas*, *asuras* took protection of Shukra's mother who was wife of *dev-rishi* Bhrigu. To help the *devas*, Vishnu had to kill Bhrigu's wife who was sheltering the *asuras*. Bhrigu was highly annoyed and cursed Vishnu that he would have to be born on earth seven times.

There are various theories as to the total number of Vishnu *avatars*; these are quoted to range from 4 to 29. However, ten *avatars* are considered the standard; the first three were fish/animal; one, half man/half lion; the remainder six were humans, out of which the last one is still to come. The ten *avatars* are described below:

1. *Matsya (Fish):* Appeared during the Manu flood as described in Chapter 8.2

2. *Kurma (Tortoise):* Vishnu asked the *devas* (gods) and *asuras* (demons) to undertake churning of the ocean with Mount Meru or Manadra as the churning stick, and serpent Sesha or Visukhi as the churning rope. Vishnu took the form of a tortoise to give support to the churning stick at its bottom end.
   Note: Before we consider the 3rd and 4th *avatars*, we need to digress a bit and introduce the reader to the 'Sarabha' form of Lord Shiva. Sarabha is a mythical eight-footed beast, mightier and stronger than anything known so far. It is jet black in color and has a sharp beak with fearsome claws; its dimensions are in thousands of feet. It would appear that the concept of Sarabha may have been introduced by Saivites to show their god Shiva as greater than Vishnu.

3. *Varaha (Boar):* Daitya (demon) Hiranyaksha had abducted the earth and taken it under water. Vishnu took the form of a boar to retrieve the earth. However, after the rescue of the earth, the boar (Vishnu) was disinclined to let go of the earth. Out of the union of the boar and the earth, three sons were produced; they took to ravaging the earth. On specific request and intervention of the gods, Vishnu agreed to give up his 'boar' form; but that was not easy. Vishnu sought help from Shiva who took the form

of Sarabha, and killed the boar and his three sons. Thus, Vishnu returned to his original form.

Note: In a different version, the boar *avatar* has been associated with Brahma, and sometimes with Prajapati. That *avatar* also rescued the earth.

4. *Narsimha (Man-Lion):* Vishnu appeared as 'man-lion' to kill the *daitya* king Hiranyakashipu (brother of Hiranyaksha) who was tormenting his own son Prahlada, a good *daitya*, and a devotee of Vishnu. Now, Hiranyakashipu was a devotee of Shiva; therefore, Shiva took the form of 'Sarabha', and attacked Narsimha. Sarabha may or may not have finally overpowered Narsimha; the position is not very clear.

   Note: Hiranyaksha and Hiranyakasipu were the first set of daityas (demons), both being sons of mother Diti, and father *dev-rishi* Kashyap. Tradition has it that the demon Hiranyakasipu was reborn as Ravan who was killed by Rama in the *Ramayana*; and later, as Shisupala — killed by Krishna in the *Mahabharta*. In short, in all the three births, the Vishnu *avatar*s killed this particular demon.

5. *Vamana (Dwarf):* There was a great and powerful *daitya* king Bali (different from the monkey King Bali of the *Ramayana*); he was the great grandson of the first *daitya* Hiranyakasipu. Bali, like Prahlad, is considered to be a good *daitya*. Bali had conquered the whole universe and usurped the earth. As a Brahmin dwarf, Vishnu asked Bali to give him as much land as he could cover in three steps; Bali agreed. Vishnu grew to a giant size and covered the heavens and the earth into two giant steps. The third step was on the head of Bali who was pushed to the under-world. Later, Vishnu allowed Bali to live on the earth.

6. *Parshurama:* His original name was Rama. He was son of rishi Jamadagni, and was thus called Rama Jamadagnya, to distinguish him from Vishnu's seventh *avatar*, also called Rama (Dasrathi). He came to be called Parshurama, after Shiva gave his '*parshu*' (axe) to him. Parshurama also inherited from his family the divine bow of Vishnu, called Vishnu-dhanu. Jamadagni suspected his wife Renuka of adultery. On orders of his father, Parshurama

beheaded his mother. Later, she was restored to life.

There was a great Kshatriya king Arjuna Kartavirya; he belonged to the powerful Haihaya tribe, an off-shoot of the Yadavas (see Chapter 8.3). Arjuna Kartavirya desecrated the hermitage of Jamadagni, and took away his divine wish-cow; therefore, Parshurama killed the king. In revenge, the sons of the king killed Jamadagni. Parshurama was greatly infuriated and vowed to wipe out the entire Kshatriya clans; and he did that not once or twice, but 21 times. However, another version says that the referred repeated killings of Kshatriyas by Parshurama, may have been widespread depredations carried out by the Haihaya hordes.

Parshurama was a very committed devotee of Shiva. Lord Rama broke the bow of Shiva called Haradhuna during the 'swyamvara' ceremony of Sita. That annoyed Pashurama, and he challenged Rama to a duel. However, Rama defeated him in the ensuing fight. Thus humbled, Parshurama retreated to his permanent abode on the Mahendra Mountain.

Tradition says that Bhisham Pitama learnt his *dharma* from Parshurama who may have also taught martial arts to Karan of *Mahabharta*. There are also some references to a conflict between Bhisham Pitama and Parshurama; the conflict may have resulted in a draw as neither was able to defeat the other. These stories might have been put into circulation to establish (or re-establish) the supremacy of the Kshatriyas (Rama and Bhisham Pitama) over the Brahmins (Parshurama). If these stories are taken as true, it would imply that Parshurama may have lived for hundreds, may be thousands, of years.

7. *Rama:* He was the son of the Suryavanshi King Dasratha. He is called Rama Dasrathi or Ramachandra, to distinguish him from Rama Jamadagnya, the 6th *avatar*. Rama is the hero of the *Ramayana* (see Chapter 5.8).

8. *Krishna:* He was the son of Vasudeva, and had a major role to play in the *Mahabharta*. Krishna is the author of the *Bhagwad Gita* (see Chapter 5.8).

Note: In all, there were three Ramas; two being Parshurama and Rama Dasrathi. The third Rama was Balarama, brother of Lord Krishna. Krishna and Balarama were twins; however, Balarama was transferred from the womb of Devaki in the seventh month, to that of Rohini, the favorite wife of Chandra, the moon. Sometimes, Krishna and Balarama are called joint *avatars* of Vishnu. Other times, Balarama is described as the *avatar* of the serpent king Sesha.

9. *Budha Gautama:* He is the founder of the Buddhist religion. Budha did not believe in the concept of God, Vedas and the cult of sacrifice. Budha was incorporated in the Hindu pantheon of gods around the 5th century AD, as the ninth *avatar* of Vishnu. That may have been against Budha's wishes, and against everything that he stood for. It was a clever step by the Brahmins, which finally helped push Buddhism out of Bharat.

   We have described in chapter 4.2, an (unbelievable) story that Buddhism was an attempt to take the demons away from the path of the Vedas and sacrifices; that was to enable the gods to overcome demons.

10. *Kalki or Kalkin:* He will come in the future as the last *avatar* of Vishnu in this cycle of the Universe, at the end of the present Kali Yuga. Kalki would reward the righteous and punish the wicked. Kalki would re-establish *dharma*, and start a new era; he may appear as a tall, handsome warrior riding a white horse.

Vasudeva Krishna is considered the most complete *avatar* of Vishnu. Vishnu is generally worshipped in his *avatars* of Rama (along with spouse Sita), and Krishna (along with Radha). Vishnu's wife Lakshmi is worshipped in her own right as the goddess of wealth and good fortune. Lakshmi *pooja* (worship) is the central activity at the time of the most holy Hindu festival of Diwali, which, otherwise, is supposed to celebrate the return of Lord Rama to Ayodhya, after his victory over the demon king Ravan.

6.10 Shiva

It is a law of nature that all created things have to die, sooner or later. So emerged the Destroyer god Lord Shiva, a fascinating god, with multifarious facets and fearsome powers. Shiva has innumerable names like Rudra, Mahesh, Shankara, Maha*deva*, Pasupati (lord of animals), Shambu, Bhava. Shiva's spouse is variously called Sati, Uma, Parvati (in the benevolent form) and Durga, Kali, Chandi (in the malevolent form). She has many other names. Shiva's mount is the bull Nandi/Nandin. His symbol is the *trishul* (trident). His battle-axe is *parshu*, which he gave to Parshurama (the 6th *avatar* of Vishnu), who used that for killing the Kshatiyas 21 times. Shiva's name as such does not appear in the Vedas. However, the Vedic god Rudra is identified with Shiva; both have many common features, some of those malevolent and highly fearsome.

Shiva, though a god of destruction, can also undertake creation and preservation. It has been expressed that after destruction, the universe sleeps in Shiva to emerge at a later date. In that sense, Brahma and Vishnu may be different aspects of Shiva. Shiva has a vast variety of many hued, colorful and fascinating natures. He is often projected as a combination of the opposites — malignant but also benevolent, active and passive, good and evil. There is also a perception of Shiva as the totality of an omnipresent, ever-changing cosmic power, ever destroying and ever creating.

Shiva believes in destruction for its own sake; he often haunts cremation grounds and other places of death, He has been called Bhairva (the terrible). The reproductive powers of Shiva are represented by his *lingam* worship (phallic worship). There is a story in the Padma Purana. Once, Rishi Bhrigu went to see Shiva. His entry was barred by a gateman saying that the Lord and the *Devi* were relaxing; Bhrigu waited for sometime. As the Lord did not appear even after a long wait, Bhrigu was furious. He cursed that thereafter, Shiva and Parvati would be worshipped as *Lingam* and *Yoni*.

There is no limit to Shiva's asceticism and austerities. He can sit motionless for centuries, with ash smeared over his (naked) body, even in the cold of the snow-covered Himalayas. Shiva has serpents around his neck and wears a garland of skulls. In the middle of his forehead, there is his third eye; this eye is normally inward looking. However, when the eye looks outward, it is a potent weapon of destruction. Skulls represent the flow of time. Shiva also exercises control over goblins, demons and spirits. At the other end, Shiva is also fond of the good things of life and is often shown inebriated. He is jovial, fun loving and sometimes may even act wild. The worshippers of Shiva in this character are the *Shakti* worshippers, called the Saktas or Shaktas.

We have referred earlier to Shiva cutting off, or burning the fifth head of Brahma. He also ordained that Brahma was not to be worshipped or worshipped only once a year. In view of such episodes, the followers of Shiva, the Shaivites express that Shiva is the greatest of the three gods of the Trimurti. They believe him to be the highest of high Lords, especially in conjunction with his spouse, the *Shakti*. Sometimes, Shiva is referred to as the Supreme Spirit. Shiva lives with his spouse Parvati on Mount Kailasha (now in China). When residing in Varanasi, Shiva is with his wife Annapoorna, who provided him food during his wanderings.

Shiva is the lord of yoga and of yogis. He is also the lord of cosmic dance, and in that form is called Nataraja. His dance called *tandava* can cause immense all-round destruction. Shiva is often shown with four, or more hands. In these hands, he holds items like the trident, *damru* (drum), a begging bowl, etc. He wears the crescent moon over his head and tiger or elephant skin over his body. Shiva is the most complex and colorful god of the Trimurti.

6.11 *Shakti*

Shakti is the female energy of God. Sometimes, *Shakti* is referred to as the primordial energy of the universal soul Purusa; it is the worship of the Supreme Lord in his female form. *Shakti* is also called *Devi*, *Mahadevi* or mother goddess.

Lakshmi has been called the *Shakti* of Vishnu. However, the spouse(s) of Shiva dominate the domain of *Shakti*. *Shakti* has two characters — one Mild or White, and the other Fierce or Black. Some of the names under the two groups are:

- **Mild or White Form**

Sati	—	Who jumped in fire
Uma	—	Of light, a type of beauty
Parvati	—	Mountaineer
Gauri	—	Yellow or Brilliant
Himmavati	—	Daughter of the Himalayas
Jagan Mata	—	Mother of the world
Annapoorna	—	Goddess of grain and food

Uma is sometimes identified with Sati; however, some texts refer to Uma as Parvati. The mild form of *Devi* has many other names.

- **Fierce or Black Form**

Durga	—	the Inaccessible
Kali	—	the Black
Bhairvi	—	the Terrible
Chamunda	—	She killed the demons Chanda and Munda.

Sati (or Uma) was the daughter of the Prajapati sage Daksha, the god of 'good order'. He was opposed to the marriage of Sati to Shiva, the mendicant wanderer. As per Daksha, Shiva had unclean habits, e.g. his body was ash-smeared; and he made rounds of cremation grounds. But marriage did take place. Daksha organized a *yajna* (fire sacrifice) to which he invited all gods, but not his own son-in-law, the great Shiva. Sati felt insulted and jumped into the sacrificial fire and died. Shiva was highly infuriated and attacked the sacrifice. Shiva took the half-burnt body of Sati (Uma), and started his destructive *tandava* dance. That was stopped on the request of other gods, especially Lord Vishnu; thus, the universe was saved from destruction. Vishnu had cut the body of Sati into its various limbs; these were scattered by Shiva's dance, all over Bharat varsha. The places where those limbs fell became centers of holy pilgrimages. Tradition has it that lord Shiva, along with the *Devi*, dwells at these places. Due to his great anger with Daksha, Shiva cut off his head. However later, Shiva replaced it with the head of a goat. The goat's head is a permanent sign of the ignorance of Daksha, in not recognizing the greatness of Shiva.

Sometimes, a different interpretation is given to the feud between

Shiva and Daksha. It is expressed that Shiva may have been a pre-Aryan god, possibly in the form of an earlier version of Rudra. There was a marked reluctance to accept a pre-Aryan god in the pantheons of the Aryan gods, perhaps dominated by Vishnu. Daksha did not invite (the pre-Aryan god) Shiva, as he did not want Shiva to get any portion of the sacrifice.

Sati was reborn as Parvati, the daughter of Himmavat, the king of the Himalayas (Some texts call her Uma). Parvati undertook extreme austerities and asceticism with a view to regain Shiva. The god of Love, Kamadeva was called upon to intervene to excite desire in Shiva's heart. For doing that, Shiva burnt Kamadeva and his wife Rati with his third eye. Later, on the intervention of the gods, the couple was restored to life by Shiva.

Finally, Parvati was united with Shiva. Parvati's sister is Ganga, which flows as the most holy river in Bharat. The Ganga river emerges out of the toe of Vishnu in heaven; Ganga's descent on the earth was softened by the matted hair of Shiva. In her human form, Ganga married king Shantanu; their son became famous as Bhisham Pitama of the *Mahabharta*.

Parvati is beautiful and gentle and full of tender feelings; her influence on her husband Shiva is always for the good of humankind. The union of Shiva and Parvati is represented in the concept of *Ardh-Narishwara* (half male and half female). Parvati and Shiva are believed to be living on Kailash mount (now in China). Parvati is the mother (though not biologically) of Ganesha and Karttikeya (also called Skanda). Once Shiva taunted Parvati on her black complexion. Parvati was highly annoyed, and retreated to the forests; there she took to hard austerities. Thereupon, Brahma granted her a boon that her color would be always golden; as such, she is called Gauri.

In her role of Durga or Kali, Shiva's consort assumes a ferocious character and role. Though still serene, Durga is fierce, sometimes even cruel, out to destroy the demons. Tradition has it that a demon Durga had become very powerful and defeated all the gods. Gods then undertook a *yajna* and surrendered all their energies and weapons to the *Shakti* of Shiva; the weapons included the trident, the disc, the sword, etc.

The *Shakti*, so empowered, started her fight with the demon Durga. The fearsome fight lasted for nine days and nine nights. It is believed that on each of those nights *Shakti* would assume a different form. Finally, the demon Durga was defeated and killed. Those nine nights are celebrated as the most holy *Navrataras* (nine nights) in Bharat, especially in Bengal. The *Shakti* in that form came to be called 'Durga', as she had slain the demon Durga; he is also called Mahishasura, as in the final fight on the ninth day Durga assumed the form of a buffalo (*mahisha* in Hindi).

In a different version of the above story, it has been stated that after the 'Ocean Churning' (chapter 9.1), there was wholesale massacre of daityas (demons). Diti, the mother of the daityas was highly dejected and agitated. She asked her husband, *dev-rishi* Kashyap for a most powerful son who would avenge the death of her many sons; her anger was specially directed against Indra, the ruler of gods. She got a most powerful son; he had body of a man and the head of a buffalo, he was called Mahishasura. He over-powered all the gods, and is reported to have captured the triple world. After due deliberations, all gods surrendered their individual powers to an entity, who emerged as the great goddess Durga, a form of *Shakti*. She finally killed Mahishasura.

Another form of *Shakti* is Kali, the 'black' manifestation of Shiva's spouse. As per one version, Kali emerged out of Durga's forehead. As Kali, the *Shakti* is fearsome and bloodthirsty; she does not hesitate to drink the blood of her victims. Tradition has it that Kali has a gaping mouth, with blazing red eyes and uncombed hair. Her tongue, red with blood, hangs out. She is generally shown with four hands, and has a garland of skulls of the giants she has slain. Kali is generous in awarding blessings to her devotees. Some other names of Kali are:

Dasa Bhuja — the ten armed
Sinha Vahini — riding on a lion she slew the demon Rakta Bija
Mahisha Mardini — killed demon Mahishasura
Chinna Mastaka — the severed-head form: slew demon Nisumba

Bhadrakali — a most severe form
Aparajit — the invincible
Chandi
Chandalika

Note: The demons mentioned above may have been killed by a form of Kali, other than the one listed above; the situation is rather flexible.

Shakti is envisaged as a savior of humankind as per the wishes of her Lord, the Supreme God. Through the worship of *Shakti*, devotees arouse the energy that is within them; with that, they try to attain oneness with Shiva, as the Supreme Spirit.

The worship of *Shakti* is connected with Tantrism in the form of rituals and magical practices (see Tantras in chapter 5.9); that is also called the Sakta movement. Worshippers of *Shakti* are called Shaktas or Saktas. Under the school of Tantra, many erotic sculptures came to be built, most notably at Khajuraho and Konarak, from about the 10th century AD.

6.12 God's Various Names

Before we proceed further, we may clarify certain similar sounding names relating to Hinduism, i.e. Brahman, Brahma, Brahmin, Brahmanas.

Brahman: Brahman may be pronounced as 'Brahmn', and is spelled as such in some texts. He is the Pre-existent, Absolute, One Universal, Supreme God of the Upanishads (Chapter 6.4). Brahman is the only Hindu God spelled with capital 'G'; He is of neuter gender. Confusion arises as in some texts He (Brahman) is spelled as 'Brahma' (with silent last 'a').

Brahma: He is the creator god of the Trimurti — the Hindu Triad (Chapter 6.6). Brahma is a male god, as opposed to the neuter Brahman. However, in some texts, Brahma is spelled as 'Brahman'; that can be the cause of confusion.

Brahmin: He is the Hindu priest. In many texts, it is spelled as Brahman (same spelling as at serial 1 above); that can cause confusion. Therefore, in our text, we use the word 'Brahmin'.

Brahmanas: The religious books containing the prose portion of the Vedas, as explained in Chapter 5.3.

In some texts, distinction between Brahman and Brahma is sought to be made by using the following nomenclature:

Brahman (neuter or n) : for Brahman
Brahman (masculine or m) : for Brahma

In spite of the above clarifications, the scope for confusion between 'Brahman' and 'Brahma' is almost endless. Ordinary people have been confused over the ages and may be confused now. Though Brahma is called the Creator god, it was Brahman who may have been involved in the first creation activity when He planted the first 'seed' in the primordial waters, which grew into the golden egg Hiranyagarbha (see Chapter 7.2). Brahma emerged out of that Egg, after one thousand years. Another way of saying the same thing is that Brahman manifested Himself as Brahma. It is an altogether different matter that sometimes Vishnu is associated with the first Creation activity, as explained in Chapter 6.7.

We may refer in passing that for most Hindu laymen, Brahman is a bit abstract concept, of which they may be only vaguely aware. Hindus generally give no direct worship to Brahman and there are no specific temples for the Supreme God. They are more aware of Brahma (as a part of the Trimurti); but even Brahma hardly gets any worship from the Hindus; there are only one or two temples for Brahma in the whole of Bharat.

The author, a normal Hindu, was not specifically aware of Brahman, before he started his studies for writing this book; he had a vague feeling about the concept. It is author's view that vast majority of Hindus perhaps fall in this category.

God's Other Names

A number of names are used in the Hindu literature for God. Some of the terms are Purusha or Purusa, Narayana, Prajapati, 'That One' and Parmatma. Those names may sometimes represent the Supreme Spirit Brahman and other times, other gods. Hundreds, perhaps thousands of pages have been written in Hindu literature to explain the concepts of Purusha, Prajapati, etc. There is no way we can attempt to cover these entities in a page or two; nevertheless, we make a feeble attempt.

Purusha (Purusa)

In simple Sanskrit/Hindi, Purusha means 'man', the male human being. His opposite gender is *Istri* — the female.

Here, Purusha is taken to mean the cosmic giant or the primeval man who may have existed before the universe was created. We have said earlier that it was the Supreme Spirit Brahman who was pre-existing. Therefore, sometimes, Purusha has been equated with the Supreme Spirit and Soul of the Universe. However, this is only a simplified version of a very complex situation, which may confuse the common man.

The Katha Upanishad says that 'Purusha is beyond the unmanifest, and beyond Purusha, there is nothing'. To a layman, that would appear to be a reference to Brahman. The Markendya Purana says as follows:

'When in the water lay the thousand-eyed, thousand-footed, thousand-headed Purusha, the golden complexioned Being beyond human perception, it is the Brahman; he is also also called Narayana'.

In the *Rig Veda*, Purusha appears in the famous Purusha Sukta, as the cosmic giant, or the primeval man. Purusha was taken up for sacrifice. From that sacrifice or dismembering of Purusha issued forth all that exists in this world, and the next; the whole universe was produced from the body of the divine victim. Who took up Purusha for the sacrifice? As per the Purusha Sukta it were the gods. However, the Creation hymn of the *Rig Veda* says that gods came after the creation of the universe. In the conventional sense, sacrifices were carried out to the gods (to please them). To whom was Purusha sacrificed? There is no simple answer. Possibly, he was sacrificed to himself; *shloka* 16 given below appears to imply that. Again, all this is a simple explanation of a very complex issue. The *Rig Veda* describes Purusha as the whole universe; 'a quarter of him is all beings, three quarters of him is immortal'.

We give below some selected verses from the Purusha Sukta of the *Rig* — Book X, Hymn 90:

1. Purusha has a thousand heads, a thousand eyes, and a thousand feet. He dominates the earth on all sides and even extends beyond it; may be as far as ten fingers.

6. When the gods performed a sacrifice with Purusha (or Prajapati) as the offering, the spring was the butter, the summer the fuel and the autumn its oblation.

12. The Brahmin came forth from Purusha's mouth, the Rajenya (Kshatriya) from his arms, the Vaishya from the thighs; and the Shudra was born of the feet.

13. The moon arose from his mind. *Indra* (the god) and *Agni* (Fire) came from the mouth and *Vayu* (wind) from the breath; from his eyes emerged the sun.

16. With that sacrifice, the gods sacrificed to the sacrifice.

In another version, it has been expressed that in the beginning, there was only Purusha. At some time, he felt lonely and split himself into two. What emerged from Purusha has been called Prakriti. If Purusha is the soul or consciousness, Prakriti is matter (or nature). Prakriti is a term used to refer to the material world, as distinct from the Supreme Spirit (Purusha). Prakriti consists of 24 elements from which Purusha is considered separate. However sometimes, the Purusha may get entangled in these elements. Liberation for Purusha lies in the ability to discriminate between the Supreme Spirit and the material form. Purusha is masculine; Prakriti is feminine. Sometimes, Prakriti has been equated with *maya* (Illusion). All this may be quite confusing; it can't be helped.

Prajapati

It appears that the concept of a Creator god might have emerged towards the end of the *Rig Veda* period; he came to be called Prajapati — Lord of Beings or Creatures. Later on, the god Prajapati appears to have merged in the Creator god Brahma of Trimurti. This again is a highly simplified account of a complex issue.

From time to time, the term Prajapati has been applied to different entities in their varying functions, viz:

- The term has often been used for the Supreme Spirit Brahman Himself, or for His manifestation.

- Prajapati has been described as the father of gods, the Creator and the Supreme Lord in the Rig.
- The *Rig Veda* applies the term Prajapati also to Hiranyagarbha (the Golden Egg), Indra, Savitre (Sun), Soma, Agni.
- In some Puranic literature, Prajapati is a name of Brahman; however, in a few Puranic texts, Prajapati is used to indicate Vishnu and Shiva.
- In Manu, the term is applied to Brahma as the active creator and supporter of the Universe.
- The term is sometimes applied to Manu Swambhuva, as the mind-born son of Brahma, and as the secondary creator of the seven *dev-rishis*.
- The most common use of Prajapatis is for the ten sages, who were the mind-born sons of Brahma (see Chapter 7.3).
- Ka (Who?) is another word sometimes used for Prajapati. Some texts mention Ka (Prajapati) as the father of Indra.

The association of Prajapati with creation is an important element; he has also been equated or clubbed with Purusha; the combination Prajapati-Purusha appears in some texts.

We regret that we may not have been able to clarify the concept of Pajapati; we can do no better.

Narayana

In simple words, '*nara*' means water; and '*yana*' means home, or 'path to move on'. Thus, Narayana means the entity, whose home was water, or who moved on water. The name Narayana has been applied both to Brahma and Vishnu. Manu applied it to Brahma. However, in present day Hinduism, it is generally applied to Vishnu.

As per one version of Creation, the Supreme Spirit Brahman planted a seed in the waters; that seed grew to the egg Hiranyagarbha. After residing for 1,000 years (some texts say one year) in the egg, the Supreme Spirit manifested Himself as Brahma; and hence the name Narayana for Brahma — having been in water for 1,000 years (or one year).

In another version of Creation, Vishnu rested in/on primordial waters, slumbering on the curled serpent Sesha. When Vishnu woke from his cosmic sleep, a lotus emerged from his navel; ensconced in the lotus was Brahma. As Vishnu was in the waters, he has been called Narayana. This is the version presently prevalent in Bharat.

Svetasuara Upanishad describes Narayana as the Supreme Deity who is Absolute and in the highest order. Mahanarayana Upanishad gives a sequence of identification from Prajapati to Brahma to Narayana, and goes on to link it up with the theme of Cosmic Sacrifice of Purusha. Sometimes, Narayana is even identified with Purusha. The Satapatha Brahmana conflates Purusha and the Creator god Prajapati, who is identified with Vishnu.

Another name suggested for Narayana is Maha-Vishnu who is identified with the Supreme Lord himself. Narayana has also been called the World Soul. The divine incarnations of Brahman have often been called Brahma, Narayana and Prajapati.

We end this account by saying that presently it is Vishnu, who in his various forms, is identified with Narayana. Sometimes, he is equated with the Supreme Spirit.

That One

It has been expressed that 'That one' or 'That which is One' (see under 'Rig Veda gods' in Chapter 6.3) is the term used in the Rig Veda for the Supreme Lord who is pre-existing and indefinable. Those two terms could be perhaps equated to the Supreme Soul Brahman of the Upanishads; that is largely by inference.

Paramatma

The term means the highest Spirit. Following names are considered inter-changeable with Paramatma — Parmeshwara, Ishwara and Bhagwan. Possibly, all of these terms refer to the Supreme Spirit.

Comments

In Hinduism, the concept of God/god is very complex, and prevails at many levels. Hundreds of books and thousands of commentaries have been written on the subject. In our above comments, we have not tried to project a scholarly point of view; that was not our intention, nor within our capability. We have tried to explain the issues in a language with which, hopefully, common laymen can relate. In that effort, we may have deviated from the straight and narrow path, and some compromises may have been made. If that be so, apologies are offered in advance. Also, we have raised a number of questions, but have not given all the answers.

6.13 Additional Gods — Ganesha and Others

In addition to the concepts of Trimuri and *Shakti*, the epics and the *Puranas* introduced a whole range and variety of gods. Amongst these, Rama and Krishna are the most important. Though they are the *avatars* (incarnations) of Vishnu, they are gods in their own right. Sometimes, Krishna is considered the most complete *avatar* of Vishnu; he has been identified with the Supreme Spirit Himself. Then, there are other gods like Ganesha, Karttikeya, Hanuman, Ayyeppa, Balaji, Jagannath and others. Ganesha and Karttikeya are the sons of Shiva and Parvati. Hanuman, a devotee and constant companion of Rama, is the monkey god. We describe below some of the more important Hindu gods.

Ganesha

Ganesha is the mind-born son of Parvati, and therefore of Shiva. As per one version, Ganesha was made by Parvati out of the scurf of her body. After creating Ganesha, Parvati put him up to guard the entrance as she was having a bath. Ganesha prevented Shiva from entering. Infuriated, Shiva cut off his head. Parvati was shocked and inconsolable; she insisted on his revival. Shiva sent his bull Nandi and his companions to get a suitable head. Nandi, not being the epitome of intelligence, got the head of Indra's elephant Airavata. Shiva grafted the elephant head on Ganesha, under instructions from Brahma. Ganesha was revived and all present blessed him, especially Brahma. Later, by the grace of Shiva, Indra got his elephant back, as it emerged during the Ocean Churning.

Ganesha has many names, viz:

Ganadhipa/Ganapati — Ruler of *ganas* (hosts)
Gajanana — The elephant-headed
Ekadantaka — One tusked; one tusk of the elephant was broken when Nandi threw his axe to get its head.
Lambodara — Pot-bellied

Vighnesa — Remover of Obstacles

Vinayak

As per Brahma's boon, Ganesha is the first god to be worshipped at all auspicious occasions. He is the god of knowledge, and the remover of all obstacles in the path of his devotees. Ganesha has a big belly, which represents the entire cosmos; the belly has a snake wound around it. Ganesha's elephant head has only one tusk; the other was cut off when Parshurama threw his axe at Ganesha, who had prevented him from meeting Shiva (this is an alternate explanation to the one given earlier). Ganesha's vehicle is the rat. He is the head of *ganas*, i.e. Shiva's personal army of spirits, demons and goblins. Ganesha had three wives, called Siddhi (spiritual power), Riddhi (prosperity) and Buddhi (intellect). Sometimes, Ganesh in represented as a bachelor.

Ganesha has four hands, each holding the following things:

Rope — To carry devotees on the path of truth
Axe — To cut devotee's attachments to wordly things
Laddoo — A sweet, as a reward to his devotees
4th Hand — Raised to bless the devotees

There are different versions as to what Ganesha holds in his hands.

Ganesha took down the dictation of the entire *Mahabharta* from Ved Vyasa. Ganesha expressed that the dictation must be continuous. Ved Vyas agreed with a counter condition that Ganesha should understand the message, and only then write it down; that gave time for Ved Vyasa to think. The annual festival of Ganesha or Vinayaka Chaturthi is celebrated with great fervor, especially in Maharashtra; the countryside reverberates with shouts of 'Ganpati Bapa Moriya'.

Karttikeya (Skanda)

Some of his other names are Murugan (the most popular), Agzhan Murugan (handsome and romantic prince), Maal Murugan (nephew of Vishnu), Kumara, Malaikilavon (Lord of the hills), Subramanya, Balasubramanya, Mahasena.

Karttikeya is considered the son of Shiva and Parvati, though he was not born of Parvati. There are various versions of the birth of Karttikeya.

There was the great demon Taraka who had become invincible for the gods. The gods went to Brahma to seek his help. Brahma expressed that Taraka could be killed only by a son born of Shiva. However, Shiva was an ascetic. Therefore, the gods requested the god of desire/love Kamadeva to excite desire in the heart of Shiva. Kamadeva did that, and Shiva was married to Parvati. Now, both Shiva and Parvati are very powerful entities, each in his/her own right. The gods feared that the universe might not be able to withstand the power of a son, born out of the union of Shiva and Parvati. They requested Shiva not to produce a son from Parvati; Shiva agreed. Parvati was furious and cursed the gods. As a result of that curse, consorts of the *devas* had to stay barren. As gods could not procreate, they are worshipped in stone.

To accommodate the gods' request, Shiva released his seed in the air. Agni, in the form of a dove, swallowed that seed; he started looking for a suitable womb for the seed. Now onwards, there are various versions:

- Six wives (except Arundhati, wife of Vasishta) of the *dev-rishis* were having a bath in the Ganga (Ganges). Seeing them in that condition, *Agni* impregnated the six wives, with the seed of Shiva (Rudra). When the wives realized their condition, they took out the six seeds/fetuses and amalgamated these into one. Taking that as adultery, the *dev-rishis* divorced their wives. That seed grew up to be the six-headed Karttikeya, nursed by the six foster mothers, i.e. wives of *dev-rishis*. In recognition of their services, Kartikkeya placed all seven wives of the *dev-rishis* in the Zodaic. They shine there as Pleiades (name of the open cluster of stars in the constellation of Tauras); they are also called the Seven Sisters. The other variants given below revolve around this basic version.
- *Agni* lodged that seed of Shiva in the Ganga; the river was unable to stand the heat of the seed and deposited it amongst the bulrushes along its banks. Six Karttikas (mythical sister-stars, abandoned wives of the *dev-rishis*) were residing on the

bank. They nursed that seed, which grew up to be the six-headed Karttikeya.

- Karttikeya was born out of the union of the fire god *Agni* (who was carrying Shiva's seed) and a minor goddess Swaha: she is sometimes referred as a daughter of prajapati Daksha. For that union, Swaha took six different forms of the six wives of the *dev-rishis* (out of the total seven). On account of her strong asceticism, Arundhati was the only exception. During *yajnas* (fire-sacrifice), Hindus utter the word '*swaha*', whilst offering *samagri* (ingredients) to *agni* (fire) of the *agni-kund* (fire vessel). It has been expressed that the word '*swaha*' represents the union of the goddess Swaha with Agni. There are also other explanations of the word *swaha*.

- As per the *Mahabharta*, gods needed a great general for their army to fight the demon Taraka. Dev-rishi Angiras's wife Swaha (also called Siva) approached Agni. Swaha assumed the shape of the six wives (except Arundhati) of the *dev-rishis* for the union. She deposited the six seeds in a golden pot among the reeds along the banks of the Ganga. Out of that pot, Swaha brought out a six-headed son; he was Karttikeya.

- In one version, the name of Uma (wife of Shiva) is mentioned in connection with the birth of Karttikeya; Uma entered Swaha for the purpose of Kartikkeya's birth.
  Note: We may recall that as per the above version, *Agni* was carrying the seed of Shiva. Even otherwise, *Agni* is sometimes identified with Shiva. The name Rudra may be considered interchangeable with Shiva in the above story.

Parvati adopted the infant Karttikeya. He killed the demon Taraka in his infancy. Karttikeya is an invincible warrior. He is the war god of the Hindus and supreme commander of their forces. Karttikeya is generally considered a bachelor. However, sometimes, names of his wives are mentioned as Valli (Spear) and Sena (Army), which match his status of a warrior.

Hanuman

Hanuman is the monkey god, and the most ardent devotee without parallel, of Lord Rama. His other names are — Pavanputra (son of *Vayu*, the Wind), Mahavira (the highly brave), Mahabali (very strong), Maruti (son of Marut, the Wind) and Anjnaya (son of Anjana). He is perhaps the most worshipped of the gods, whose temples are probably the most numerous, all over the countryside.

The story of the birth of Hanuman is as follows. There was a female angel Punjiksthala in the ashram of rishi Brhaspati. Due to some wrong doing, she was cursed to take birth in the form of a monkey, to be released only by an incarnate of Shiva. She was born as a female monkey named Anjana, and was married to Keshari, monkey king of Kishindhya. Anjana got incidentally impregnated by a touch of *Vayu* (the Wind god); Hanuman was born, and Anjana got her release. That would appear to suggest that Hanuman may have been the Shiva incarnate. Hanuman was very mischievous as a child. Once he thought that the rising sun (being all red) was a toy and grabbed it; the earth was plunged into darkness. Indra had to fire his thunderbolt on Hanuman to retrieve the sun. The weapon hit the left jaw of Hanuman, and it was broken. Perhaps, that is how Hanuman got his name, as hanuman means 'broken jaw'.

Hanuman rendered invaluable service to Lord Rama during the war with Ravan. He is the one who traced and met Sita when she was imprisoned in Lanka. During the war, Lakshman was critically injured. Hanuman flew all the way to the Himalayas, and returned with a part of the mountain full of herbs, out of which *sanjivini* was plucked to cure Lakshman.

Hanuman had magical powers. He could become as big as a mountain, and as small as a rat, or a fly. He could kill demons and *rakshasas*. Hanuman could fly through the air and leap over oceans. In his childhood, Hanuman played some pranks with the *rishis*. He was cursed by them that he would forget his powers when needed. However, these would come back when reminded of the same powers. Hanuman is the embodiment of a lot of attributes like devotion, strength, knowledge and dedication. Devotees pray to Hanuman to get rid of their suffering, and to get all types of boons. Sometimes Hanuman is considered to be one of the ten people who were blessed with immortality.

Lord Ayyappa

He is believed to be born out of the union of Shiva, and Vishnu, when the latter was in the form of a damsel named Mohini; the child was abandoned at birth. South India's King Rajashekhara (of the Pandian dynasty) had no issue of his own; he found the infant child and adopted him. Later, a son was born to the King. Thereafter, the queen became jealous of Ayyappa, and did not want him to ascend the throne. She feigned sickness and asked Ayyappa to get tiger's milk. Ayyappa went to the forests. He returned with not only tiger's milk, but a whole pride of docile tigers. The king could detect the divine nature of Ayyappa who asked the king to make a temple for him at Sabrimala north of the holy river Pamba in Kerala.

Ayyappa is also called Sastha or Manikhantan — with a golden bell at the neck. He is also referred as Harihara, since he is son of Hari (Vishnu, but in form of a damsel) and Hara (Shiva). He is worshipped as a major deity in the South.

Triputi Tirumala Balaji

His other name is Lord Sri Venkateswara, an incarnation of Vishnu. His most ancient temple is located on the seventh peak Venkatachala of the Tirupati hill, and lies on the southern banks of Sri Swami Pushkarini in Andhra Pradesh. He is also called the Lord of the Seven Hills. The Tirupati Tirumala temple is reputed to be the richest temple of the world.

Lord Venkatswara bestows boons on his devotees. As per the *Puranas, mukti* (salvation) in Kali Yuga can only be achieved by worshipping Lord Venkateswara.

Kamadeva

He is the god of love and desire. It is his duty to ensure that the universe continues to procreate. For that, he acts in consort with the female energy. Kamadev's spouse is Rati, the embodiment of sexual desire.

On the urgings of the gods, Kamadeva undertook the hazardous task of arousing desire in Shiva's heart, when the latter was in deep meditation. Kamadeva was successful in his task, and Shiva was united with Parvati. But, Shiva was furious on being disturbed, and burnt up Kamadev and Rati with his third eye. However, on the intervention of the gods, the couple was restored to life.

General Comments

To say that the vast numbers of their gods have the Hindus confused would be an unwelcome statement to most devout Hindus. To them, in the very multiplicity and variety of gods lies the true spirit and beauty of Hinduism. You take away these gods and goddesses, a very insipid Hinduism would be left.

6.1 World Religions — Comparison

We give below a comparative table of salient features of Hinduism, Buddhism and the three Judaic religions (Judaism, Christianity and Islam).

		Judaic Religions	Hinduism	Buddhism
1	God	One God of each religion	Multi-gods	No God
2	Soul			
	Man	Yes	Yes	No
	Animal	No	Yes	No
3	Rebirth	No	Yes	Yes
	Soul Trans-Migration	No	Yes	No
4	Day of Judgment	Yes	No	No
5	Divinity of Prophets/ *Avataras*	Abraham — Human Christ — Divine Muhammad — Human	Divine *Avataras*	Budha — Human
6	Paradise & Hell	Yes (All three)	Yes	No
7	One Single Book	One Book for each Religion	No	No

8	Origin	Judaism — 2000BC (estimation) Christianity — 30 AD Islam — 622 AD	4000/3000 BC (Estimate)	6th century BC

Notes

1. Hindus consider the soul of man and animal as equal.
2. In Judaic religions, only man being higher is considered to have a soul.
3. Buddhism believes in rebirth, but without soul and its trans-migration.
4. Christians believe Christ was the Son of God and Divine (a part of God).

7
HINDU CREATION CONCEPTS

7.1 Hindu Concept of Time
Hinduism has a cyclic concept of Time. The universe is considered *anadi* (without beginning) and *anant* (without end); it goes through cycles of manifestation and destruction. The universe is destroyed at the end of each Day of Brahma (called Kalpa); that is followed by a Night of Brahma. The universe is recreated when Brahma's night ends. There is the concept of *pralaya* (deluge) and *maha-pralaya* (major deluge) to destroy the universe.

Four Yugas
In Hindu mythology, 360 human years are equal to one 'year of gods', or celestial year. There is the concept of four Yugas — Krita (or *Satya, Dharma*), Treta, Dvapar and Kali, in that order. The Creation starts with the Krita Yuga with a duration of 4,800 'years of gods', or 1,728,000 human years. Duration of the each subsequent Yuga decreases by ¼th of the duration of Kritya Yuga. The position can be summarized as follows:

Yuga	Years of Gods	Human Years
Krita — Age of Winning	4,800	1,728,000
Treta — Age of Trey	3,600	1,296,000
Dvapar — Age of Duece	2,400	864,000
Kali — Age of Losing	1,200	432,000
Total — MahaYuga	12,000	4,320,000

It would appear that the Yuga theory came into being somewhere between the age of the *Rig Veda* and *Mahabharta*. There is no mention of this theory in the Vedas, or in the Upanishads. Tradition has it that the four Yugas may have been described for the first time during the *Mahabharta* age. That event took place when one of the Pandava brothers, Bhima in his journey to collect the flowers of paradise, happened to run into an old and feeble monkey; he turned out to be the great god Hanuman of the *Ramayana*. Hanuman told Bhima that they were half-brothers, both being the sons of *Vayu* (Wind god), from different mothers. The account of the four Yugas given by Hanuman runs somewhat along the following lines:

Krita Yuga: It was all *satya* (truth) and happiness. There was morality, knowledge and abundance of good things. Human life may have been in hundreds of years. There was complete absence of malice, deceit, cruelty, etc. Man did not have to seek or yearn for anything more, as the facility of every type was freely available to him. Though Krita was all happiness and goodness, there is talk of Daitya (demon) king Bali ruling during the Krita Yuga.

Treta Yuga: In this Yuga, the good qualities decreased by ¼th. The *Ramayana* was in the Treta Yuga. In the *Ramayana satya* (truth) dominated; however, we do find some (very few) deviations here and there.

Dvapar Yuga: At the beginning of the Dvapar Yuga, good and bad qualities were half-and-half, i.e. Good and Evil were in equal proportion. As the Yuga elapsed, the good qualities decreased and bad qualities increased. The *Mahabharta* was towards the end of the Dvapar Yuga. We find many cases of *asatya* (untruth) and 'war code' violations dominating the scene during the *Mahabharta*.

Kali Yuga: At the start of the (present) Kali Yuga, 'evil' was three times the 'good'. As the Yuga elapses, evil increases until towards the end of the Kali Yuga, all evil will prevail. As per one version, the Kali Yuga started in 3102 BC, when Lord Krishna was killed by a hunter's arrow hitting his heel. Tradition says that in the Kali Yuga, the stock and prestige of Brahmins and Kshatriyas will fall, and that of Vaiyshas and Shudras will rise. Power will pass from the former to the latter, and may be from men to women. (Do we see a glimpse of that in the phenominal rise of Mayavati?)

The Kalpa

The period of the sum of the four Yugas, i.e. 12,000 'years of gods', or 4.32 million human years, constitutes One Mahayuga, i.e. four Yugas combined; it is also called Chatur-Yuga. One thousand Mahayugas make one Kalpa. Summarized:

One Kalpa   = One Day of Brahma = 1,000 Mahayugas
                = 12 million 'years of gods' or celestial years
                = 4.32 billion human years.

Tradition has it that Brahma destroys the universe at the end of each Kalpa. Then, Brahma sleeps for a Kalpa night, equal in duration to one Kalpa day. At the end of the Kalpa night, Brahma recreates the universe. 360 Kalpas make one year of Brahma. Life of Brahma is 100 such years; at the end of that, a new Brahma emerges. Tradition has it that presently we are in the 51st year of Brahma.

The Practical Yugas

The length of various Yugas given above is pure mythology. It is highly unlikely that the Yugas would have been in millions of years. If, indeed, there were the four Yugas, their length is likely to have been in hundreds or maximum in thousands of years. Thus, on a practical note, the following durations have been suggested for the four Yugas:

Krita Yuga: It started with Manu and ended with King Bahu, who was 40th generation of the Suryavanshi kings. Tradition says that an off-shoot tribe of the Yadavas, called Haihayas carried out major depredations all over North India. They deposed King Bahu (aka Asita). Haihayas had support from foreign tribes like Yavanas (Greeks), Pahlavas (Iranians) and Sakas. A view has been expressed that Parshurama's repeated exterminations of Kshatriyas, actually refers to depredations by the Haihayas. Likely duration of the Krita Yuga would have been 800 years @ 20 years per generation.

Treta Yuga: King Sagara (son of Bahu) and of the 41st Solar generation, snatched power from the Haihayas, and drove them out. He re-established the Suryravanshi line; that was the start of the Treta Yuga. Lord Rama, seventh *avatar* of Vishnu, lived in this age. Treta ended with Rama's victory over Ravana, and the destruction of the

Rakshasas. Treta lasted 25 generations (65th of Rama — 41st of Sagara); that would be about 500 years.

Dvapar Yuga: This Yuga started with coronation of Lord Rama, and ended with the death of Lord Krishna, the eighth *avatar* of Vishnu, at the end of the *Mahabharta* war. The Yuga lasted 30 generations (94th of Krishna — 65th of Rama); that would be about 600 years.

Kali Yuga: This current Yuga started with the death of Lord Krishna, which is sometimes placed in 3102 BC. The Kali Yuga will end at some future date, when Kalki, the tenth *avatar* of Vishnu, would appear.

The practically possible Yugas, in tabular form are as follows:

YUGA	START	END
KRITA (40 generations)	Manu	40th Solar king Bahu was defeated by the Haihaya king Talajangha
TRETA (25 generations)	41st Solar generation: King Sagara snatches back power from Haihayas	65th Solar generation: Lord Rama defeats and kills Ravana
DVAPAR (30 generations)	65th Solar generation: Lord Rama's coronation	94th Solar generation: Death of Lord Krishna
KALI	94th Solar generation: Death of Lord Krishna (3102 BC)	Current

Manvantras

The period of one Kalpa is ruled by 14 Manus, who are specially designated by Brahma for this purpose. The period of rule of one Manu is called a Manvantara, which is equal to 71 Mahayugas, or about 300 million human years. It is believed that presently we are in the rule of the seventh Manu, called Manu Vaivaswat (Chapter 8.1). Some texts refer to the destruction of the universe at the end of each Manvantara, and not Kalpa as referred above.

7.2 Creation of the Universe

In Hinduism, there are many accounts of the creation of the Universe. Manusmriti gives the following account:

"The whole universe was enveloped in darkness, without being perceived, and undiscoverable as it was in sleep. Then the self-existent and indiscernible Lord, causing this universe with its five elements and other things, to become discernible, was manifested, dispelling the all prevading gloom. The Supreme Lord, who is beyond all senses, and essence of all beings, shone forth Himself. Thereafter, desiring to produce a variety of creations, He first created the Waters and deposited a seed in it. That seed became the golden egg Hiranyagarbha, shining as the sun, in which the Supreme Lord manifested Himself as Brahma, the progenitor of the universe. He continued one year in the egg; thereafter, he divided the egg into two parts by his thought. With these two parts, Brahma formed the heavens and the earth, and in the middle he placed the skies, the eight regions and the eternal abode of the waters."

The Vishnu Purana describes the wonderful egg Hiranyagarbha as follows:

"Its womb vast as the mountain Meru was composed of high mountains and the great oceans which filled all its cavity. In that egg were the gods, the demons, the human beings, the oceans, the continents, the mountains, the planets and all other divisions of the universe. The egg after the Creator had inhabited it for a thousand years burst open; Brahma issuing forth by his meditation, commenced the work of creation."

However, the *Mahabharta* and some *Puranas* give a somewhat different account of the emergence of Brahma. Lord Vishnu in the form of Narayana rested in the waters, reclining on the coiled serpent Sesha or Ananta. When Narayana woke from his cosmic sleep, a lotus emerged from his navel; ensconced in the lotus was Brahma, who then set about the task of creation.

Rishi Vasishta in the *Ramayana* gives still another account of the creation of the world:

"Initially all was water in which earth was formed. Thereafter rose Brahma, the self-existent along with all the deities. Then becoming a boar, he raised up the earth and created the whole world with the *dev-rishis*, his sons."

Note 1: Here, the 'boar' has been associated with Brahma. However, in the more conventional version, 'boar' was the 3rd *avatar* of Vishnu, who recovered the earth from down-under.

Note 2: The stay of Brahman/Brahma in the egg Hiranyagarbha has been variously given as one year, or 1000 years. The one-year period could be a reference to some sort of a 'year of gods', which we have, however, earlier defined as 360 human years.

The word Hiranyagharbha literally means 'golden womb'; however, over time it came to be called golden egg. It has also been called Brahmanda — the egg of Brahma. The egg split into two, forming the heaven and earth. Mount Meru was in the centre of the earth, with rings of continents and seas surrounding it. Seven heavenly realms (called Lokas) where gods reside are above the earth. Seven nether worlds (called Talas) are below the earth; these are the residence of various types of beings like demons. yakshas, *nagas* (serpents), and others. In addition, there are seven hells, where people with bad *karma* (deeds) have to reside.

7.3 Primeval Rishis (Heavenly Sages)

Brahma the Creator god has unlimited powers of creation. At beginning of the universe, Brahma, through his mind power, created primeval or mythical *rishis* (sages). There are various versions of their creation, as well the number of the *rishis* so created. The numbers vary from 7 to 17 in various texts like Manusmriti, Satpatha Brahmana, *Mahabharta* and *Vayu* Purana. For our text, we will adopt the figure 17. The primeval *rishis* are further sub-divided into *dev-rishis* (7 numbers) and prajapatis (10 numbers); that sub-division is a bit arbitrary.

Before we go on to describe the primeval *rishis*, some points need to be noted:

- There are various versions of stories/events given below. We give only one of those versions, and that too, in a highly simplified form.

- We have called *dev-rishis* and prajapatis as mind-born sons of Brahma. However, there are alternate versions of creation/ birth of many of these primeval *rishis*. That may be a cause of some confusion.

- Some of these *rishis*, especially Vasishta and Vishvamitra, are reported to have lived during periods and events, which were separated from each other by hundreds, or even thousands of years, e.g. same rishi being present during *Rig Veda* as well as during the *Mahabharta* and/or *Ramayana*. There could be two possible explanations for the same:

  — Vasishta and Vishvamitra were not names of individual *rishis*, but were family names, sometimes called *gotra*. That would suggest that there were many Vasishtas and Vishvamitras, who would then have their respective first names. We do come across such first names here and there.

  — Such exalted *rishis* may have lived for hundreds, or even thousands of years. That may not be as big an impossibility, as it appears to us today. There are references to very long lives during the Krita Yuga.

The above argument would be equally applicable in the case of other *dev-rishis* like Angiras, Bhardwaj, etc.

- One possible way to describe the primeval *rishis* may be to call them as semi-celestial beings–neither fully men, nor fully gods. They perhaps had equal access to earth and heaven. Narada was the most frequent traveler between the two domains.

- In a rather arbitrary manner, we list the following seven as *dev-rishis*: Bhrigu, Angiras, Vasishta, Vishvamitra, Kashyap, Atri, Pulastya. In an alternate list, Bhrigu and Angiras get replaced by Gautama and Jamadagni. Generally, Bhrigu is considered the senior-most rishi, he was the first to emerge. It is difficult to explain his non-inclusion in any list of *dev-rishis*.

- The 10 remaining *rishis* not listed as *dev-rishis* above may be considered Prajapatis.

- In 1.35 of Manusmiriti, Manu lists the following ten as the primeval *rishis*; Marichi, Atri, Angiras, Pulastya, Pulaha, Kratu, Pracetas, Vasishta, Bhrigu and Narada. For some strange reason, two of the important *rishis*, i.e. Kashyap and Daksha (a prajapati) are missing from this list; these two were primarily responsible for populating the world. Even Vishvamitra does not find a place; rather unimportant *rishis* like Kratu and Pracetas have been included.

- Many of the primeval *rishis* are associated with the hymns of the *Rig Veda*. The name of the particular rishi appears along with that hymn.

Dev-Rishis and Prajapatis

Tradition says that Brahma, through his mind-power and/or means of *yajnas* (fire-sacrifices), created the following:

- Four Sanat Kumars: Pre-adolescent boys, unaware of sex
- Seven Dev-Rishis: Heavenly sages, sons of Brahma
- Ten Prajapatis: Another set of heavenly sages

The *dev-rishis* and prajapatis make a total of 17 primeval *rishis*. There is no unanimity of opinion as to which *rishis* come under the *dev-rishis* and which are the prajapatis. We list the 17 *rishis* below.

1. Bhrigu

Tradition says Bhrigu was born of a *yajna* (fire-sacrifice), conducted by Brahma himself. Bhrigu founded the Bhargava line, whose family tree is as follows:

Bhargavas

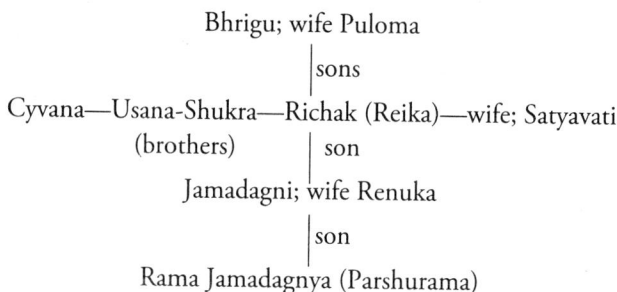

Bhrigu; wife Puloma
|sons
Cyvana—Usana-Shukra—Richak (Reika)—wife; Satyavati
(brothers) | son
Jamadagni; wife Renuka
|son
Rama Jamadagnya (Parshurama)

Richak's wife Satyavati was the sister of *dev-rishi* Vishvamitra, who was thus the maternal uncle of Jamadagni and granduncle of Parshurama. In one version, Jamadagni had two brothers named Sunassepa and Sunapucha. However, in a second version, Jamadagni had a brother called Ajigarta, whose two sons (out of the three) were named Sunassepa (middle son) and Sunapucha. Sunassepa became a great sage, whose story we give a few paragraphs later. Usanas-Shukra or Shukracharya, the guru of the *asuras* (demons) was also a son of Bhrigu, hence a Bhargava. Through a second wife Khyati, Bhrigu had a daughter Lakshmi; she became the wife of Vishnu.

Bhrigu wanted to give a boon to his daughter-in-law Satyavati. He made two potions, one for Satyavati and another for her mother who was Kshatriya. But the potions got interchanged. As a result —

- Satyavati's mother gave birth to Vishvamitra (nee Vishvavartha), who though born a Kshatriya, finally became a Brahmin.
- Satyavati's son Jamadagni was set to get Kshatriya characteristics. However, on the specific request of Satyavati, Kshatriya characteristics were passed over to Satyavati's grandson Parshurama.

Bhrigu was the patriarch of a Vedic priestly family. He married a daughter of prajapati Daksha and officiated as a priest at the *yajna* (fire-sacrifice) organized by Daksha. That gave offence to Shiva, who pulled Bhrigu's beard. Once Bhrigu was deputed to confirm as to which of the three gods of the Trimurti was the greatest. Tradition says that after visiting the three gods, Bhrigu declared Vishnu to be the greatest amongst the three.

Bhrigu's wife was Puloma, who was abducted by a *rakshas*, also named Puloma; she was recovered later. Brighu is considered the most senior *dev-rishi*. In the *Bhagwad Gita*, Lord Krishna called himself Bhrigu amongst the *dev-rishis*.

2. Kashyap

He is one of the important *dev-rishis*, though sometimes, he is not listed among the *dev-rishis*; some texts refer to him as a prajapati. Sometimes, he has been called the son of Marichi.

After Brahma, Kashyap had the major task of creation. He was a mind-born son of Brahma, who married 13 daughters of prajapati Daksha. Demons, gods, animals, serpents, and birds emerged out of the union of Kashyap with his various wives. Later, humankind emerged from one of the gods. The more important wives of Kashyap (all daughters of Daksha) were:

- *Aditi:* She gave birth to adityas and other *deva*s (gods) and has been called *deva-matri.*
- *Diti:* Gave birth to Daityas (demons). She was the mother of daityas Hiranyaksha and Hiranyakashipu, father of Prahlad.
- *Danu:* Mother of *Danavas* (demons)
- *Vinita:* Gave birth to garuda (eagles) and birds
- *Kadru:* Mother of *nagas* (serpents)

Of the above, Aditi is the most important. We take her up for detailed study in the following paragraphs.

Aditi

Aditi has been called the Vast Expanse, or Infinite Universe. In all, Aditi gave birth to 33 sons; amongst these were 12 (or 8) Adityas (sun gods), eight Vasus (assistants of Indra) and 11 Rudras (Maruts- storm gods, sons of Rudra).

As per an earlier version, eight adityas were born to Aditi. She took seven adityas (Varuna, Mitra, Aryaman, Bhaga, Daksha, Indra and Ansu) to the gods. In some form or the other, these are names of Surya, the sun. Even Indra is identified with the blazing, mid-day sun; of the seven, Indra was perhaps the strongest aditya. The eighth aditya called Martanda (another name of Surya) was cast away. Here, we quote two verses from a *Rig Veda* hymn (*RV,* X, 72):

Verse 8: 'Eight sons were there of Aditi, who were born of her body. With seven, she went forth among the gods, but she threw away Martanda, the sun.'

Verse 9: 'With seven sons, Aditi went forth into the earlier ages. But she bore Martanda so that he would in time beget his off-springs and then die.'

Originally, Martanda meant 'born of an egg'. Later, it came to mean 'dead in the egg', or even miscarriage. Martanda was unformed and unshaped; other adityas shaped him and made him into the sun. Another name of Martanda is Vivaswat, which means the shining one, and is the name of the sun in its creator form. On growing up, Vivaswat (Martanda) fathered Manu, who is, therefore, called Manu Vaivaswat — the seventh Manu (see Chapter 8.1).

As per a later version, Aditi gave birth to 12 adityas. Names of the additional four adityas are Dhatr, Amsa, Pusan and Tvastr (or Vishvakarma). In another version, Tvastr is called the father of Vishvakarma. The twelve adityas are sometimes identified with the 12 aspects or phases of the sun. Another popular name for Surya, the sun is Savitre or Savitri; it represents the invisible (night time) or setting sun. Savitre is also an aditya. The word aditya (in singular) is used for the sun.

Sometimes, Vishnu and Prajanya (rain-cloud) are mentioned as adityas; this may not cause much surprise, as Vishnu of the Vedas was a sun god. Tradition also says that Vishnu as a dwarf was born to Aditi; Vishnu is reported to have called Aditi his mother.

We may note that Daksha is listed above as the son of Aditi. We started our story with Aditi being a daughter of Daksha. Here we quote a *Rig Veda* hymn (*RV*, X, 72, 4):

'The earth was born from her who crouched with legs wide spread; and from earth, were born the quarters of sky. From Aditi, Daksha was born. From Daksha was born Aditi.'

Thus, *dev-rishi* Kashyap was the creator of all living beings, i.e. gods, demons, birds, serpents, etc; however, some alternate versions of their creation/birth are given later. Humans emerged from a son of Kashyap, named Vivasvat, the sun. Vivasvat's son was Manu Vaivasvat, who is the progenitor of humankind.

3. Vasishta
Vasishta was a mind-born son of Brahma. However, due to a curse, he was later born as son of Varuna and Mitra, through the

nymph Urvashi; thus, he is called 'Maitravaruni'. The family tree of Vashistas is:

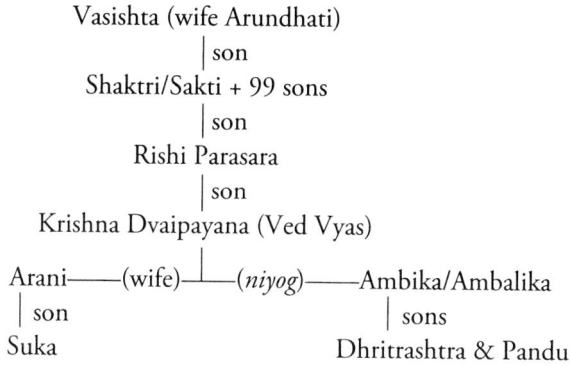

<pre>
                 Vasishta (wife Arundhati)
                          │ son
                 Shaktri/Sakti + 99 sons
                          │ son
                    Rishi Parasara
                          │ son
            Krishna Dvaipayana (Ved Vyas)
                          │
    Arani────(wife)──┴──(niyog)────Ambika/Ambalika
      │ son   │ sons
    Suka   Dhritrashtra & Pandu
</pre>

Rishi Parasara was born of his mother Adrsyanti, after the death of his father Sakti; Parasara became a great saint. Parasara's son Ved Vyas impregnated the widowed Paurava queens Amba and Ambalika, through the rite of *niyog*; for more details, see *Mahabharta* in Chapter 5.8.

Along with Vishvamitra, Vasishta was one of the most important *dev-rishis*. In addition to being the family priest of *Rig Veda* King Sudas, he was the family priest of the Suryavanshi clan of Ikshvaku, son of Manu. Vasishta served many Suryavanshi kings; from the very early times, through Satyavrata Trishanku and Harishchandra, on to Dasrath and Rama. Sometimes, he has been given the name Devaraj Vasishta.

Vasishta had the wondrous wish-cow of plenty, Nandini (off-spring of the cow Surabhi which had emerged during the Ocean Churning); Vishvamira (as the Kshatriya king Visvaratha) wanted to posses the cow. That led to enmity between the two. Vasishta, with his one breath reduced the hundred sons of Vishvamitra to ashes. Vishvamitra also destroyed the 100 sons of Vasishta.

Vasishta was present at events, which were separated by hundreds (may be thousands) of years. This has led to the controversy whether there was one Vasishta who lived for hundreds of years, or there were many Vasishtas. Without taking a specific position on that, we list below the seven (+1) times when Vasishta is reported to have been present. Where possible, the respective first name of Vasishta has also been given. By listing the seven (+1) Vasishtas, we are not saying that there were, in fact, seven (+1) Vasishta. The first Vasishta was the primeval rishi, who was the mind-born son of Brahma.

Vasishtas

Vasishta's Sl. No. & First Name	Solar Kings	Solar Generations	Remarks
1 Dev-Rishi	Brahma's son	N/A	Primeval Rishi
2* Devraj	Satyavrata Harishchandra Arjuna (Haihaya)	31 32 31	
3 Apava (Varuni) or Arthavanidhi I	Bahu Sagara	40 41	
4 Sethabhaj	Kalmasapada	54	100 sons, eldest Sakti
5 Arthavanidhi II	Dilipa II	60	
6	Dasratha Rama	64 65	
7	Sudas (Rig Veda)	68	
8	Mount Abu Yajna		Indeterminate Date (Rajput Times?)

Columns in the above table are as follows:

Col. 1: Possible serial number of Vasishta, with his first name
Col. 2: Names of kings when the respective Vasishta lived
Col. 3: Serial number of Solar kings in col. 2

The following points are made in respect to the above table:

- We have shown 100 sons of Vasishta against serial 4. They could be shown against some other serial number, especially serial 7.

- In the earlier Vasishtas, there is a difference of 10 to 15 generations; that could mean 200 to 300 years @ 20 years per generation.
- The eighth Vasishta has been tentatively included. Tradition says that he conducted a *yajna* (fire-sacrifice) at Mount Abu, at an indeterminate date in the past. The folk-lore says that four Rajput mythical figures emerged out of that *yajna*. Those figures went on to start Rajput clans like Parmara, Chauhan, etc.

4. Vishvamitra

Manu's daughter Ila had a son Pururavas. A son of Pururavas, Amavasu founded a Kshatriya dynasty at Kanyakubja (present Kannauj). In that dynasty, there was King Kush or Kusa, who is listed as the 12th king of that dynasty. Kusa had synchronism with the 28th king of the Solar line which ran without a break. King Kusa's family tree is as follows:

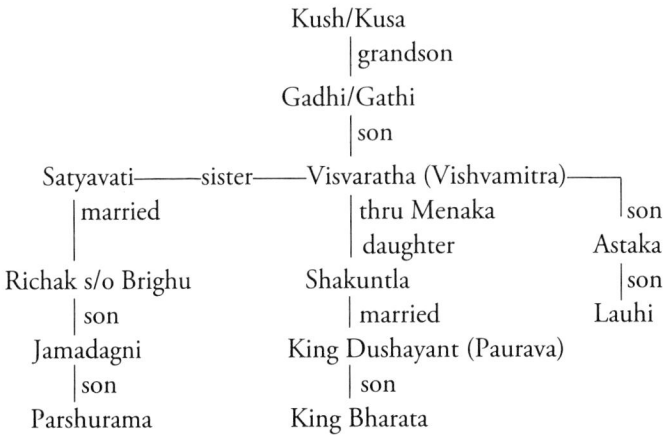

```
                        Kush/Kusa
                          │ grandson
                        Gadhi/Gathi
                          │ son
Satyavati────sister────Visvaratha (Vishvamitra)────┐
  │ married              │ thru Menaka              │ son
  │                      │ daughter              Astaka
Richak s/o Brighu       Shakuntla                  │ son
  │ son                  │ married               Lauhi
Jamadagni              King Dushayant (Paurava)
  │ son                  │ son
Parshurama            King Bharata
```

The above means that Vishvamitra had synchronism with the 32nd king of the Solar line, who was King Harishchandra. As Vishvamitra was descended from Kush/Kusa, the family is called Kaushik, or Kusika. The birth of Vishvamitra as a Kshatriya, and his above genealogy may raise a question mark about him being a mind-born son of Brahma. This is often explained by saying that there may have been many Vishvamitras.

Once, King Vishvratha (later Vishvamitra) visited the hermitage of Vasishta. Vishvaratha was very impressed with all the services provided by the wondrous wish-cow Nandini of Vasishta. He wanted to possess the cow, but Vasishta refused. That resulted in a fight. The poor Brahmin Vasishta succeeded in defeating the great Kshatriya King Vishvaratha. In that fight, Nandini helped Vasishta, by producing warriors from her body. Vishvaratha realized that he must raise himself to be a Brahmin, and took to severe *tapasya* (austerities and penance) for that purpose.

There are references that Vishvamitra's austerities might have continued for hundreds of years; however, another version says that these lasted 12 years. At some stage, Visvaratha changed his name to Vishvamitra. Indra sent celestial nymphs, first Menaka and then Rambha to disrupt the austerities of Vishvamitra. Whilst Menaka succeeded, Rambha failed. Out of the union of Vishvamitra and Menaka was born the girl Shakuntala; she became the wife of King Dushyant, and mother of king Bharat after whom our country Bharat is named. Thus, king Bharat was a grandson of Vishvamitra, through his daughter Shakuntla. After a lot of twists and turns, Brahma finally raised Vishvamitra to be a Brahmin. This tale appears to show that it was very difficult (almost impossible) for a Kshatriya to become a Brahmin. From a Brahmin to a Kshatriya may have been an easier process. There are tales of the mixing of Brahmin and Kshatriya blood, and the term 'Kshatriya-Brahmins' was in use.

Vishvamitra and Vasishta, were in turn, the family priests of King Sudas of the *Rig Veda*. There was severe rivalry and enmity between the two; the *Rig Veda* also refers to their famous rivalry. Each is held responsible for the death of 100 sons of the other. There was a King Kalmasapada of Ayodhya; Vishvamitra turned him into a *rakshas*, who ate the hundred sons of Vasishta. Kalmasapada's insanity lasted 12 years. There is also a reference to a fight between Vishvamitra and Vasishta, with both the *dev-rishis* being in the form of birds. Brahma had to intervene to stop that fight. Manusmriti 10.108 records as follows: 'When Vishvamitra, who knew the difference between right and wrong, was distressed by hunger, he set out to eat the hindquarters of a dog, which he received at the hands of a Fierce Untouchable'.

There was the Suryavanshi king Trayyaruna (30th generation from Manu). He sent his son Satyavrata in exile, and handed over the reigns of his kingdom to his family priest Vasishta. Vishvamitra helped Satyavrata regain his throne after 12 years. At a later stage, Satyavrata wanted to ascend to heaven in his bodily form, which was against the very law of nature. He approached his family priest Vasishta and his 100 sons for their help; however, they refused. In some texts, this particular Vasishta's full name has been given as Devaraj Vasishta. King Satyavrata then turned to Vishvamitra who agreed to help, to spite Vasishta. By his special powers, Vishvamitra sent Satyavrata to heaven. Whilst Vishvamitra was pushing Satyavrata to heaven, Indra was pushing him downwards. Vishvamitra even threatened to create a separate heaven for Satyavrata; that unnerved Indra. Finally, in that conflict situation, a compromise was reached. Satyavrata was suspended head downwards, in between heaven and earth. Since then, Satyavrata has stayed in that position, and is called Tri-Shanku.

Satyavrata's son was the famous King Harishchandra (32nd generation from Manu). Once his family priest, Vasishta happened to praise the king. That annoyed Vishvamitra, who decided to reduce Harishchandra to penury. In the guise of getting alms, Vishvamitra divested Harishchandra of his empire and all riches. So much so, that Harishchandra had to sell his wife and son Rohitashwa, and finally himself. He had to take the job of a *chandal* (a fierce untouchable), to burn corpses. Finally, Vishvamitra was pleased with the sacrifices made by Harishchandra, and restored his empire, riches and family.

We list below six possible Vishvamitras. Columns in the table below indicate the following:

Column 1: Serial number of the respective Vishvamitra
Column 2: Names of the kings, Solar or otherwise, in synchronism

Column 3: Serial number of synchronous Solar kings

Vishvamitras

Serial No	Solar King	King's Generation	Remarks
1	Dev-Rishi	0	Brahma's mind-born son
2	Satyavrata	31	Kshatriya Visvaratha becomes
	Harishchandra	32	Brahmin Vishvamitra
3	Bharata (Lunar king)	44	Vishvamitra was grandfather of Bharata
4	Kalmasapada	54	
5	Rama	65	
6	Sudas (N Panchala king)	68	Of *Rig Veda*

We make the following comments on the above table:

- By listing six Vishvamitras, we are not saying that there were, in fact, six Vishvamitras. There may have been only one, the original mind-born *dev-rishi*, who might have lived for hundreds of years.
- Unlike Vasishta, we do not come across the first names of the six Vishvamitras.

Vishvamitra was the maternal uncle of Jamadagni, and grand-uncle of Parshurama. He was also uncle or grand-uncle of rishi Sunassepa, whom he adopted as a son, and renamed him Devarata (see his story later in the chapter). That move was not liked by some of his 100 sons; Vishvamitra cursed them to become dog-eaters. One of Vishvamitra's sons, Astaka ascended the Kankyakubja throne; the rest remained Brahmins.

5. Jamadagni

Jamadagni, whilst being the mind-born son of Brahma, is also shown as the grandson of Bhrigu (see above under Bhrigu). He is thus a Bhargava. Jamadagni was husband of Renuka, a princess of Ayodhya. They had four sons, of which Parshurama, the sixth *avatar* of Vishnu, was the youngest. Jamadagni suspected Renuka of adultery and got her beheaded by Parshurama. However, she was later restored to life.

A Kshatriya, Haihaya (an offshoot of the Yadavas) king Arjuna Kartavirya misused the hospitality of Jamadagni and carried away his sacred wish-cow. Parshurama killed the king. In revenge, the sons of Arjuna killed Jamadagni. That infuriated Parshurama and he wiped out entire every Kshatriya clans 21 times (However, another version says that repeated killings of Kshatriyas by Parshurama, in fact, refer to the depredations carried out by the Haihaya hordes).

6. Atri

The Brahmin family founded by Atri is called the Atreyas. Atri's wife was Ansuya, a daughter of prajapati Daksha. There is a story that once Brahma, Vishnu and Shiva came as mendicants to Ansuya's house. They expressed that they would accept the alms only if Ansuya appeared 'in the buff'. Ansuya turned the three gods into toddlers, and hence had no problem in meeting with their condition.

Ansuya then expressed that she would like to have sons like the three gods. Thereupon, Ansuya became mother to three sons, i.e. Soma/ Chandra (like Brahma), Datt-atreya (like Vishnu), and Durvasa (like Shiva). In some cases Datt-atreya has been called an *avatar* (incarnation) of Vishnu. In the *Ramayana*, there is an account of Lord Rama and Sita visiting the hermitage of Atri and Ansuya.

As per another version, the three gods, under the influence of Kama, were full of desire for Ansuya, and asked to be accommodated. On being refused, they threatened to use force. That annoyed Ansuya, and she cursed the gods, impinging upon their worship. She added that they would be free of the curse only after they were born as her sons. Therefore, the three gods were indeed born as her sons, as per the names in the preceding paragraph.

7. Gautama

He was husband of the most beauteous Ahelya. She was seduced by Indra — by deceit as per one version, by her consent as per another. Gautama cursed Ahelya and turned her into stone. Lord Rama in the Treta Yuga removed the curse. Gautama also made Indra suffer for that adultery; Indra was emasculated and his scrotum was detached.

8. Angiras

Sometimes Angiras and Bhrigu have been called the most senior *dev-rishis*. Tradition says that Angiras emerged from the 'angaras' (burning coal). Rishi Angiras was adopted by Agni; as such, the family has been called Agneya; it is also called Angirasa or Atharvan. Angirases was priest to the kings of the Vaisali dynasty; Marutta Aviksita was a famous king of that dynasty. Angiras had three sons Brahspati, Samvarta and Utathya (Ucathya).

The following belong to the Angirasa family:

Brhaspati: son of Angiras, and acharya/guru of the *devas* (gods)
Utathya or Ucathya: brother of Brhaspati
Bhardwaj and Raca: sons of Brhaspati
Krpacharya: priest of the Kauravas, and a great warrior (*Mahabharta*)
Dronacharya: guru of the Kauravas and Pandavas (*Mahabharta*)
Ashothama: son of Dronacharya, and a great warrior

Dronacharya has also been listed as son of Bhardwaj, who could be a later Bhardwaj. The story of Raca is given a few paragraphs later under 'Shukracharya'.

9. Bhardwaj

As Angiras emerged from the '*angaras*' (burning coal), Bhardwaj emerged from quenched '*angaras*', when these were re-ignited. Bhardwaj is an Angirasa, and the son of Brhaspati; hence, he has been called Bhardwaj Brhaspatya. Bhardwaj was a major contributor to Book 6 of the *Rig Veda*, where 57 hymns are attributed to him.

There is a fanciful tale told about the birth of Bhardwaj. Utathya (Ucathya) and Brhaspati were brothers. Utathya's wife was Mamata, daughter of Soma, the moon. As it happened, Mamata got simultaneously pregnant, both by her husband (resulted in the birth of son Dhirgatamas, born blind) and Brhaspati (result, son Bhardwaj). Dirghatamas kicked his half-brother Bhardwaj out of the womb, before his time.

Tradition says that Bhardwaj was adopted as a son by the great Paurva King Bharata; his full name was Vidathin Bhardwaj, who may have been a descendent of the original Bhardwaj. However, for all we know, he might have been the original Bhardwaj himself. Ultimately, Vidathin Bhardwaj's son Vitatha succeeded to the Paurva throne, after the death of Bharata. That indicates induction of the Brahmin blood in the great Paurva/Puru line, which goes down to the *Mahabharata* dynasties.

10. Narada

Tradition says that Narada sprang from the forehead of Brahma. However, he became the son of *dev-rishi* Kashyap, due to a curse by Daksha. *Rig Veda* lists him as belonging to the Kanwa family.

Narada is wise and elegant and a musician of extra-ordinary caliber; he invented the musical instrument *'vina'* (lute). He has a reputation of being a gossip, a strife-maker and a meddlesome person. Narada is a devotee of Vishnu and keeps on reciting the name 'Narayana'. Narada visited *Patala* (the nether worlds), and was very happy with what he saw there.

Narada dissuaded the *Dev-rishis* and Prajapatis from populating the world. For that, Brahma cursed Narada; the curse included that Narada would be born of a slave girl; the curse may have been later recalled. In turn, Narada cursed Brahma (his father) that he would not be worshipped for three eons.

11. Marichi

He is one of the primeval *rishis*. Sometimes, he is projected as the father of Kashyap. However, we have listed Kashyap above as a *dev-rishi* in his own right, and not as the son of Marichi.

12. Daksha

He is regarded as the chief of the prajapatis. Sometimes, he is identified with Prajapati, the Creator himself. Daksha had, by various counts, 24, 50 or 60 daughters. According to Manu, he gave 10 of his daughters to Dharma (god of Justice); 13 of them to *dev-rishi* Kashyap, who became mothers of gods, demons, birds, serpents and all living beings. 27 daughters were given in marriage to Chandra, the moon; these are often projected as the 27 lunar *nakshatras* (asterisms — lunar mansions).

One of the daughters of Daksha, Sati (aka Uma) married Lord Shiva, against Daksha's wish. Daksha organized a major *yajna* (fire-sacrifice). He invited all the gods, but not his own son-in-law, the great Shiva. Sati felt insulted and jumped into the sacrificial fire. Shiva was infuriated and attacked the sacrifice; he also cut off Daksha's head. Later, he replaced it with the head of a goat as a perpetual sign of Daksha's ignorance in not recognizing the majesty of Shiva.

Sometimes, the feud between Shiva and Daksha is represented as the reluctance of (presumed) Aryans to accept a pre-Aryan god that Shiva might have been. Shiva himself told Sati that other gods were not prepared to share the 'sacrifice' with him; that is why he had not been invited.

Daksha is recognized as a lawgiver and is reckoned among the 18 writers of the Dharam-Shastras.

13. Pulastya
He received the Vishnu Purana from Brahma and communicated it to rishi Parasara (grandson of rishi Vasishta), who made it known to humankind. Pulastya was the father of Visravas, whose sons were Ravan, the demon king of Lanka, and his half brother Kubera, the god of wealth and king of yakshas. Pulastya is also considered to be the progenitor of all rakshasas (earthly demons). Sometimes, *daityas* and *danavas* are also traced to Pulastya. Sons of Pulastya are called Poulastyas. Pulastya may also be the creator of vanaras (monkeys), yakshas and kinnaras.

14. Pulaha
He is the progenitor of goblins, tigers, lions and other animals.

15. Kratu
He perhaps has no descendents.

Kanva (an offshoot of Angiras) and Prachetas are mentioned as the two remaining *rishis*, making a total of 17 primeval *rishis*.

Dev-rishis' Residence
Out of the 17 *rishis* listed above, seven are Dev-Rishis. There is no unanimity on their names; we have given tentative lists above. The seven

dev-rishis reside in the Zodiac as seven stars of the Great Bear, called *sapt-rishis* in Hindi.

The six wives of the *dev-rishis* (except Arundhati, wife of Vasishta) became foster mothers to *Karttikeya* (Skanda), son of Shiva. As a reward, *Karttikeya* put all seven wives of the *dev-rishis* in the Zodiac, where they shine as Pleiades, in the constellation of Ursa Major; they are also called the Seven-Sisters.

Additional Sages
At this stage, in addition to *Dev-Rishis* and Prajapatis, we would like to cover some other ancient sages, whose names appear in the scriptures. We know some (but not all) of these to be actual historic figures. Due to constraint of space, we are listing only a few names.

Panini
Panini was a famous Sanskrit grammarian, who lived between sixth to fourth century BC. His grammar Ashtdhyayi (eight books) with about 4,000 *sutras* is an unparalleled linguistic achievement; it describes the science of grammar with great clarity. We have listed two of his books Siksha and Vyakaran under 'Vedangas' (Chapter 5.14). His exceptional skill in the subject of grammar has won the admiration of numerous western linguists. He is the standard authority on Sanskrit grammar; tradition says that for that work of his, he might have got inspiration from Shiva himself.

Patanjali
He perhaps lived in the 2nd century BC, and is founder of the Yoga philosophy. He wrote the Yoga Sutra, an influential text on Raja Yoga. Patanjali is also the author of the text called Maha-bhashya, which is a recognized commentary on the grammar of Panini, and supports his work.

Markandeya
As a boy, he was given a life of only 16 years and was destined to die on his 16th birthday. Markandeya took to severe austerities to Shiva. On the appointed day, he clung to Shiva's *lingam*, chanting the name of the great Lord. Yama, the god of death came and snatched him away from the lingam. Shiva was infuriated at his devotee being treated in

that manner; he burnt up Yama with his third eye. That upset the very balance of the cosmos. Other gods went running to Shiva with the plea to restore life to Yama. Finally, Shiva relented and revived Yama.

No harm came to Markandeya; Shiva granted him (near) immortality. Markandeya lived to a very old age and is called 'Dhrig-ayu' (long-life). Children are blessed to have the age of Markandeya. He is reputed to be the author of the Markandeya Purana.

Parasara

He was a great saint and grandson of *dev-rishi* Vasishta, and a disciple of rishi Kapila. He was born of his mother Adrsyanti, after the death of his father Sakti, who was reportedly killed by Vishvamitra. Parasara was the father of Krishna Dvaipayana (Ved Vyas), through the fisherwoman Matsyaghangi, who later became famous as Queen Satyawati.

Parasara received the Vishnu Purana from *dev-rishi* Pulastya; he taught it to his disciple Maitreya. Parasara is also the author of some of the Dharam Shastras (law codes).

Yajnavalkeya

He was one of the greatest *rishis* (sage), whose name is associated with the authorship of the White Yajur Veda, the Satapatha Brahmana and the Brihad Aranyaka. He also wrote the Dharam Shastras on the social and legal codes of the Hindus. He is counted amongst the top authors; he stands out as a colossus.

Various accounts make him to be present during the *Mahabharta* age. His name is also associated with the Yoga doctrines. He had two wives, Maitreyi and Katyayani. There are many accounts of dialogues between Yajnavalkeya and his wife Maitreyi.

The present Hindu legal system in Bharat is based on Yajnavalkeya's Dharam Shastra (law code), and not so much on that of Manu's Law Code, as is the common impression.

Sunassepa

Sunassepa (Sunahsepha) was a very famous Rishi (seer). There are two versions of his legend. In both the versions, Sunassepa, before he became a rishi, got involved in a case of (planned) human sacrifice.

First, we give the version in the Taittirya Yajur, Aitareya and Kaushtaki Brahmanas. In the clan of Ishvaku, there was the Suryavanshi King Harishchandra, as 32nd generation from Manu; he was childless. As advised by Rishi Narada, the king approached god Varuna for a son, promising to give that son in sacrifice to the great Varuna. A son was born to Harishchandra; he was named Rohita (or Rohitshwa). However, the king kept on delaying the sacrifice on one pretext or the other, until Rohita was 16 years old. Learning of his impending doom, Rohita slipped to the forests where he roamed for 6–7 years. Then, he met an impoverished Brahmin Ajigarta (brother of rishi Jamadagni, and nephew of Vishvamitra) who had three sons. It appears that Ajigarta was on the point of killing his middle son named Sunassepa. Manusmriti 10.105 records as follows: "Ajigarta, famished, stepped forward to kill his own son, but was not smeared with evil, for he was acting to remedy his hunger." Rohita arrived there at that moment and bought Sunassepa for the sacrifice, with a price of 1000 cows. Rohita came to his father along with Sunassepa. The sacrifice was prepared and Sunassepa was tied to the yupa, the sacrificial pillar. Dev-*rishis* Vasishta, Vishvamitra and Jamadagni were present at the sacrifice. At that stage, Vishvamitra gave a *shloka* (verse) to Sunassepa to sing in praise of Varuna. The god was pleased and freed Sunassepa.

As per the *Ramayana* version, Sunaseppa was brother of Jamadagni, son of Richak (Reika), grandson of Bhrigu, and nephew of Vishwamitra. King Ambrisha of Ayodhya was conducting a *yajna* (fire-sacrifice) at Pushkar; Indra stole the sacrificial animal(s) from the *yajna*. The priest expressed that it was a serious development, which called for human sacrifice. King Ambrisha bought Sunassepa for the purpose. Sunassepa appealed to his uncle Vishvamitra for help. Vishvamitra gave him *shloka*s (verses) in praise of *Agni* (god of fire), Indra and Vishnu; the gods were highly pleased. *Agni* saved Sunassepa from harm when he was put in the sacrificial fire. Ambrisha could complete his *yajna* with the blessings of Indra.

In both the above versions, Vishvamitra, who already had 100 sons, adopted Sunassepa as his son; he was renamed Devarata. Some of the sons refused to accept this action of Vishvamitra; he cursed them to become dog-eaters. Sunassepa went on to become a great seer.

Sunassepa also appears in the *Rig Veda*. Hymn RV 5.2.7 addressed to *Agni* says as follows:

"When Sunassepa was bound for a thousand*, you (meaning Agni) set him free from the stake, for he sacrificed with fervour—"

Agastya

He was a very powerful rishi (sage) of the Vedic times. He was the brother of Vasishta, both being sons of Varuna and Mitra, through the nymph Urvashi. Another version says that Agastya was the son of Pulastya.

He made the Vindhya Mountain bow its head as it was interfering in the journey of Surya, the sun. Once, Agastya drank the seas, as demons used to hide in it after disturbing the gods. Agastya, along with many other sages, came to congratulate Lord Rama on his coronation. Agastya narrated many stories to Rama at that time.

Shukracharya

His full name was Shukra Usanas, he is also called Kavya or Kavi (poet). Shukra was the son of *dev-rishi* Bhrigu; hence, he was a Bhargava. His brothers were Cyvana and Richak (Reika); the latter was the father of *dev-rishi* Jamadagni and grandfather of Parshurama. Shukra was the guru, teacher and preceptor of *Daityas*, *Danavas* and *Asuras* (all demons).

Shukra had the unique knowledge of the 'mrytu-sanjivini' *mantra*, which enabled him to raise the dead back to life. He picked up the *mantra* from Lord Shiva himself: he used it with telling effect to help the demons in their fight against the gods (see chapter 4.2 for details). As soon as the gods killed the demons, they were restored to life by Shukra. That was the major reasons for the demons gaining an upper hand vis-á-vis the gods. At one stage, worried by Shukra's austerities, Indra sent his daughter Jayanti to disturb him. However, Shukra could complete his penance, and then married Jayanti.

---

*   We have noted above that Sunassepa was bought for 1000 cows.

Worried by the power of Shukra, the gods approached Raca (Racha), son of their preceptor Brhaspati. They requested him to pick up the 'mrytu-sanjivini' *mantra* from Shukra, by hook or by crook; Raca agreed to do that. He went to Shukra and became his pupil for 1000 years. Shukra's daughter Devyani fell in love with Raca. When the demons came to know of this arrangement, they managed to kill Raca; but Shukra brought him back to life twice. Next time, when the demons killed Raca, they burned him up and mixed his ashes in wine. That wine was given to Shukra; he drank it without knowing its contents. Thus, (the dead) Raca got lodged inside Shukra. If Raca was to be revived from such a situation, Shukra would have to die. Therefore, Shukra taught the 'mrytu-sanjivini' *mantra* to Raca, so that he could revive Shukra, when he would die as Raca would burst out of Shukra. And Raca did revive Shukra. Shukra was indulgent towards Raca only for the sake of his daughter.

Having learnt the *mantra*, Raca decided to depart. He turned down the request of Devyani for marriage on the plea that being his guru's daughter, she was like a sister to him. The knowledge of the *mantra* by Raca helped the gods to finally defeat the demons. Later, Shukra's daughter Devyani was married to the great lunar king Yayati. However, Yayati was unfaithful to Devyani; Shukra cursed Yayati of premature old age. That curse was taken on by Yayati's youngest son Puru on behalf of his father. Later, Puru became the progenitor of the Puru dynasty which finally led to the Kauravas and Pandavas of the *Mahabharta*.

Brhaspati

Brhaspati was the son of *dev-rishi* Angiras. He was the guru, teacher and preceptor of gods. Thus, he was in direct confrontation with Usanas Kavya Shukra, preceptor of demons. Unlike Shukra, Brhaspati did not know the *mantra* for reviving the dead. That was a major disadvantage for the gods, and the main reason for their losing to the demons. Brhaspati had two sons, Bhardwaj and Raca. As brought out in the preceding paragraphs, Raca learnt the revival *mantra*, and helped overcome the great disadvantage of the gods (Also see chapter 4.2 for further details).

Rishis — Likely Periods

There is lot of debate as to when the various *rishis* appeared and how long they lived. One view is that some of the *rishis* might have lived for hundreds, even thousands of years. The other view is that there were many Vishvamitras, Vasishtas, Bhardwajas, etc, each living his normal span of life. We list in the table below the group of kings, when the various *rishis* might have lived —

Solar Generation and King's Name	Rishis
0 Near Start of Universe	1 *Dev-rishi* Vasistha (1)—mind-born 2 *Dev-rishi* Vishvamitra (1)—mind-born
30 Trayyaruna (Solar) 31 Satyavrata Trishanku (Solar) Arjun Kartavirya (Haihaya) 32 Harishchandra (Solar)	1 Vasishta II 2 Vishvamitra II—uncle of Jamadagni 3 Jamadagni—Parshurama—Sunaseppa 4 Bhrigu—grandfather of Jamadagni
40 Bahu (Solar) Talajangha (Haihaya) 41 Sagara (Solar) 43 Dushyant (Lunar) 44 Bharata (Lunar) Dilipa I (Solar)	1 (Apava/Atharvanidhi I) Vasishta III 2 Brahaspati—Bhardwaja 3 Vishvamitra III—grandfather of Bharata 4 (Vidathin) Bhardwaja—adopted as son by Bharata
54 Kalmasapada (Solar)	1 (Sresthabhaj) Vasishta IV 2 Vishvamitra IV 3 Bhardwaja
60 Dilipa II (Solar)	1 (Atharvanidhi II) Vasishta V
64 Dasratha (Solar) 65 Rama (Solar)	1 Vasishta VI 2 Vishvamitra V
68 Sudas (N Panchala)— *Rig Veda*	1 (Sakti) Vasishta VII 2 Vishvamitra VI
94 *Mahabharta* Age (Lunar)	1 Parasara—grandson of Vasishta 2 Ved Vyas — son of Parasara

The two columns of the above table are as follows:

Column 1: Names of cohesive group of kings (Solar and others) covering 2 to 3 (maximum 4) touching generations of kings.
Column 2: Names of Rishis who might have lived in the times of kings in Column 1.

We offer the following comments on the above table:

- Though we have listed 6 Visvamitras and 7 Vasishtas, it does not mean that there were, in fact, such different people; it just offers a possible option.
- Wherever possible, we have tried to give the first name of the respective rishi. It has been possible to do so especially in case of rishi Vasishta; no such names could be traced for Vishvamitra.
- For all we know, there might have been just one Vishvamitra and one Vasishta, the mind-born primeval *rishis*. They might have lived for hundreds, even thousands, of years.
- In the first group, we have included four kings of 30th, 31st and 32nd generations; three are Solar kings, and one Haihaya — an off-shoot of the Yadavas. We have listed four (+2) *rishis*, who could have possibly lived during their regime of say 60 years.
- The second group is clustered around 41st generation. There is a difference of 10 generations between the first and second groups; that could mean a gap of about 200 years @ 20 years per generation.
- There is a gap of about 37 (68–31) generations between the second and seventh Vasishtas; that would be about 750 years.

8
MANU

8.1 Origin of Manu

Manu is the mystery man of Hinduism. He is the first man in Hindu mythology — equivalent to Adam of the Jews and Christians. From the word Manu is derived the Sanskrit word '*mannav*' which means human being. Thus, in Hinduism, all are considered children of Manu. There are various versions of the origin and birth of Manu, and many stories are woven around that. In order not to complicate matters, we have selected the more important versions. There could be other versions, some equally valid.

As stated in Chapter 7.1, one *Kalpa* of Brahma lasts 4.32 billion years. During the period of one *Kalpa*, 14 Manus rule over the Universe; they are specially appointed by Brahma for the purpose. Names of the 14 Manus appear in the Puranas. Tradition says that presently we are in the rule of the seventh Manu.

However, some texts refer to only seven Manus. Manusmriti 1, 61 and 62 says as follows:

> (61) When Manu had spoken to *dev-rishi* Bhrigu in this way, Bhrigu's soul rejoiced and he addressed all the sages.
> (62) There are six other Manus in the dynasty of that Manu who was born of the pre-existent Brahma; they all have great souls and great energy and each has emitted his own progeny.

We go along with the 14 Manus version. For ease of reference, we take up only the first and seventh Manus for our study; we give below simplified versions of their birth.

The First Manu

The great Lord Brahma, after creating the first Manu divulged the mystery of creation to him. Later, on specific request of *dev-rishi* Bhrigu, Manu described his own birth to the primeval *rishis*, for onward transmission to humanity. Manu states as follows in the first chapter of Manusmriti (an approximate translation):

> (32) He (Brahma) divided his own self into two halves; one half was man, the other half a woman. In her, the Lord emitted the male Viraj. (33) Viraj generated heat and then emitted me (Manu), the best of the twice-born and creator of the universe. (34) Because I (Manu) wanted to emit creatures, I too generated heat, that is very hard to produce; thereafter, I emitted the ten *dev-rishis*. (35) Marichi, Atri, Angiras, Pulastya, Pulaha, Kratu, Pracetas, Vasishta, Bhrigu, Narada.

The Manu so created was the first Manu called Manu Swayambhuva (meaning self-existent). Manu then created *dev-rishis*, prajapatis and others.

The Seventh Manu

The creator god Brahma, at the start of the universe, by his mind power created *dev-rishis* and prajapatis (Chapter 7.3). One of the *dev-rishis* so created was Kashyap; he married 13 daughters of the prajapati Daksha. Manu is descended from *dev-rishi* Kashyap; there are two main versions:

First Version

In the first version, Kashyap's wife was Dakshayani (a daughter of Daksha); their son was Vivaswat (Surya, the sun in its creator aspect). Vivaswat had two wives:

- *Samjna (also called Sarayu):* She was the daughter of Tvastr/Vishvakarma. She had the following children — Manu; twins Yama and Yami; river Yamuna/Tapti
- *Chhaya (means shadow):* A double of Samjna created by Samjna herself. She had the following children — Sarvani (of the same kind) Manu and Shani

Manu is the progenitor of the human race and rules over this world. His brother Yama rules over the next world (of the dead, or of the fathers). Sometimes, Yama and Yami, though a brother-sister duo, are referred as the first human pair, from whom humanity descended. In the *Rig Veda*, there is Hymn X.10 of 14 *shloka*s (verses) containing rather extensive dialogue between Yama and Yami, on the subject of their possible union. Often Yami has been identified with the river Yamuna, which is replaced with Tapti in some texts. Yamuna/Tapti cleans humanity of its sins. Tradition says that Tapti or Yamuna was married to the Lunar king Samvarana. That may not cause surprise, as in Hindu mythology, rivers take on human form and marry humans, mostly kings. Shani is considered a malevolent planet. It punishes the unrighteous, but helps the righteous who have nothing to fear from Shani. Thus, Shani is a much misunderstood planet.

After the children were born, Samjna left Vivaswat (Surya), as she was not able to withstand his brilliance. She created Chhaya in her own exact form and left her behind. Vivasvat could not detect the difference and produced two children with Chhaya. Samjna went to her father's house, and started her austerities. At some stage, Vivasvat detected that Chhaya was not Samjna, and went after the latter. Samjna took the form of a mare, and went into the jungle. Vivasvat took the form of a horse and went after her. When they met, they touched their mouths, and Samjna (as a mare) got pregnant. Two sons Nastya and Dasra, the two Ashwins, were produced (from the nose or mouth); they were excellent physicians.

. Later, Chhaya also left Vivasvat, due to his brilliance. Therefore, Samjna's father Tvastr (Vishvakarma) took up Vivaswat for grinding on his lathe to reduce his brilliance; 15 parts out of 16 were removed. Some fragments fell in the process — out of those fragments, emerged 12 adityas. A further ⅛ part of Vivaswat (Surya) fell on the earth. Vishvakarma (Tvastr) used that portion to make Vishnu's Sudarshan Chakra, Shiva's Trishul (trident) and weapons for other gods and goddesses.

Second Version

In the second version, Kashyap's wife Aditi, also a daughter of

Daksha, gave birth to the eight adityas. She took seven to the gods, but cast off the eighth named Martanda (literally dead in the egg). Martanda is a name of the sun, also called Vivaswat. Manu is the son of Vivaswat. Another version talks of the birth of 12 adityas to Aditi (See 'Kashyap' in Chapter 7.3).

In both the above versions, Manu is the son of Vivaswat; hence, he is called Manu Vaivaswat; this is the standard name for the seventh Manu. Manu Vaivaswat could be considered as the great grandson of Brahma, through Brahma — Kashyap — Vivaswat — Manu.

Another Version

In an entirely different version, there is reference to a self-created daughter of Brahma called Shatarupa (literally meaning hundred forms). Of this, there are two variants;

- Brahma split himself into two parts, male Manu and female Shatarupa. Manu then took Shatarupa as his wife.
- Manu sprang from one side of Brahma's mind and Shatarupa from the other side. Manu married Shatarupa.

Humankind emerged out of the union of Manu and Shatarupa.

We offer the following clarifications:

- We have done the above division between the first and seventh Manu for ease of understanding; this need not be taken too literally.
- The first Manu Swayambhuva was born before the *dev-rishis*: in fact, he created the great sages.
- The seventh Manu Vaivaswat was born after the *dev-rishis*; he was a descendent of one of them, i.e. Kashyap. As we are now in his reign, we will take it as the standard version for our text.
- Names of *dev-rishis* and prajapatis are largely common in both the cases.
- The 'Another Version' given above could be considered applicable to the first Manu.
- There are two versions of the birth of adityas. In the first version, 12 adityas emerged out of fragments when Vivaswat was

taken up for grinding by Tvastr. In the second version, 8 or 12 adityas were born to Aditi.

- Both Dakshayani and Aditi were daughters of prajapati Daksha. However, Daksha sometimes appears as an aditya — either as the son of Aditi; or he, who emerged out of the fragments.
- Vivaswat is considered to be the creator of humans, super humans, sub humans and natural phenomena.

Manvantra

One Kalpa (4.32 billion years) has the rule of 14 Manus. The period of rule of one Manu is a Manvantra, which is equal to 71 Mahayugas, or nearly 300 million human years. Presently, we are in the seventh Manvantra of the seventh Manu, called Manu Vaivaswat. Tradition says that we are in the 51st year (out of a total of 100 years) of Brahma. Thus, we are in the middle of time of the 7th (out of 14) Manvantra and 51st (out of 100) year of Brahma.

Whilst the generally accepted version is that the Universe is destroyed at the end of each Kalpa, some texts refer to the destruction of the universe at the end of each Manvantra.

8.2 The Manu Flood

The highly interesting legend of the great deluge/flood is connected with Manu Vaivaswat, the seventh Manu. The first account is found in the Satapatha Brahmana, which runs as follows:

> One day, the seventh Manu was practicing asceticism on the sea-shore. A small fish approached him. Being small, the fish feared for her life and asked Manu to take her under his protection; Manu did that. That fish was released in the sea when fully grown. One day, the same fish approached Manu and warned him of an approaching big deluge. She asked him to build a boat for himself. When the flood came, the fish guided the boat to the top of Mount Meru. Thus, Manu was saved, whilst everyone and everything else was destroyed.

As per another version, Manu took the seven *dev-rishis* and seeds of all creation elements with him. After the flood receded, Manu restarted the human race, other living creatures and everything else. The fish which guided Manu's boat was Lord Vishnu himself in his first *avatar* (incarnation) of Matsya (fish). However, that fish is referred to as Brahma's *avatar* too sometimes.

There are stories of similar floods in the Epic of Gilgemish of Babylonia and the Jewish tale of Noah's Ark in the Old Testament. It is difficult to say as to who borrowed the story from whom, if at all.

Whilst talking of Manvantra (one Manu period) we have referred to millions of years. However, at a more practical level, the year of Manu's flood has been placed at 3102 BC (see Chapter 9.5).

8.3 Progeny of Manu

As per Hindu mythology, all humankind is descended from Manu Vaivasvat who was the first man (equivalent to the Jewish Adam). In all, Manu had nine sons and one daughter. We give below the broad outlines of the major dynasties founded by Manu's progeny, separately for Manu's sons, and daughter Ila.

Manu's Sons

Names of Manu's sons are Ikshvaku, Nabhaga, Saryati, Primsu, Karusa, Dhrsta, Narisyanta, Nabhanedistha and Pradha. Manu divided Bharat varsha of those days, amongst his nine sons and daughter. Names of areas assigned to them are given in the texts; some names suggest even foreign locations, e.g. Balikha (Balkh?). Most of the sons established ruling Kshatriya families. However, some may have headed Brahmin families. One son Pradha had to become a Shudra, as he killed the cow of his guru.

Ikshvaku was Manu Vaivaswat's eldest son. He headed the main branch of the dynasty that ruled in Kosala, with the capital at Ayodhya. Another son of Manu, Nabhane-dishta founded the dynasty at Vaisali (North Bihar). The remaining seven sons ruled over their respective kingdoms.

There are references to Ikshvaku having 100 sons. However, names of only two sons, Vikukshi and Nimi are in general circulation. Vikukshi inherited the main line of the dynasty at Ayodhya. This dynasty has been called the Suryavanshi or Solar Line, after Manu Vaivaswat, who was son of Vivaswat, Surya, the sun.

Nimi inherited the dynasty at Videha (literally 'bodyless'), with its capital at Mithila. Kings of this dynasty were called 'Janaks'. Sirudhvaja Janak ruled Videha as a contemporary of King Dasratha of the *Ramayana*. His daughter was Sita, who became Rama's wife. The remaining 98 sons might have had their own kingdoms. The main dynasties were as follows:

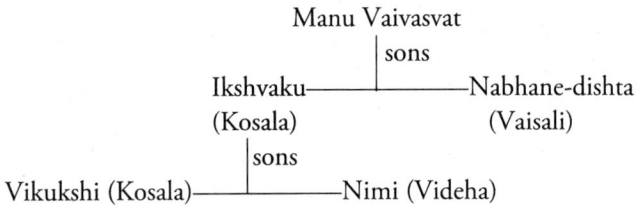

```
                      Manu Vaivasvat
                            |  sons
            Ikshvaku————————┴————————Nabhane-dishta
            (Kosala)                   (Vaisali)
                 | sons
Vikukshi (Kosala)————————┴————————Nimi (Videha)
```

Some of the famous kings of the Suryavanshi dynasty were as follows (numbers indicate the generation from Manu):

1 Manu	41 Sagara	60 Dilipa II
2 Ikshvaku	45 Bhagirtha	62 Raghu
3 Vikukshi	47 Nabhaga	63 Aja
31 Satyavrata	48 Ambrisha	64 Dasrath
32 Harishchandra	54 Kalmaspada	65 Rama

Manu's Daughter

The story of Manu's daughter Ila is more colorful. Manu Vaivaswat, before he had a son, conducted a sacrifice to Mitra and Varuna to obtain a son. The priest mismanaged and a daughter was born, called Ila. With the favor of Mitra and Varuna, Ila's sex was changed and she became a man, Su-dyumna (sometime called Il). Shiva again turned Su-dyumna into a woman. At a later stage, under the favor of Vishnu, Ila again became the man Su-dyumna, and was father of three sons. There are other versions of Ila's sex change. As per one such version, she was a woman one month and a man the next month, as ordained by Shiva/Parvati.

Ila, when a woman, married Budha, son of Chandra, the moon. A son Pururavas was born to Ila. Ila set up a ruling dynasty in Madhyadesh, with its capital at Pratisthana (present Allahabad).

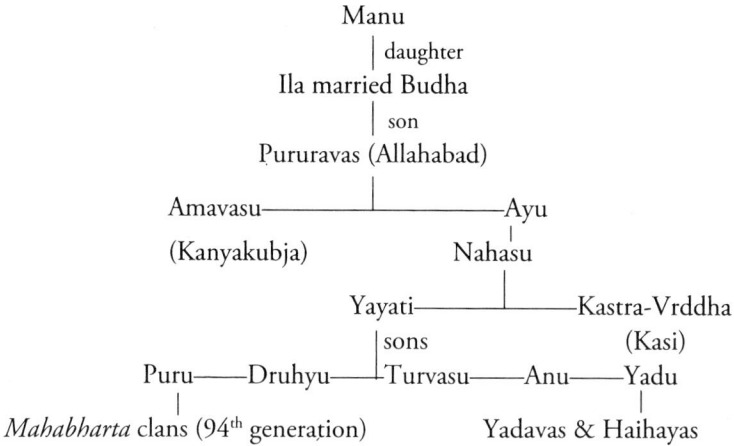

```
                        Manu
                         | daughter
                  Ila married Budha
                         | son
                  Pururavas (Allahabad)
                         |
   Amavasu───────────────┴───────────────Ayu
   (Kanyakubja)                        Nahasu
                                         |
                      Yayati─────────────┴──────Kastra-Vrddha
                         | sons                    (Kasi)
   Puru────Druhyu────┴Turvasu─────Anu─────Yadu
      |                                      |
Mahabharta clans (94ᵗʰ generation)    Yadavas & Haihayas
```

The above dynasties have been called Ailas, after Ila. They are also called Paurvas, Chandravanshi or Lunar Line. Pururavas had six sons, out of which two, Amavasu and Ayu were important; the latter was more famous than the former.

Amavasu founded a ruling dynasty at Kankyakubhja (modern Kannauj). To that dynasty belonged King Gadhi, corresponding to the 30th generation of Solar kings. His son was Vishvavartha, who became a Brahmin, the famous *dev-rishi* Vishvamitra. The birth of Vishvamitra in synchronism with the 31st Solar king would cast a doubt on Vishvamitra being the mind-born son of Brahma. That anomaly is explained by saying that there may have been many Vishvamitras, which may have been a *gotra* (family) name.

Ayu had a grandson Yayati, who was a great conqueror and cakravartin (universal) ruler. In turn, Yayati had five sons; they inherited/established five kingdoms, the dynasties being named after their own respective names, as given in the above chart. These five Aila kingdoms, along with the earlier two of Kankyakubhja and Kasi, were spread all over Bharat varsha, from Punjab to East Bihar, may be also in the South. We may record that Suryavanshi clans were also ruling simultaneously. The dynasties founded by Yayati's two sons Puru and Yadu, were more important.

Puru Line: The line founded by Puru became the main Aila Line, which ultimately led to the Kaurva and Pandava clans of the *Mahabharta*. It came to be called the Chandravanshi or Lunar line, after Chandra, the moon, the father of Budha, Ila's husband. There was a break in the Puru line after 20 generations. King Dushyant, who was in synchronism with the 43rd Solar king, revived the dynasty. Dushyant married Shakuntla, who was born out of the union of rishi Vishvamitra and *apsara* Maneka. Their son was king Bharata, after whom our country Bharat is named. 91st generation King Shantanu was father of Bhisham Pitama and grandfather of Dhritrashtra and Pandu of the *Mahabharta*. The latter corresponded to the 93rd Solar generation of kings. For more information, see *Mahabharta* in Chapter 5.8.

Some of the famous kings of the Puru Line were:

1 Manu	43 Dushyant	70 Kuru
2 Ila	44 Bharata	86 Dilipa II
3 Pururavas	45 Vitatha (Brahmin)	91 Shantanu
6 Yayati	51 Hastin	93 Dhritrashtra
7 Puru	52 Ajamidha	94 Kauravas
20 Tamsu	63 Rksa I	95 Prikshit

The following points need to be noticed in respect to the above table:

- Unlike the Solar line which ran continuously, there were major breaks in the Lunar line. The first break was for 22 generations after Tamsu (serial 20). The dynasty was revived by Dushyant (serial 43). There was another break of 11 generations after Ajamidha (serial 52).

- In view of the breaks, it is not easy to give serial numbers to the Lunar kings. Hence, the above serial numbers relate to the correspding Solar kings who were in synchronism with that Lunar king. For example, serial number 44 for the Lunar king Bharata means that Solar king serial number 44 (Dilipa I) was ruling at that time.

Yadu Line

In the Yadu line were the Yadavas and Haihayas. The great Lord Krishna was born in the main Yadava line as the 94th generation.

Haihayas

Haihayas were an off-shoot of the Yadavas. The word Haihaya is a derivative of 'haya', meaning horse; Haihayas were excellent horsemen.

Haihayas have been associated with foreign tribes like the Sakas (Scythians), Pahlavas/Parthians (Iranians) and Yavanas (Greeks); they may have even originated from these tribes. The tribes helped Haihayas in their all-round conquests. The possible presence of those foreign tribes at that stage of pre-history is most baffling. We know from the recorded history of India that these tribes came to India in the closing centuries of BCE. Manusmriti records these tribes as 'lapsed Kshatriyas', meaning degraded Kshatriyas. Is their some link between these tribes and Rajputs? We will study this aspect in 'Chapter 23 — Rajputs'.

In the 30th and 31st generation of the Haihaya dynasty were the famous kings Kartavirya and his son, Arjuna Kartavirya. Haihaya king Talajangha, the great grandson of Arjuna Kartavirya, drove out the powerful Suryavanshi King Bahu (40th generation), with the help of foreign tribes mentioned in the preceding paragraph. Haihayas also conquered the kingdom of Kashi (Benaras) from King Divodasa. Tradition says that at that stage, Haihayas hordes carried out major depredations over North Bharat varsha of those days. Those all-round Haihaya raids have been sometimes projected as the repeated killings of Kshatriyas by Parshurama.

Summary: The sons and grandsons of Manu, and the progeny of his daughter Ila, set up innumerable kingdoms all over Bharat varsha of those days. These could have included some kingdoms in the South. There are some hints of rule, even in areas presently not in India/Pakistan. Of the various kingdoms/dynasties, the more famous were as follows:

By Sons/Grandsons of Manu (Solar Line):

- Kosala: Capital Ayodhya; clan Suryavanshis: Famous kings were Sagara, Bhagritha, Dasrath and Rama; Chief Priests — Vasishtas — from the beginning to the end.

- Videha: Capital Mithila; Famous king Janak, father of Sita.
- Vaisali: Nothern Bihar; Famous kings were Marutta Aviksita and Trnabindhu; Priests Angirasas/Bhardwajas.

By Manu's Daughter Ila & her Progeny (Ailas/Lunar Line/Pauravas):
- Paurvas (Mainline): Capital first at Pratisthana (Allahabad), then Hastinapur; Chandravanshis or Lunar Line: Innumerable famous kings: Chief priest — Angirasas/Bhardwajas.
- Kanyakubhja (Kannauj): Famous kings Jahnu and Gadhi, father of Vishvartha, who became *dev-rishi* Vishvamitra. The dynasty was extinct after Lauhi, grandson of Vishvamitra.
- Kasi (Varanasi): Famous kings Divodasa II and Pratadana.
- North Pachala: Located somewhere in UP, north of Ganga, Dynasty of King Sudas of the *Rig Veda*. There was also South Panchala ruled by King Draupad, at the time of the *Mahabharta* war.
- Yadavas: A continuous chain of kings ruled, right down to Lord Krishna. Following kings ruled during the *Mahabharta* age:

    Jarasandh: Ruler of Magadh; Father-in-law of Kamsa (Kans) ruler of Mathura and maternal uncle of Krishna.

    Sishupala: He was the son of King Dama-ghosha of Chedi. Sishupala was the cousin and foe of Krishna who was forced to kill Sishupala.

- Haiyayas: We have covered them earlier.

Manu Dynasties — Tabular Form

At this stage, we give below in a tabular form some of the more important dynasties emerging out of the sons and daughter of Manu. The Ikshvaku (Suryavanshi or Solar) dynasty was continuous without a break. Hence, even the serial number of the Paurvas (Chandravanshi or Lunar) kings listed in the table below, relate to the corresponding (synchronous) Ikshvaku kings. For example, Dushyant has been listed as serial 43. Actually, he was the 21st king in the Paurva line. As there was a break of 22 generations in the Paurva line, he is listed as serial 43 (21+22).

Survanshis/Ikshvakus	Chandravanshis/Ailas/Pautvas		
	Mainline-Paurvas	Yadavas	Kanyakubhjas
1 Manu	1 Manu	1 Manu	1 Manu
2 Ikshvaku	2 Ila	2 Ila	2 Ila
3 Vikukshi	3 Pururavas	3 Pururavas	3 Pururavas
	4 Ayu	4 Ayu	4 Amavasu
	6 Yayati	6 Yayati	
	7 Puru	7 Yadu	
	20 Tamsu	*Haihayas*	28 Kusa
30 Trayyaruna		30 Kartavirya	30 Gadhi
31 Satyavrata		31 Arjuna	31 Vishvamitra
32 Harishchandra	*(Dynasty Break- 22 generations)*	Kartavirya	32 Astaka
		34 Talajangha	33 Lauhi
			(Ends)
40 Bahu			
41 Sagara	43 Dushyant	*Yadavas*	*North Panchala*
44 Dilipa I	44 Bharata	Continued	(Start)
45 Bhagirtha	45 Vitatha Bhardwaj		
47 Nabhaga			
48 Ambrisha			
	51 Hastin		
	53 Ajamidha		54 Nila
	(Dynasty break)		
60 Dilipa II			60 Mugdala
62 Raghu			
63 Aja	63 Rksa I		63 Divodass
64 Dasrath			
65 Rama			
66 Luv/Kush			67 Cyvana
	70 Kuru		68 Sudas
	86 Dilipa II		71 Jantu
	91 Shantanu		91 Prasta
		92 Ugrasen	
	93 Dhritrashtra	93 Kansa	93 Drona
	94 Kaurvas/ Pandavas	94 Krishna s/o Vasu*deva*	94 Asothama
	95 Prikshit		
	96 Janmejaya		

8.4 Manusmriti

Manu is the author of Mannav Dharam Shastra, also called Manusmriti, or Manu's Law Code. Tradition has it that the Law Code was given by Brahma himself to Manu, may be at the very beginning of humankind. It appears that originally Manusmriti had some one lakh entries, which were reduced to about 12,000 entries by Narada; these were further abridged to 4,000. Currently, there are some 2,685 entries.

In the very first chapter of Manusmriti, Manu, describes the creation of the universe by Brahma, and his own birth. He does this at the request of the *dev-rishis* (sages). The text expounds the source of *dharma* and enumerates the duties of man from birth to death, divided intro four periods, viz.:

Brahmcharya	Student	0 to 25 years
Grihsth	Householder	26 to 50 years
Van-prast	Forest Dwelling	51 to 75 years
Sanyas	Renunciation	76 to 100 years

Manusmriti covers every possible aspect of human conduct and administration of justice; it lays down graded punishment according to caste (jati/varna) of the offender. Manusmriti prescribes duties of the king and all its related topics. The text divides human society into four castes; Brahmin (Priest), Kshatriya (Warrior), Vaishya (Trader/Agriculturist), Shudra (Service Class).

It has been expressed that Manusmriti was written in its present form in the closing centuries of BCE, or in early ADs. That would appear to suggest that the Law Code may have been carried forward in the oral tradition for thousands of years. Alternately, the Manusmriti might have been compiled in the closing centuries of BCE; Manu was projected as its author to give it ancient authenticity. We face this dilemma in respect of most Hindu literature. It is generally difficult to say when any text was evolved, and when was it reduced to the written form. Once reduced to writing, the texts might have undergone a number of re-writings and updating.

Chapter X, entries 43 to 45 of the Manusmriti refer to the following classes/tribes as having been reduced to the 'level of servants' (or degraded/lapsed Kshatriyas), as they did not undertake the proper rituals, or sought blessings of the priests: Cholas, Greeks, Scythians, Persians, Moutaineers, Kambojas, etc. The entries also say that these classes have been traditionally regarded as 'aliens'.

Now, we know that Bactrian Greeks, Persians (Iranians) and Scythians (Sakas) came to India after the 2nd century BCE (see Chapter 18). That would appear to show that Manusmriti may have been written in the closing centuries of BCE or early ADs. It is also possible that these entries may be a later addition. However, there is a small possibility that these foreign tribes might have come to India at a much earlier date in history, rather mythology. There are references to a branch of the Yadavas, called Haihayas existing 25 generations before the *Ramayana*. It is recorded that a Haihaya King Talajangha took help of the above quoted foreign tribes to wrest power from the Suryavanshi King Bahu (see Chapter 8.3).

Manusmriti has come in for severe criticism for its bias against women and Shudras. Some sample extracts in respect of women:

Manusmriti, Chapter 9

(3) Her father guards her in childhood, her husband does that in her youth, her sons would look after her in her old age. A woman need not be given total independence.

(10) Men cannot guard women by force; however, they can be guarded by using the following means.

(11) He should keep her busy in hoarding money, and also in spending it; she should be kept engaged in all types of purification rites, and attending to household duties like cooking food and cleaning the furniture.

(15) By their desire for men, by their fickle minds, by their lack of love(?), the women may not remain faithful to their husbands even when well guarded.

(17) The bed, lot of jewelery, some lust and anger, traces of crookedness, an uncertain nature and bad conduct are the attributes what Manu assigned to women.

(18) There are no rituals with Vedic verses for women; that is a firmly established principle of law.

Manusmriti, Chapter 5

(147) A girl, a young woman, or even an old woman should not do anything independently, in (her own) house.

(148) In childhood a woman should be under the supervision of her father. In youth under that of her husband's. When her husband dies, she should be under the supervision of her sons.

(150) A woman should always be happy and cheerful; she should be an expert at her household work. She should keep her utensils clean, and not have a free hand in spending money.

Following types of statements also appear in Manusmriti:

- Even if destitute of virtue and devoid of good qualities, a husband should be worshipped as a god by a faithful wife; if a wife continues to obey her husband, she will be exalted in heaven.
- Through their passion for men, their mutable temper and their heartlessness, they (women) become disloyal towards their husbands, howsoever carefully they may be guarded.
- For the sake of marriage, Manu lists forced abduction of the bride as one of the alternatives. For the Kshatriyas, even the murder of the bride's family may be an option. For women, there is no separate sacrifice, no vow, not even fasting; she may just obey her husband.

As against the above, the Manusmriti also contains following types of laudatory passages for women:

- Where women are honored, there the gods are pleased; but where they are not honored, no sacred rites yield rewards.
- Where female relations live in grief, the family soon perishes; however, the family where women are happy always prospers.

While the statements derogatory to women are innumerable, lauding them are few and far between.

Manusmriti's harsh treatment of Shudras is covered later in chapter 11 — 'Hindu Untouchability'.

9

HINDUISM
Miscellaneous

9.1 The Ocean Churning (*Sumandra Manthan*)

Sumandra Manthan (Churning of the Ocean of Milk) is an important event in Hindu mythology, involving the second *avatar* (incarnation) of Vishnu, i.e. Kurma or Tortoise. As per one version, we have placed the first *avatar* of Vishnu, *matysa* (fish) at the time of Manu's flood in 3102 BC. The churning of the ocean could be placed sometimes after that.

Tradition has it that in the ancient past, there was a major conflict between the *deva*s (gods) and the *asura*s (demons), for control of the universe. The conflict might have lasted one celestial year (360 human years). The *asura*s got the upper hand, and the *deva*s lost most of their clout. The gods were feeling frustrated; they went to Lord Vishnu and requested him to rescue them. Vishnu expressed that numerous precious things had been lost during Manu's flood and that the ocean should be churned to get those back. Special emphasis was to get the 'Water of Life', i.e. *amrita* (ambrosia), which would impart immortality to the gods.

Vishnu advised that Mandara or Meru, the mountain should be taken as the churning stick, and the serpent Sheshanaga (aka Vasuki, Shesha, Adi Shesha, Ananta) as the churning rope. Both gods and demons got on the job of churning, with Vishnu as a tortoise, providing the base for the churning stick Mandara. Gods held the tail end of the snake rope and the demons the mouth-end. Lord Shiva was the mediator. The churning continued for one thousand years (need not be taken literally).

The following emerged out of the ocean, in some flexible order:

1. *Surabhi:* She was the wish-cow of plenty, which gave huge quantities of milk and butter, and many other things. She is also called Kama-dhenu or Savala. *Surabhi* gave unlimited quantity of *ghee* (clarified butter); this was the main ingredient for the *yajnas* (fire-sacrifice) conducted to please the gods. Therefore, the Brahmin priests took Surabhi. Thereafter, they enjoyed a big advantage, in pleasing the gods, through *yajnas*. Nandini was an off-spring of Surabhi; she was owned by Vasishta, and was the cause of trouble between him and Vishvamitra. Dev-rishi Jamadagni also owned an off-spring of Surabhi.

2. *Varuni:* Daughter of Varuna and goddess of wine; also called Sura. Gods drank it; hence, they came to be called Suras. Demons did not drink it; hence, they were called a-*suras* (in Sanskrit, addition of 'a' to any word indicates its opposite).

3. *Parijata:* The tree of paradise with its blossoms.

4. *Apsaras:* Celestial nymphs, matchless in their beauty. They adorned Indra's court in heaven, called Indra-Lok. Some texts talk of the number of *apsaras* in millions.

5. *Chandra:* The (crescent) moon; it was taken by Maha*deva* (Shiva), who put it on his matted hair.

6. *Airvata:* The elephant — claimed by Indra

7. *Uchchaih-sravas:* This was the king of horses, of white color; Indra took it.

8. *Vish (Poison):* Nagas, the snake-gods, claimed it. However, Shiva drank the poison to save the universe (see comments below).

9. *Lakshmi or Sri:* Demons desired to possess her. After some skirmishes between gods and demons, Lakshmi became Vishnu's wife. As per one version, she was daughter of *dev-rishi* Bhrigu.

10. *Kaustabha:* The Jewel.

11. *Sankha:* Conch.

12. *Dhanus:* A matchless bow.

13. *Amrita:* This was the 'Nectar of Life' or ambrosia, held in a bowl by the great physician Dhanwantri (see comments below).

Tradition says that at some stage, the serpent Sheshanaga gave out the deadly poison known as halahala or kalakata. As that poison would have consumed the universe, the great lord Shiva in his form of Mahakala decided to take it within himself. As Shiva-Mahakala gulped the poison, Parvati realized the danger; she clenched Shiva's throat. The poison got stuck there and Shiva's throat acquired a blue color in perpetuity; hence, Shiva is called Neelkanthan — the blue-throated.

Before the churning could begin, it was agreed that both gods and demons would each get half portion of the *amrita* (the life giving draught of nectar); but, as usual, the gods played foul. As the bowl of amrita emerged, both gods and demons lunged at it. At that stage, Vishnu appeared as the beauteous damsel Mohini; the demons went after her. In the ensuing confusion, Vishnu (as Mohini) handed over the bowl of *amrita* to the gods who drank it, thus getting immortality. Shiva was the great renunciator, who did not hanker after the riches emerging out of the ocean.

As the *amrit* (nectar) was being distributed amongst the gods, one of the demons Rahu sat among the gods, in disguise. He managed to get a sip of *amrit* on his tongue. But, before he could swallow it, Surya, the Sun and Chandra, the Moon recognized him. They informed Vishnu who attacked Rahu, splitting him into two, called Rahu and Ketu. Vishnu cut off the head of Ketu. However, as the head had already tasted *amrit*, it became immortal and was placed in or near the heavenly abode. The lifeless body fell on the earth. Rahu wreaks his vengeance on the Sun and the Moon, by occasionally swallowing these (i.e., eclipses). The duo of Rahu and Ketu, are two malefic planets.

9.2 Ved Vyas

Ved Vyas is one of the most important persons of Hindu mythology. In various texts, he has been credited with the following:

— Author of the *Mahabharta*
— Editor or Re-arranger (not author) of the Vedas
— Author or Arranger of the Puranas

Who was Ved Vyas?

There was the famous *dev-rishi* Vasishta, who had a grandson Rishi

Parasara. One day, Parasara wanted to cross a stream. The fisherman sent his adopted daughter named Matsyaghangi (she has many other names), to row the Rishi across. Matsyaghangi was the daughter of Uparichara, king of Chedi and an apsara Adhriki who was cursed to live as a fish. Being born of a fish, Matsyaghangi had a fishy smell. However, she was extremely beautiful. Rishi Parasara could not resist her charm. Out of that union, was born a son in nine seconds, on the theory of 'Rishi Garbham'. The virginity of the girl was restored and the fishy smell replaced by perfume. The son had a dark complexion, and was born on an island (Dvipa); thus, he was named Krishna Dvaipayana. Later, he came to be called Ved Vyas. Keeping in view the imperishable nature of his work, Ved Vyas has also been called Saswatas (Immortal).

Now, the Chandervanshi king Shantanu was ruling at that time and was rather advanced in years. Shantanu was married to Ganga, otherwise a river. Through her, he had a son Bhisham Pitama, the greatest warrior of all times. Shantanu happened to see Matsyaghangi; he could not resist her charm and proposed marriage. Her father laid the condition that after Shantanu, his daughter's progeny should ascend the throne. When Bhisham Pitama heard of this condition, he on his own volition renounced all his rights to the throne. He also took a vow of celibacy so that no progeny of his could claim the throne in the future.

King Shantanu was married to Matsyaghangi, who became famous as Queen Satyavati. She had two sons, who died childless, leaving behind two young widows. As no heir to the throne was left, Queen Satyawati asked Bhisham Pitama to impregnate the young widows through the rite of 'niyog'. Bhisham Pitama refused, citing his oath of celibacy. In that situation, Satyavati called upon her earlier son Krishna Dvaipayana (Ved Vyas) to do the needful. The result was birth of Dhritrashtra, who was blind and Pandu, who was pale. A third child Vidur was also produced through a maid; he went on to become the Chief Minister in the court of Dhritrashtra. Ved Vyas had another (regular) son called Suka.

As per our above story, Ved Vyas was born during the *Mahabharta* age and was the biological grandfather of the Kaurava and Pandava brothers. As Ved Vyas lived in the age of the *Mahabharta*, there is no doubt that he was indeed the author of the same (It is another matter that we have expressed earlier that the epics got written around the 4th–3rd century BC).

Tradition also has it that Ved Vyas may have classified the original one Veda into three or four Vedas. As for Ved Vyas's authorship of the *Puranas*, we would have to stretch our imagination a bit. As brought out by us earlier, *Puranas* were written much later and over a period stretching more than 1,000 years (say from 200 BC to 1000 AD). It may be difficult to attribute their authorship to a single person. It is possible that the authors of the *Puranas* associated the name of Ved Vyas to give ancient authenticity to those texts. Alternately, there could be many Vyasas, as Vyas may have been a common name those days. Whatever, Ved Vyas is generally credited with the authorship/editorship of the *Puranas*.

Notwithstanding the above comments, tradition says that Brahma received the Puranic knowledge from the Supreme Spirit Brahman. Brahma passed that knowledge to his mind-born sons, the four Sanat Kumars. They in turn gave that knowledge to Narada, who transmitted that to Ved Vyas.

9.3 The *Gayatri Mantra*

OM BHOOR, BHUVAHA, SWAHA
TAT SAVITRE VARENIYAM
BHARGO DEVASYADHEEMAHI
DHIYO YO NAHA PRACHODAYAT

The Gayatri is the most holy *mantra* (hymn) of the Hindus and the most sacred verse of the *Rig Veda*. As per one version, the Gayatri *mantra* is addressed to *Surya* (Sun), in its form of Savitre or Savitri, the invisible nighttime, or setting Sun. There are various translations in existence; we give below a few:

- We meditate in the adorable light of the divine Savitri that he may raise our thoughts (Max Muller).
- Earth, sky, heaven; let us meditate on the most excellent light and power of that generous, sportive and resplendent Sun, praying that it may guide our intellect (Colebrooke).

- We may meditate on that desirable light of the divine Svitre, who influences our pious rites (Wilson).
- God, thou art the giver of life, the remover of pain and sorrow, the bestower of happiness; O Creator of the Universe, may we receive thy sin destroying light; may thou guide our intellect in the right direction.
- May the Almighty God illuminate our intellect, which may lead us to the righteous path.
- May we receive the glorious brightness of this, the generator, of the god who shall prosper our works.
- May we attain the excellent glory of the divine Vivifier; so that we may enlighten or stimulate our understanding.
- Aware of the supreme sublimity of the Kindly Light, we solemnly pray unto the Glorious Almighty; inspire and fire our hearts to take the right initiative and act accordingly.

We may not be wrong in taking the translation by Max Muller as the most authentic, without in any way reducing the importance of other translations.

Each word in the Gayatri *mantra* may have a different meaning, as is common in ancient Sanskrit. We give below one version of the meanings:

Om	—	The Supreme Spirit, the Almighty God
Bhoor	—	Spiritual Energy
Bhuvaha	—	Eliminator of all Suffering
Swaha	—	Happiness
Tat	—	That — the Almighty God
Svitre	—	Surya, the Sun, perhaps the setting Sun
Vareniyum	—	The High and Mighty
Bhargo	—	Destroyer of all sins
*Devas*ya	—	Divine
Deemahi	—	May receive
Dhio	—	Intellect
Yo	—	Who
Maha	—	Our
Prochodyat	—	Inspire

Ga means 'to sing', *yatri* means 'protection'; therefore, Gayatri means 'Those who worship get protected'. The Gayatri *mantra* is a scientifically composed verse for the invocation of the Supreme God, who always uplifts and inspires. It is a combination of letters, which are an arrangement of words with hidden spiritual meaning. The real power of the *mantra* is in the form of syllables, which can be awakened by uttering the *mantra* again and again. The *mantra* has deep philosophical meaning. Its singing repeatedly benefits both physical and spiritual health. A possible explanation of the Gayatri —

'All this Creation and Cosmos is a physical manifestation of the Almighty; He pervades in each and every particle of it. I see Him at all places and everywhere. I shall remain away from bad thoughts and deeds; and shall perform the true worship of God. I will promote true happiness and peace in this universe, which is His creation. I am assimilating God within myself. By such assimilation, I am becoming more and more virtuous. These virtues are developing in all parts of my mind and body. I am becoming one with God; may He lead me to the righteous path.'

The Skanda Purana says —

"Nothing in the Vedas can be superior to the Gayatri. No invocation is equal to it, as no city is equal to Varanasi. The Gayatri is the mother of the Vedas. By repeating it, one is always saved. There is nothing that cannot be affected by the Gayatri; for the Gayatri is Vishnu, Brahma and Shiva, and the three Vedas."

The sentiments expressed in the Gayatri are powerful, righteous and elevating. One should contemplate these feelings daily. By doing that, the Gayatri would be fully assimilated in every pore of the person. The result is that in a few days, the mind is diverted from evil deeds and one starts righteous thinking. The Gayatri does not belong to any community; it is a universal truth.

9.4 *Aum (Om)*

Aum (Om) is the most sacred syllable in Hinduism. It is sometimes referred to as The '*Udgitha*' or '*Pranava mantra*' (primordial hymn) because it is the primeval sound.

Aum is seen as the manifestation of the Brahman that resulted in the creation of the whole universe. All cosmos stem from the vibrations of the word Aum. It is considered so sacred that almost all Hindu mantras start with it. It is the most representative symbol of Hinduism.

Aum, written in Hindi has its left part like the Hindi figure '3' representing the merger of three into one, i.e.:

— The Trimurti of Brahma, Vishnu and Shiva into Brahman
— Rajas (activity, heat, fire), Sattva (purity, light, *shanti*), Tamas (dullness, ignorance, darkness), into Brahman
— Body, speech and mind into Oneness

In the scriptures, it is stated as follows, "One who chants AUM, which is the closest to Brahman, approaches Brahman. That liberates him from the fear of the material issues; therefore, it is known us *tarak* Brahman".

Aum takes the form of the Gayatri, then the Veda and Vedanta Sutra; then it takes the shape of the Srimad Bhagvatam. The Chandogya Upanishad states 'The Udighita (Aum) is the highest of all essences.'

Aum can also be seen as Sri Ganesh, whose figure represents the shape of Aum; Ganesha is thus known as Aumkar (shape of Aum). Nataraja, the dancing Shiva is also considered to have the shape of Aum. Aum is said to be the perfect approximation of the cosmic existence within time and space, and therefore, the sound closest to *Satya* (Truth). The first word '*Aum*' is called '*Pranav*' because its sound emanates from the *Prana* (vital vibration), which makes the Universe.

When you pronounce Aum, A — emerges from the navel, U — rolls over the tongue, M — ends on the lips.

Aum is the sum and substance of all the words that can emanate from the human throat. It is the primordial fundamental symbol of the Universe. Aum is the standard sign for Hinduism.

Gods and goddesses are sometimes referred to as Aumkar, which means the form of Aum; thus implying they are limitless, the vibration whole of the cosmic. '*Ek Omkar*' meaning one God, is the central tenant of the Sikh religious philosophy. In Hindu metaphysics, the closest approximation to the name and form of the universe is Aum, since all existence is fundamentally composed of vibrations.

9.5 3102 BC & 1400 BC

The reader would have noticed that almost all periods and dates in Hindu mythology cum history are generally rough approximations. However, two dates are precisely mentioned to which specific events are related. These are 3102 BC and 1400 BC; the former is more or less mythology, but the latter is recorded history. We study these dates below.

3102 BC

The year 3102 BC has a special significance in the Hindu calendar. Two events totally separate in space and time have been associated with this date.

One version is that Manu's flood occurred in 3102 BC; that involved the first *avatar* (incarnation) of Vishnu in the form of *matsya* (fish). Being the date of the first *avatar*, 3102 BC could be considered as very near to the creation or re-creation of the universe, rather humankind. That would place the Vedas, Upanishads and the two epics later than 3102 BC. In this version, the date of the *Mahabharta* (actual war) could be around 1400 BC, or 900 BC, as we have referred to earlier. The *Ramayana* could be 500 to 1,000 years earlier than the *Mahabharta*.

In the second version, 3102 BC is the age of *Mahabharta* war. After the war, Lord Krishna was killed by a hunter's arrow piecing his heel (shades of Achilles' heel story). That is considered to be the end of the Dvapar Yuga and start of the (present) Kali Yuga. In this version, the Vedas could be considered to have been evolved say 1000 years before, i.e. around 4000 BC.

Now, only one version of the above can be true; or both can be untrue. That is the beauty of mythology. It tells you a story, but does not force you to believe in it; you may form your own view.

We may record here that recognized historians like Max Muller would disagree with both the above versions. Max Muller considers the evolution of the *Rig Veda* around 1500–1200 BC. In all probability, historians may not give much credence to Manu's flood story. There is also a divergence of opinion on the historicity of the *Mahabharta* and the *Ramayana*.

1400 BC

There is one date which helps establish a sort of anchor point of Hindu history — 1400 BC.

In the ancient past, in the area around the present Iraq/Syria border, there were two kingdoms of Hittite and Mittani; they had a war. The warring parties agreed on a truce, and signed it in the form of a clay tablet. Hindu gods Indra, Varuna, Mitra, etc. were quoted as witnesses to that truce agreement.

The original clay tablet is available. Its date has been scientifically established around 1400 BC. That would imply that by that time, Hindu gods were in circulation. There is no easy answer as to what they were doing in that area of West Asia.

9.6 *Tulsi* (Basil Plant)

Tulsi, the Basil shrub is the most holy plant (shrub) in India. It grows all over India and receives a lot of worship. Tulsi is often worshipped, in conjunction with the shaligram stone — a fossil ammonite, primarily by followers of Vishnu. Shaligram is a special type of stone found in a village named Shalagrami, situated near the origin of River Ghandhak in Nepal. There are various stories relating to the origin and worship of tulsi, viz.:

- A woman named Tulsi engaged in religious austerities for a thousand years, as she wanted to marry Vishnu. Hearing of that, Lakshmi, wife of Vishnu, cursed Tulsi and converted her into a plant. However, Vishnu assured Tulsi that he would assume the form of shaligram and continue to be near her.
- Vishnu was fascinated with the beauty of one damsel Vrinda, wife of Jalandra. To distract Vishnu, Lakshmi, Gauri and Swadha gave seeds to sow, from which trees/shrubs grew with which Vishnu got enchanted; one of the shrub was tulsi. Thus, Vishnu was saved from the wiles of Vrinda.

- Krishna was in love with Radha who was otherwise a married woman and thus, could not marry Krishna. As a symbolic gesture, Krishna was married to a tulsi in place of Radha.
- Saraswati and Lakshmi had a quarrel, and Saraswati cursed Lakshmi. The curse transformed Lakshmi into a tulsi plant, which had to live on earth.

As most of the above relationships were outside marriage, tulsi has to stay in the courtyard and cannot enter the inner household. Vishnu or Krishna, in the representation of shaligram, is married to tulsi annually in the month of October/November, in parts of North India.

The same honors are paid to tulsi as to idols of gods. Salvation is assured to anyone who carefully tends to the tulsi. Young women worship tulsi to get a good husband, married couples for children, old people for space in heaven and widows for their salvation. It is a common practice to place a sprig of tulsi at the head of a dying person. There are two types of tulsis — Rama tulsi with green leaves and Krishna tulsi with dark green leaves. Tulsi leaves have medicinal value. Sometimes, tulsi is considred to be a product of the 'Ocean Churning' episode.

9.7 The Hindu *Purusharthas* (Goals)

The Hindu religion sets four Purusharthas or Goals for the Hindus, i.e. —

Dharma
Artha
Kama
Moksha

- *Dharma:* Generally, *Dharma* is taken to mean religion; however, it has a much wider connotation. *Dharma* is to realize the underlying universal truth and to build one's life around it. *Dharma* includes concepts like morality, good behavior, righteousness, and justice. A path of spiritualism, which leads to the Supreme Spirit, is also called *dharma*.
- *Artha: Artha* is meant to achieve the well-being of an individual; it means both money and power. In other words, Hinduism stresses on economic power, including possession of material

things. However, these are not to be gained through immoral or wrong means.

- *Kama:* *Kama* means sensual enjoyment. If God has given this sense, it must be used for the enjoyment and pleasure of life. This would appear to imply that Hindus should neither be shy nor apologetic about sex. (The actual prevalent situation may be quite different).
- *Moksha:* The final goal of a Hindu is to achieve *moksha*, i.e. freedom from *samsara* — the continuous cycle of birth and death. *Moksha* is self-liberation of the soul and its merger with the Outside Reality Brahman. Many means of attaining *moksha* are prescribed in the scriptures.

In addition to the four Purusharthas, there are other subjects in groups of four each, i.e. —

- The Four Varnas or Castes — Brahmin, Kshatriya, Vaishya, Shudra
- The Four Vedas — Rig, Yajur, Sama, Atharva
- The Four Ashrams or Stages of Life — Brahmcharya, Gristha, Vanaprastha and Sanyas
- The four paths of Yogas — *Jnana* (knowledge), *Bhakti* (devotion), *Karma* (selfless action), *Raja* (physical and psychic control)

9.8 Hindu *Navagrahs* (Nine Planets)

Hindu astrology recognizes Navagrahas (nine planets), which are supposed to have an influence on human life. Out of the nine, seven are actual planets; the remaining two being shadow planets, as these do not exist in physical form. Names of the seven planets, along with days of the week corresponding to each planet as per the Hindu system are given alongside:

Sun	Ravi, Surya	Raviwar	Sunday
Moon	Soma, Chandra	Somwar	Monday
Mars	Mangal	Mangalwar	Tuesday
Mercury	Budha	Budhwar	Wednesday
Jupiter	Brhaspati	Brhaspat	Thursday
Venus	Shukra	Shukrwar	Friday
Saturn	Shani	Shaniwar	Saturday

The two shadow planets are —

Rahu — represents ascending node of the Moon
Ketu — represents descending node of the Moon

- Surya is an aditya born to mother Aditi and father *dev-rishi* Kashyapa; he has more than 12 other names.
- Chandra/Soma was the son of *dev-rishi* Atri and his wife Ansuya; he had the characteristics of Brahma.
- Budha was the son of Chandra/Soma through Tara, who was otherwise the wife of Brhaspati; Chandra had abducted her.
- Brhaspati was the son of *dev-rishi* Angiras; his brother was Samvarta. Brhaspati was *acharya* (priest) and preceptor of *deva*s (gods) in their fight against the *asura*s (demons). His wife's name was Tara, who was abducted by Chandra/Soma. He is lord of the inspired or sacred speech. Rishi Bhardwaj was a son of Brhaspati. It would be reasonable to distinguish between Brhaspati as a planet, and Brhaspati as the acharya of *deva*s. The planet might have been named after the Acharya.
- Shukra or Usanas-Shukra was son of *dev-rishi* Bhrigu; hence he was a Bhargava. He was *acharya* (priest) and preceptor of the *asura*s, *daitya*s and *danava*s (all demons), in their fight against the *deva*s (gods); he is also called Shukracharya. He is thus an opponent of Brhaspati. Shukra had a *mantra*, by which he could revive the demons, killed in war; he had learnt the *mantra* from Lord Shiva himself. Brhaspati had no such *mantra*; that was a great advantage for the demons. However, Raca, son of Brhaspati picked up that *mantra* from Shukra: that turned the tables against the demons, and the gods could emerge victorious. Shukra as a planet and Shukra as an acharya may be considered as separate entities — the planet being named after the Acharya.
- Shani was born of Vivaswat (Surya, the Sun) and his wife Chhaya (a double of his first wife Samjhna). He was the half-brother of Manu. The common perception is that Shani has a baleful influence on humankind. However, an alternate view is that Shani is only for the righteous. He is not a subject to flattery, and would never help people who are not on the righteous path.

Rahu and Ketu are inauspicious and the most dreaded planets; Rahu is dragon's head and Ketu its tail. Rahu is considered to cause eclipses, by occasionally swallowing the Sun and the Moon. (We know otherwise).

9.9 Hindu Kshatriyas

As our subject relates to military issues of the Hindus, we may spend some time in studying the Hindu martial class, the Kshatriyas. Their primary duty was to engage in warfare to protect the motherland, and its people; they were also to ensure the rule of *dharma* (righteousness). However, the Kshatriyas did not have a smooth sailing and had to face a lot of vicissitudes. We study the progression of Kshatriyas over the ages in the following paragraphs.

Parshurama

Very near to the beginning of humankind, the great lord Vishnu appeared in his sixth *avatar* as Ram (before Rama son of Dasrath). Lord Shiva gave his *parshu* (axe) to Ram, and he became Parshu-Ram. He was the son of *dev-rishi* Jamadagni, and the grandnephew of Vishvamitra. A Kshatriya king Arjuna Kartavirya (some texts say Sahastrabhu) deprived Jamadagni of his divine cow. Parshurama killed the king, whose sons thereupon killed Jamadagni. Infuriated by that, Parshurama decided to kill all Kshatriyas. Using Shiva's parshu, Parshurama wiped out the entire Kshatriya clans, not once or twice, but a full 21 times, (there are different versions of this episode).

That is a lot of punishment for digression by one king and his sons. Parshurama was the Vishnu *avatar*, who used Shiva's parshu (axe) to exterminate Kshayriyas. That would appear to show that Lords Vishnu and Shiva may not have been very favorably inclined towards the Kshatriyas.

Ramayana

In his seventh *avatar*, Lord Vishnu was born as Rama, son of King Dasrath. Tradition says that the four sons of Dasrath were born by divine intervention, as a result of a *yajna* (fire-sacrifice); a special kind of mixture was produced, which was consumed by the three queens.

The *Ramayana* story revolves around the Great War that was fought.

On one side was the demon king Ravan and his force of rakshasas. Ravan, his brother Kumbhkaran, and son Meghnad, were all blue-blooded Brahmins; the first two were grandsons of *dev-rishi* Pulastya. It would be reasonable to presume that the other rakshas forces had no particular caste system. Thus, there were no Kshatriyas on the side of Ravan; the main fighters were Brahmins.

Lord Rama's forces consisted of monkeys and bears; presumably, these did not have any castes either. It would thus appear that in that mother of all battles, there were only two Kshatriyas, i.e. the great Lord himself and his brother Lakshmana. That is not a very good advertisement for the Kshatriyas, whose primary role was to engage in combat.

Ravan was not only a high class Brahmin, but was also well versed in all the scriptures, including the four Vedas. Through severe austerities, he had obtained boons from both Brahma and Shiva. The killing of such a person by Lord Rama amounted to *brahm-hatya* (Brahmin killing), a heinous crime in Hinduism. It would appear that Rama might have had to suffer for that, as witnessed by the subsequent events:

- Rama had to spend the remainder of his life without his beloved wife Sita. He had to live without his two sons for some 14 years
- The Suryavanshi Line after Rama appears to have hit a roadblock. Though the names of several later Solar kings are mentioned in the Puranas, nothing much is known of their rule or exploits.

Note: We may remind ourselves here that though Lord Rama was an *avatar*, he decided to live the life of an ordinary mortal.

Mahabharta

There was a break in the Chandravanshi line of the *Mahabharta* after the 20th generation. The break lasted 22 generations; the line was re-established by King Dushyanta. His son was the great King Bharata, after whom our country Bharat is named. Sons of King Bharata were not worthy, and he killed them all. Later, Bharata adopted rishi Bhardwaj (a blue-blooded Brahmin) as his son. On the death of King Bharata, a son of Bhardwaj named Vitatha, ascended the Paurava throne. If true, this tale suggests the induction of Brahmin blood in the Chandravanshi line.

We next shift our gaze to King Shantanu, the 91st generation in the Chandravanshi Line. As outlined in the preceding paragraph, there was hardly any Kshatriya blood left in the Chandravanshi kings after King Bharata. Even that low percentage was further diluted. Shantanu's eldest son Bhisham Pitama took a vow of celibacy. Shantanu's other two sons, though married, died before they could produce any children. Hence, none of the three sons could contribute towards the propagation of the dynasty. Again, Brahmin blood had to intervene to continue the great Puru dynasty.

A practicing Brahmin, Ved Vyas was called in to render his services, so that the dynasty could continue. He impregnated the two widowed queens (wives of the deceased sons of Shantanu), through the rite of niyog. Now, Ved Vyas was born from the out-of-wedlock union of the great Rishi Parasara (grandson of *dev-rishi* Vasishta) with a low caste fisherwoman called Matsyaghangi. Thus, Ved Vyas had a mixture of Brahmin and low-caste blood. Dhritrashtra and Pandu were biological sons of Ved Vyas. Thus, these two had no Kshatriya blood whatsoever from their father's side. They might have had some Kshatriya blood from their mother's side; that does not count for much in patrilineal societies. We may note that in any case there was no Chandravanshi blood in the veins of Dhritrashtra and Pandu.

Later, the fisherwoman Matsyaghangi married King Shantanu, and became famous as Queen Satyawati. Her two sons were married to the mothers of Dhritrashtra and Pandu; but both died childless (as referred above). Thus, from both biological and non-biological angles, the grandmother of Dhritrashtra and Pandu was a fisherwoman.

Matters did not end there. The five Pandava brothers were not biological sons of their father Pandu who, in any case, had hardly any Kshatriya blood. They were all born with the help of gods Dharamraj, *Vayu*, Indra and the Ashwins. We are not aware if gods had any caste. If at all, they could perhaps be Brahmins, with the possible exception of Indra who was a warrior. Thus, it would appear that from their father's side, the five Pandava brothers had no Kshatriya blood in their veins.

From the foregoing, it is difficult to avoid reaching the following conclusions:

- There was very limited to nil Kshatriya blood in the Kauravas and Pandavas; the former had a shade more than the latter.
- Both Kauravas and Pandavas had no Chandravanshi blood.

What did the *Mahabharta* war do?

- It wiped out the Kaurava clan in its entirety.
- A little after the *Mahabharta* war, the Dwarka Yadavas (Kshatriyas) were destroyed in an internecine war.
- The Pandavas also relinquished their rule after a few years. One version says that the last king in that dynasty was Janmejeya, son of Parikshat who was the grandson of Arjuna. However, another version says that the dynasty may have continued for many generations in some diluted form.

In the preceding paragraphs, we have laid a lot of emphasis on the purity of the Kshatriya (and by implication Brahmin) blood. Our emphasis may be, somewhat, misplaced. In actual fact, there was a lot of mixing of Kshatriya and Brahmin blood. There are number of cases of even *dev-rishis* (the most blue-blooded Brahmins) marrying Kshatriya princesses. Satyavati, daughter of Kankyakubhja Kshatriya king Gathi, was married to Richik (Reika), son of *dev-rishi* Bhrigu. Satyavati's son was *dev-rishi* Jamadagni whose wife Renuka was a princess of Ayodhya. Many Kshatriya kings took Brahmin wives. Even low-castes (Shudras) wives were not ruled out. Tradition has it that Sumitra, wife of the Suryavanshi king Dasrath and mother of Lakshmana, might have been low-case. The great Chandravanshi king Shantanu married the low-caste Matsyaghangi.

It would appear that after the *Mahabharta*, Kshatriyas may have been severely weakened. Was there a divine design in that? There is a conspiracy theory which says that *Mahabharta* war was fought to finish off/weaken the Kshatriyas. We may recall from the story of Parshurama that Lords Vishnu and Shiva were not favorably inclined towards Kshatriyas.

Though we have concentrated on the Chandravansi dynasty, we may keep in view that there were many other Kshatriya dynasties ruling in different parts of Bharat-varsha of those days (see chapter 8.3: Manu's Progeny). Even those dynasties appear to have gradually melted away. We do not find any dominance of Kshatriyas, on eve of the recorded period of Indian history, starting around the 7th century BC. (see Chapter 16).

After the *Mahabharta* war, the history of Bharat again goes into oblivion, until we come to the recorded period of Indian history. As a very rough approximation, that twilight period might have been around 500 to 1000 years.

Recorded History Period

Now we come to the period of the actual recorded history of Bharat, i.e. around the 7th century BC onwards. Knowing the designated role of Kshatriyas, one would expect major Kshatriya empires flourishing at the dawn of history in Bharat. However, Kshatriyas are found to be largely missing. Rather, the major Hindu empires were non-Kshatriya, as shown below.

7th–6th century BC

At the dawn of history in Bharat around 7th century BC, we find Sisunaga dynasty ruling Magadh located in the eastern parts of Bharat. The general belief is that the Sisunaga dynasty was low-caste; the ending word 'naga' indicated an aboriginal origin. Brahmin records of those days speak of the Sisunaga kings in rather derogatory terms.

4th century BC

Around 400 BC, a low-caste upstart by the name of Mahapadma Nanda snatched power in Magadh, from the last Sisunaga king. He founded the Nanda dynasty. Mahapadma was a great conqueror; he undertook all round expansions. His empire became a major one, dominating the whole of North India, excepting the western areas. The Nandas ruled for about 80 years. From various accounts, it appears that Mahapadma Nanda treated Kshatriyas rather harshly. One version goes to the extent of saying that Mahapadma tried to exterminate the Kshatriyas.

In the late 4th century BC, Porurava (Porus to the Greeks) was ruling in the Punjab (NW India); in all probability, he was a Kshatriya, who claimed descent from the Puru clan of the *Mahabharta*.

4th to 2nd centuries BC

Around 320 BC, a nowhere man named Chandragupta Maurya defeated the Greek general Seleucus Nikater and the last Nanda king. He established his rule over the whole of North India. He set up the first pan-Indian Hindu empire, called the Mauryan Empire. It is believed that Chandragupta Maurya was the son of a (low-caste) Nanda king, through a low-caste woman, either named Mauri, or belonging to a tribe of that name. The Mauryan Empire was the biggest ever to rule Bharat from 320 to 183 BC.

2nd to 1st centuries BC

The Maurayan Empire collapsed in 183 BC. One would expect that at least by then, Kshatriyas would establish their rule. But, we are again disappointed. A Brahmin Commander of the Mauryas seized power, and established the Brahmin Shunga dynasty, ruling over the Eastern parts of Bharat. When the Shungas lost power, another Brahmin dynasty Kanva took control in the East of Bharat.

In the North West of India, barbarian tribes seized power. Hindus referred to them as *mlecchas* — lowest of the low caste. A Kushan Ruler Kaniksha established perhaps the greatest ever empire, covering whole of North India; it extended beyond the borders of India up to Central Asia, from where the Kushans had come. The Kshatriyas were again largely missing from action. If indeed there were any (brave) Kshatriyas left, they would have protected this holy land from the depredations of the barbarian tribes.

4th to 5th centuries AD

Barbarian tribes ruled over North West India for some 500 years, up to the 3rd century AD. Who snatched power from the barbarian tribes? Kshatriyas? No way; they were again nowhere on the action scene. A Vaishya, Chandragupta I emerged on the scene from nowhere. He established the great Gupta dynasty, which went on to rule for some 200 years from 320 AD to 520 AD.

It appears from the foregoing that there were hardly any major Kshatriya states, during the first 1000 years of India's recorded history, i.e. from 700 BC to 700 AD. The Kshatriyas might have been ruling some small states.

Rajputs

Around the 8th century AD, a new class emerged on the Indian scene; they called themselves Rajputs. They had their base in the south of North West India, called Rajputana, rechristened Rajasthan now. Rajputs claimed that they were Kshatriyas of the highest class. They trace their ancestry to:

- Suryavanshis of the *Ramayana*
- Chandravanshis of the *Mahabharta*
- Agnivanshis who emerged out of a *yajna* (fire-sacrifice) conducted at Mount Abu by *dev-rishi* Vasishta, at an undetermined date in the past.

If Rajputs (Kshatriyas) were on-going people from the *Mahabharta* time, the center of power at the dawn of recorded history in India (7th century BC) should have been in and around Rajputana, where Rajputs abound. Far from it, all power was concentrated in the Eastern corner of Bharat, away from Rajputana. We start hearing of Rajputs a full one thousand years after start of the recorded history in Bharat. In the late 4th century BC, when Alexander attacked, we find powerful kings in the East (Magadh Empire), Punjab (Porus) and Gandharva (Afghanistan). Rajputs and Rajputana were hardly in the picture.

From 2nd century BC onwards, Barbarian tribes started their invasions of India right in the backyard of Rajputana, the homeland of the Rajputs. We do not find records of any great resistance that may have been offered by the Rajputs to these barbarians. That would lead us to the conclusion that there was no great presence of Rajputs those days, i.e. from the 2nd century BC to the 3rd century AD, and perhaps even the following four or five centuries. In view of the totality of circumstances, it is difficult to accept the claim of Rajputs of their lineage to the Suryavanshi, Chandravanshi and Agnivanshi Lines. We have discussed in detail the issue of Rajput ancestry in Chapter 23.

Great Hindu Generals

The four essential ingredients of a 'world-class' General are:

- An Offensive Mindset, resulting in a series of offensive actions
- Record of Victory after Victory — No need to advance excuses for defeats
- Repeated Expansion of Empire — including (successful) efforts to unite Bharat under one flag; at least partly, if not fully.
- Some offensive forays beyond Indian borders (in earlier times, Afghanistan was mostly part of Bharat).

Going by the above criteria, no Hindu General qualifies for inclusion amongst the 'World's Great Generals'. Though, if such a study were to be conducted, among the Hindu Generals, the following (in chronological order) would rank at the top:

> Bimbisara and his son Ajatasatru — low caste
> Mahapadma Nanda — low caste
> Chandragupta Maurya — low caste
> Chandragupta I & II, and Samudragupta — Vaishyas
> Shivaji and Bajirao I — Marathas

Without doubt, amongst the above, Chandragupta Maurya would get the top billing. He built a pan-India empire from absolute scratch. Ashoka's name does not appear in the list, as his military exploits, compared to Chandragupta Maurya, were quite modest. It is the present day Hindus who have attributed great military exploits to Ashoka. The second rank would go to Chandragupta I of the Gupta dynasty.

We cannot help noticing that there is no Rajput name in the above list, though they produced two top class Generals in Rana Sangha and Maharana Pratap later on. The Rajput Generals fail to qualify for the above list as they:

- Lacked Offensive Spirit — they fought only when attacked. Rajput Generals did not march on Delhi even when there was a total power vacuum there during the Muslim rule (that happened many a time — see Chapter 22).

- Far from uniting the whole of Bharat, Rajputs failed to bring even Rajputana under one flag. They were generally satisfied with their small or medium sized states.
- They did not attack foreign lands, even when upstarts from those lands became their tormentors, i.e. Ghazni and Ghauri.
- They failed to see the danger as it manifested itself repeatedly at the Khyber Pass, and beyond.

Names of some generals of South India could have been included in the above list — names like Pushkin II, Rajaraja I and Rajendra I would be strong candidates. Though these generals showed offensive spirit and made many conquests, they remained largely confined to the South. They did not expand their empire to the North, the heartland of Bharat those days. They also failed to take on the Muslims, as the Muslims were establishing their rule over Bharat, or otherwise plundering it. They had the duty to take on these invaders as soon as they crossed the Khyber Pass, or even before that; some of the generals of the South had the military muscle to do that. A few Southern generals undertook military expeditions to South East Asia, and made conquests. However, authentic details about their conquests are not available.

Conclusions

From the preceding, we draw the following conclusions:

- The two biggest and the most powerful empires to ever rule over India were non-Kshatriya, i.e.
  — Low-caste Mauryan Empire (320 to 183 BC)
  — Mllechha (low-caste) empire of Kushan King Kanishka (78 to 114 AD)
- The longest ruling Hindu empire (200 years, from 320 to 520 AD) was the Gupta Empire, which was Vaishya. That has been called the golden period of Indian history.
- At the dawn of history in Bharat, around 7th century BC, the

low-caste Sisunaga dynasty was ruling (Eastern) Bharat; its rule lasted some 200 years (6th & 5th centuries BC).

- For about 1000 years after the dawn of history in the 7th century BC, Kshatriyas are not found to be ruling the major states of Bharat; they might have been heading some smaller states.

In an overall analysis, it is difficult to avoid the impression that the downfall of the Kshatriyas might have started from the *Mahabharta* days. They were pushed to the background, and appear to have gradually melted away, yielding place to other castes. Was that, in some sort of way, a result of divine intervention? We have no answer. We may just recall the Parshurama episode — the great Lords Vishnu and Shiva had contributed to the extermination of the Kshatriyas.

Present Position

Let us survey the caste scene in present day India. It is easy to identify a Brahmin by the last name of a person, i.e. Trivedi, Diwedi, Vajpayee, Sharma, Kaul and innumerable such names. It is also the case with the Vaishyas, i.e. Gupta, Bansal, Aggarwal, Goenka, and others. Of course, the Shudras in India are always easily identifiable by their caste names and even otherwise. Other than Rajputs, it is difficult to identify caste names of Kshatriyas easily. As brought out earlier, we are not very sure about the Kshatriya origin of the Rajputs. Most of the present-day martial classes are either low caste, or want to be counted amongst the OBCs; some examples:

Low Caste: Majhbi Sikhs, Mahars of Maharashtra.
OBCs: Yadavas, Jats, etc.
We may state here that at the start, the Yadavas were the Kshatriyas of the Chandravanshi Line.

Section C
Some Issues of Hindus

10
POLYTHEISM VS MONOTHEISM
The Confusion

Christianity and Islam dominate the present day world; both are fiercely monotheistic religions, the latter even more than the former. Each believes in One Single Supreme God. Another popular religion, i.e. Buddhism has no need of God. Hinduism is the only formal religion, which believes in multi-gods; though once in a while, Hindus also talk of One Single God.

In Bharat, Muslims are a dominant minority. The Hindu–Muslim relations are hostage to their disparate religious belief systems, 600 years of master-slave equation, and 1000 years of composite history, mostly of conflict. The partition of India on communal lines made the situation worse. The recent rise in Islamic terrorism has added another twist to the tale. In this largely monotheistic world, Hindus are often criticized and badgered for their polytheism. To get out of the tight situation, (pseudo) Hindu leaders are often heard making the following types of statements (mostly on TV channels):

Firstly, the Hindu leaders say that there is but One Universal God, though the wise call Him by different names; and that there may be different ways to worship Him, and all these ways may be acceptable. The Hindu leaders tend to imply that this Universal God is common to all religions, though they do not say so in these words. During TV debates, the Muslim leaders, being a minority in Bharat, let this statement pass, though taking care to avoid any hint of endorsing it. Secondly, the Hindu leaders say that the multi-gods are but a manifestation of the One Universal God; so, there is no harm in having multi-gods.

In making the above types of statements, Hindu leaders have two aims. Firstly, by talking of the One Universal God, Hindu leaders seek to put Hinduism in the company of the monotheistic religions which dominate the world. Secondly, Hindu leaders think that the talk of a common God may help reduce the chasm between the Hindus and Muslims, and bring them closer.

Let us try to see if there could be a concept of a God, which is common to all religions. We start our search with the three Judaic religions, which otherwise, have a lot of commonality, i.e. common roots and some common prophets. If there is a possibility of a common God, chances are He would be amongst the Judaic religions. The three Judaic Gods are:

Judaism   — Yahweh
Christianity — The Father, the Son and the Holy Spirit
Islam   — Allah

Yahweh, by his own declaration, is God only of the Jews and no other people. He was openly partial to Jews, and promised them many bounties. Yahweh is a jealous and intolerant God who told Jews that He would punish them if they worship any other God, and He did that, indeed.

The concept of the Christian God of 'The Father, the Son and the Holy Spirit' is unique and exclusive to Christianity. Jews refused to recognize Christ even as a messiah, leave alone as God. Islam though recognizing Christ as an earlier prophet does not recognize Christ as the 'Son of God'.

Allah is so exclusive to Islam that non-Muslims cannot even begin to fathom His exclusivity. It is blasphemous for Muslims to think of a God other than Allah; the question of any commonality would not even arise.

Each of the three Judaic Gods own a small piece of real-estate in the holy city of Jerusalem, viz.;

Jews   — The Wailing Wall

Christians — Church of the Holy Sepulcher
Islam — Dome of the Rock and Al-Aqsa mosque

The above pieces of real estate have been the cause of severe discord and violent conflicts. These threaten world peace, with possibilities of a nuclear war. If, indeed, there were any degree of commonality between the three Gods, matters would have been resolved long time back. If the issue ever gets resolved, it would be due to a future President of the USA, and not due to any common God.

Tradition has it and folklore confirms that each of the three Judaic Gods always acted in the interests of His own people; no signs of universality are seen. As stated earlier, Yawveh is openly pro-Jewish. Allah, at all times and under all circumstances, helped Muslim armies win in the battlefield. The Christian God helped His people rule over almost the entire world. Islamic, Hindu and Buddhist civilizations lay prostrate before the Christians during the 18th to 20th Century AD. The Christian God helped Christians discover an entire new continent and dominate it.

The occasions when the Hindu God might have helped Hindu armies in the battlefield can be counted, if at all, on the fingers of one hand. Hindu warriors were always careful to shout 'Jai Bhawani' or 'Har Har Mahadev' (hail/salutation to war goddess/war god). That does not appear to have helped them. Does that convey a message? Perhaps yes, may be no.

There is no dearth of occasions in history when one Judaic community indulged in the mass slaughter of another Judaic community in the name of its God. During most part of history, the three communities have been at each other's throats, mostly in the name of God. During the first Crusade, Christians captured the city of Jerusalem in 1099 AD. There was such slaughter of Muslims and Jews that it is reported that in some streets of Jerusalem, blood flowed up to the level of horses' reins (not to be taken literally). During the 1940s, Hitler carried out mass slaughter of a few million Jews. Could all that have happened if there was even a small amount of commonality between the three Judaic Gods?

Muslims would never accept the thesis that their Allah could have any degree of commonality with the Brahman or Brahma of Hindus. Muslims are clear that Allah is exclusive to Islam and there is no question of sharing Him with any other religion, least of all with the (idolatrous) Hindus. Even the thought of sharing is blasphemous to Muslims. Therefore, Hindu leaders should proceed with caution in hinting about One Universal God being common to all religions. Such a formulation, far from unifying the two communities, may drive them apart.

When the Hindu leaders see that their thesis of One Universal God does not cut much ice, they adopt another stratagem. They take the position that the multi-gods of the Hindus are but a manifestation of the One Universal God. As per them, this is necessary so that the Hindu masses can relate to the One God, through the intermediate gods. The concept of God, by its very nature, imparts Him with such majesty and grandeur, that none can have any difficulty in relating to him directly. That is why the dominant and successful religions of the world, i.e. Christianity and Islam, insist on direct relationship with God. They frown upon the need of any intermediaries, i.e. the lesser gods.

The need for intermediaries between God and man can arise only under one of the two following conditions:

- The Supreme Universal God is somewhat weak and not capable of direct interaction.
- The people are dim-witted and intellectually under-developed, who cannot see the majesty and grandeur of the Supreme One, without the help of the intermediaries.

None of the above condition is applicable to Hindus. Their Supreme God Brahman is all-powerful and majestic. He was there before 'time and space' and would be there after 'time and space' cease to be. He actually resides in individual consciousness as *atma* (soul). Therefore, there is no question of the Brahman having any difficulty in direct interaction with man; and vice-versa. Also, Hindus were and are a set of most prescient and intelligent people. They were the first to discover and realize the intricacies of nature, and to develop the very concept of

God. So, there should be no question of Hindus having any difficulty in relating to the One Universal God. However, regretfully, in actual practice, things have not worked out that way.

In studying the Hindu leaders' manifestation theory, we start at the very beginning, i.e. with the Vedas. With all the sublimity and purity of the Vedas, the main emphasis of the *Rig Veda* is on multi-gods, 33 of them. Indra is selected for special deification. Hymn after hymn is addressed to Indra, invoking him for various benedictions. Out of the 1028 hymns in the *Rig Veda*, some 250 are addressed to Indra, and another 200 to Agni. The Vedic gods appear to be gods in their own right and not as proxies (or manifestation) of the Supreme One, who gets mentioned only in passing in the Rig, and that too in a later book.

Before the concept of the One Supreme God developed by the Upanishads had the time to settle down, two new religions emerged in Bharat, i.e. Buddhism and Jainism. These religions did away with the very concept of God, single or multi. Buddhism flourished in Bharat for some 800 years, which could be considered to be period of almost 'No-God'. When the Hindus drove out Buddhism, they did not revert to the Upanishads and their concept of One Supreme God. Rather, they went in for the Puranic Hinduism, which introduced a new set of multi-gods. Seeing the awesome powers of Lord Shiva, it is difficult to see him as proxy (or manifestation). He is a god in his own right. So is Lord Vishnu, who appears from time to time in his *avatars* (incarnations), including the most supreme and powerful ones, i.e. Rama and Krishna.

It is ordained that under all circumstances, Ganesha is the god to be worshipped first. If Ganesha was a proxy (or manifestation) of the Supreme One, he should be worshipped after the Supreme One. But, in the *Puranas,* there is no great stress on the worship of the Supreme One. With the exception of Arya Samaj mandirs, there are no Hindu temples for the Supreme One. All temples are dedicated to one or the other god, or a clutch of them. As such, Hindus are generally at a loss to understand where and how to pay their respects to the Supreme One. As against that, each Christian church and each Muslim mosque is dedicated exclusively to their respective Supreme God. Perhaps, in that lies a major reason for the success of Christianity and Islam.

No it's in
'jahalat'

Rig Veda lists mundane (in the spiritual sense) entities like *Vayu* (wind) and *Soma* (an intoxicating drink), as gods; they could hardly be the manifestation of the Supreme One. Diwali is a major festival of the Hindus. It is supposed to celebrate the return of Lord Rama after his famous victory of 'good over evil'. On this most holy occasion, Hindus worship money, giving it the name of goddess Lakshmi; even Lord Rama is pushed to the background.

The manifestation theory of modern Hindu leaders appears to be a non-starter. If one's belief in the Supreme God is firm and strong, there can be no difficulty in relating to Him directly. Only if the belief system is weak and wavering, one starts looking for intermediaries, i.e. the lesser gods. It is worth examining as to how and when the 'manifestation theory' emerged on the Hindu scene.

The genesis of emergence of the manifestation theory appears to lie in the advent of Islam in India. Islam is a fiercely monotheistic religion. Muslims found the Hindu religious practices appalling. They offered Hindus a way out by conversion; Hindus rejected that option. Being the ruling class, Muslims took to taunting and tormenting Hindus for their religious belief in multi-gods. To come out of the tight corner, Hindus developed a counter strategy. They started arguing that, in fact, Hindus also have One Supreme God, and that their multi-gods were but a manifestation of the Supreme One. In practice, that argument did not cut much ice; neither with the Muslims, nor even with the Hindus. Most Hindus, especially the common folk in the rural hinterland consider their multi-gods as gods in their own right. *too much choice!*

Swamy Dayanand of the Arya Samaj is one of the most exalted leaders of Hindus is modern times. He did not favor the manifestation theory. He provided Hindus the direct route to One Supreme God. Regretfully, Hindus adopted his route in a half-hearted way, and only in parts of Bharat. Even that commitment appears to be weakening now.

11
HINDU UNTOUCHABILITY
A Curse

It would appear that untouchability has been part of Hindu society from the very beginning.

- *Rig Veda*: In the Rig, the creation of humankind has been narrated as the dismemberment of the Cosmic or Primeval Man Purusha (Prajapati), offered in total sacrifice. The *Rig* Hymn X.90.12 says —
  From his mouth was the Brahmin
  From his arms was fashioned the Rajenya (Kshatriya)
  His thighs, the Vaishya
  Shudra was born from his feet.
- Manusmriti — Chapter 1, Entry 31 says — Then, so that the world and its people may prosper and increase in numbers, he (Brahma) created the Brahmin (priest) from his mouth, Kshatriya (ruler) from the arms; from his thigh was born the Vaishya (trader/agriculturist), and from his feet the Shudra (service class).

Manu was the very first man of creation; Manusmriti lays down the foundation of the caste system. It divides society into four castes, viz: Brahmin, Kshatriya, Vaishya and Shudra. The first three castes have been called the 'twice-born'; this word appears again and again in the Manusmriti. Most of the activities which are permitted to the 'twice-born' are prohibited to the Shudras. A few sample entries from Manusmriti:

Chapter 3, Entry 329 — An Untouchable, a pig, a cock, a dog, a menstruating woman or an impotent man, should not watch the priest eat.

Chapter 10: (51) The dwellings of untouchables and dog-cookers should be outside the village boundary; they should use discarded broken bowls; dogs and donkeys should constitute their wealth (52) Their clothes should be those of the dead, and their food should be taken in broken dishes (55) When they move about in the day to do their work, they should be recognizable by some special marks, as may be prescribed by the ruler; they should have the duty to carry out the corpses of people who have no close relatives.

Chapter 11, Entry 176 — If a priest, unknowingly has sex with an untouchable or low caste woman, eats, or accepts, he falls; if knowingly, he becomes their equal.

As for mobility amongst the castes, Chapter 10, Entry 64 says:

'If someone born from a union of a priest with a low-caste woman, produces a child with someone of a higher caste, the lower caste will reach the status of the higher caste after seven generations.'

This entry would appear to suggest that mobility amongst castes was almost impossible, except in some exceptional circumstances.

In a thought process unprecedented in human history, Manu decried that the very touch of a Shudra would pollute the other three castes. God was not available to the Shudra; his entry to temples was barred. A Shudra may not pass in the vicinity of a temple, as he might accidentally hear Vedic *shlokas* (verses); if that was to happen, lead was to be poured in his ears. If a Shudra chants a Vedic *shloka*, he may lose his tongue. Shudras were to stay duly segregated in their own clusters; generally outside the village boundary. They may carry a clapper or bell, so that higher castes are informed of their approach. It would appear that dalits (new name for Shudras) have been the most oppressed portion of society anywhere in the world, in the entire human history of say 5000 years.

Shudras were not to enter the village before say 10 o'clock or after

3 o'clock, when they would caste longer shadows — the sun being low on the horizon. Such a shadow may accidentally touch the higher castes and pollute them. Shudras had to have their own well and could not draw water from the common village well. The list of discriminations against the Shudras is long and disgusting. For their originality, sheer asinity and gross depravity, there is no parallel to the taboos prescribed for the Shudras. As per the Hindu Upanishadic philosophy, the Outside Reality, the Supreme Spirit Brahman resides inside every man as *Atma* (soul); so does Brahman reside within the Shudra. How can that Brahman be an untouchable? Not many Hindu scholars or saints have tried to explain this great dichotomy in a style or language which a layman Hindu can understand.

For all the taboos against the Shudra, he was still a part of the Hindu society, though at the rock bottom. Hindus did not stop at that; they went on to prescribe even lower categories like Chandalas, Mllechha, Nishadas, Doms, etc. They were out of the very domain of society. These non-castes begged to be given a category even lower than the Shudra; but the Hindus would not agree. For example, Chandalas were required to wear the dresses of the dead, as the disposal of dead bodies was their main task; they were forced to live in and around cremation grounds. The abhorrent treatment given to these non-castes was and is beyond human imagination. Never in history have human beings (in whom God Himself resides) been so degraded. God punished Hindus for that, by making them slaves for some 1300 long years, out of the last 2300 years — that is a dubious world record.

There is the Eklavya story in the *Mahabharta*. Eklavya was a low caste aboriginal. He approached the Brahmin, Guru Dronacharya for lessons in archery. He was turned away due to this low caste. Nevertheless, Eklavya achieved mastery in archery on his own. The Brahmin guru came to know of his achievement and feared that he might outshine the high caste prince Arjuna — a pupil of Dronacharya. The guru now claimed that he was in fact, the guru of Eklavya and asked Eklavya's right hand thumb as *guru-dakshina* — guru's fees. Eklavya cut his thumb and gave it to the guru. Being an aboriginal, Eklavya lived by hunting animals; with his thumb gone, he might have starved.

The moral of the story is that the low castes must give their limb, livelihood and life for the high castes; that is the *dharma* (duty) of low castes. There can be no greater case of hypocrisy, depravity and sheer high degree of oppressive injustice. Even the present day upper caste Hindus narrate the Eklavya story with great gusto and pride, and tout it as the hallmark of a great civilization. Attitudes have not changed even in the 21st century. In 2007 AD, one of the most progressive film directors made a movie named *Eklavya*, to convey exactly the same message, i.e. the lower castes must give their limb and life for the higher castes. Unexpectedly, but fortunately, the movie bombed at the box-office.

There are two stories in the *Ramayana*. One relates to Lord Rama eating the bitten and tasted 'bers' (Indian berries) of a Shudra woman called Shabri. It indicates the disapproval of Lord Ram of untouchability.

There is another story in the *Ramayana*. Son of a Brahmin died in the days of Rama Rajya — Rama's rule. The Brahmin went to King Rama; he expressed that Rama's reign must be evil, because a son had died before the father. Rama called his counsellors. After investigation, the counsellors informed king Rama that the cosmos balance had been disturbed, as one Shudra named Shambuk was practicing asceticism, which was not permitted to the Shudras. King Rama proceeded and cut off the head of Shambuk. As per the legend, that restored the cosmic order.

In the 16th Century AD, a most learned scholar and renowned author Tulsidas wrote the *Ramayana* in Hindi called *Ramchritmanas* — a profound book by any standards. About the Shudras, he writes as follows:

Shudra, gwar, pashu aur nari,
Hain sab taran ke adhikari
or,
Shudra, rustic, animal and woman — need to be beaten up

There have been leaders who have tried to ameliorate the status of Shudras. Among them were Mahatma Gandhi and BR Ambedkar.

Gandhi coined the word harijan (sons of God) for the Shudras; but they did not particularly like it. Gandhi tried, but was only marginally successful. BR Ambedkar perhaps achieved more success. Ambedkar, along with a large number of his followers, converted to Buddhism. That was his manner of protest against Hinduism's oppression of Shudras. The British made their contribution to ameliorate the condition of the Shudras; they coined the word 'scheduled castes', which had greater acceptability. The British set up exclusive regiments for the scheduled castes in the Army, i.e. Mahar Regiment, Sikh Light Infantry (for *majhbi* Sikhs). At some stage, the word 'Dalit' came into usage for the Shudras; this is considered acceptable and is in common use these days.

In the present, untouchability has reduced amongst the educated classes in cities and larger towns. There is also the emergence of a creamy layer among the dalits; it is, however, limited to say 20% or so. In the villages, untouchability is prevalent in varying intensity, from area to area. Still, a million humiliations are inflicted on dalits every day, in many parts of Bharat.

The dalit woman is an easy game for the village 'thakur'; police normally do not register such a case. In 2006, there was the case of mass rape and murder of dalit women in Khairlingi in Maharashtra. Police Officers were involved in the cover up till the case incidentally came to media notice. Dalits have to still stay in their own clusters in one corner of the village. In many places, they still cannot enter a temple, or draw water from the common village well. In some cases, a dalit groom may not be allowed to ride a horse to the bride's house. The whole system is against the dalits their poverty only making it worse.

Mercifully, by the grace of God, things started changing in the 1990s. This happened due to the emergence on the Dalit scene of a man-woman duo. Both were nowhere people with very modest backgrounds, modest education, no means and no pedigree. One was a government servant, and the other a schoolteacher. Their names are Kanshiram (now deceased) and Mayawati — rather uninspiring names by the standards of modernized Hindus. Though the latter was the pupil of the former, she soon outshone him. Mayawati started succeeding where much bigger leaders had failed.

Mayawati has been successful in achieving a modicum of respectability for the dalits. However, caste Hindus like to concentrate on her alleged corruption. She is dubbed as erratic, temperamental and irresponsible. The truth is that caste Hindus, including the moderate and modernists, are not able to stomach her success. She has out-foxed the old foxes of well-established political parties. She has stolen their thunder and vote-bank, not only of the dalits but also of the Brahmins, from right under their feet. Mayawati won a major victory in the UP elections of 2007. Established and pedigreed political leaders were left looking on in stunned silence. They could give out only incoherent excuses for their shattering defeat. Defying all predictions, Mayawati found herself ensconced in the chair of the Chief Minister of the biggest state of India, for the long-term. That was one of the greatest moments of history of the Hindus in general, and dalits in particular. However, not many (caste) Hindus would see it that way; they would like to predict dire consequences. Some would go to the extent of saying that cosmos' balance is likely to be disturbed, as a dalit (Shudra) has 'grabbed' power.

All the actions and (so-called) antics of Mayawati are well-rehearsed and for a set aim. If she spends a crore (ten million) or two on her birthday, it is to send a message. If she erects Ambedkar statues, it is for a set objective. She is one of the very few effective administrators that independent India has seen; ask any IAS/IPS officer who might have served in UP. The caste Hindus' ire is aroused as to how a no-where dalit woman from a humble background, with no lineage, could single-handedly outwit them all, including the most pedigreed. Hence, there is hostile reaction to Mayawati and refusal to recognize her obvious merit. Even if (as falsely alleged), she has made some money on the side (which she has not), it is no big deal. Political parties need funds to run; no Indian moneybag will give money to a dalit party. In any case, caste Hindus have been making their billions from times immemorial. It is about time a dalit also has his/her hand in the till.

Mayawati's Dalit/Brahmin combination is a political masterstroke, without parallel in Indian (Hindu) history. It is the first successful attack on the Manu citadel and a challenge to the 5,000-year-old Hindu society. Manu and Mayawati are at war, and Manu is crying 'foul', perhaps just crying. In the meantime, the higher classes are leaving no stone unturned to 'demonize' Mayawati; the media is at their beck and call. However, it is certain that she will have the last laugh. With her sitting in the Chief Minister's chair, police SHOs in UP are recording cases of dalit rape, even before these are reported to them.

Non-dalits cannot even begin to imagine what it means to be a dalit in Bharat — How helpless! How humiliating! How horrifying! No non-dalit can represent repressions and aspirations of the dalits; they must do it themselves. Towards that end, Mayawati needs to be encouraged all the way. Perhaps, she is in no need of any encouragement, and may do it entirely on her own steam.

12
HINDU AHIMSA
The Undoing

In Hinduism, there is a saying *'Ahimsa Parmo Dharma'*; translated, it means 'Non-violence is the first duty'. This slogan has confused the Hindus no end and is responsible for many of their ills. No other civilization talks of non-violence in that manner, or with such reverence. Actually, not many civilizations even have this concept of *ahimsa*, as the Hindus have. There may not be an equivalent word for *ahimsa* in most languages. Even in English, it is a derivative word from the word 'Violence'.

Adopting any yardstick and considered from any angle, the first (and the second and the third) duty of a man is to his motherland, to defend her honor. He has to keep the flame of liberty burning, for a slave society is a dead society. This is a truth, which can be ignored only at great peril to national honor and freedom. After the motherland come family, community and a host of such entities. *Ahimsa*, if at all, would not appear even at serial 20 of duties. Actually, *ahimsa* cannot be a duty; at best, it can be an option. It cannot be even an option for people who cherish and value their liberty and wish to live as free people. It could only be an option for those who wish to live in bondage and slavery. The *shloka* is not of Hindu origin and has been adapted from Buddhism and Jainism. The emphasis of these two religions was on other-worldly things, divorced from day to day issues of life, living and liberty.

The Vedic people were of martial nature. The Vedic religion requires the Hindu to be powerful and assertive. If the Hindu is weak, his *dharma* is not only dubious, but distasteful. Hindus were worshippers of *Shakti* and could not be believers in *ahimsa*. The creed of *ahimsa* is foreign to the Vedic religion; it was dreamt up by those, who look for excuses to enjoy cowardice. (Mahatma Gandhi was an honorable exception; he devised *ahimsa* to cope with a specific situation. See comments in a later paragraph).

Even if there is an oblique reference to *ahimsa* in some scripture (which is doubtful), Hindus have no tradition of *ahimsa*. Their holy book, the *Gita* is delivered right in the midst of a battle; it calls for war — To kill, or be killed. The delivery of the *Gita's* message is followed by *himsa* (violence) of gigantic proportions, in which the whole Kaurva clan is wiped out, and thousands are killed.

Lord Vishnu took the *avatar* as Parshurama to indulge in the greatest bouts of *himsa*, during which he slaughtered the entire Kshatriya clan 21 times, without adequate or even reasonable cause (as it appears to us now). After all, just one Kshatriya king had done the digression — why punish the entire clans, and that too 21 times. It would appear there was a lot of avoidable *himsa* (violence), and for no particular purpose. Can there be any greater advertisement for the place of *himsa* in human affairs? In his *avatars* of Rama and Krishna, the great lord Vishnu had to resort to repeated *himsa* to fulfill the roles for which he descended in this world. Rama killed Bali, Ravan and others; Krishna killed Kansa and Sishupal. The mythological history of Bharat is full of *himsic* wars. The great lords Vishnu and Shiva play roles in those wars. If gods were so committed to and fascinated by wars and *himsa*, it does not behove mere mortals like us to shun these. War and *himsa* are a part of the humans' and gods' DNA.

Now, coming to actual history. We had the great Chanakya Kautilya in the 4th century BC. He was an unabashed admirer of *himsa*. In his *Arthshastra*, Chanakya prescribes that if a king has a weak neighboring state, he must attack it and capture it. Chandragupta Maurya, a friend of Chanakya, followed his dictum and established the first pan-Indian Hindu empire, covering the whole of North India, including Afghanistan. Chandragupta's grandson Ashoka decided to ignore Chanakya and adopted *ahimsa* as his creed. The result was that the great Mauryan Empire was blown to bits. Due to neglect of the army, Ashoka's descendents were unable to defend Bharat against the barbarian tribes. These tribes poured along the North West borders; their invasions and conquests lasted for some 500 years. Those years have been called the Dark Ages of Bharat and were the direct result of Ashoka's *ahimsa*. After Ashoka, the concept of *ahimsa* got lodged in the Hindu sub-conscious. That was, in no small measure, responsible for the subsequent bouts of Hindu slavery, including some 800 years in the second millennium AD.

Pages of history consist mostly of exploits of generals and of wars, which is but another name for *himsa*. If a king is strong, he must attack his neighbors, as all famous generals did. If the king is weak, he will invite attack. When a king dies, there is *himsa* to establish succession. When the king ascends to the throne, there is *himsa* as contenders have to be eliminated. Many a times, the king has to kill his own brothers and other kin to safeguard his throne. Emperor Ashoka did that; later, he became an apostle of *ahimsa* (shades of hypocrisy there).

No essay on *ahimsa* is complete without reference to the modern apostle of *ahimsa*, the one and only Mahatma Gandhi. He was a master strategist; the great Mahatma saw that British had a vice-like military grip over India. The prospects of success of an armed rebellion against the British were bleak; the past military performance of the Hindus did not inspire much confidence. Therefore, showing rare foresight, he devised his weapon of *ahimsa*. The point to note is that it was a specific weapon for a specific situation. In fact, in 1942, Mahatma Gandhi gave a call for 'Do or Die' or '*Karo ya Maro*'. To us laymen, that would appear to be a call for *himsa*. Of course, apologists of *ahimsa* would say that small minds like ours cannot understand Gandhi; we plead guilty — we cannot.

It would not be fair to attribute the Indian independence entirely and solely to Gandhi's *ahimsa*. Many other factors were at work; not the least of these being Hitler's *himsa*, which drained Britain of all energy, both moral and physical. By the end of World War II, Britain was no longer in a position to hold on to its overseas possessions. In addition, there was unremitting pressure from the emerging super power, the USA. The role of *himsa* of Bhagat Singh and other martyrs also needs to be acknowledged with gratitude. Also, there was the INA of Bose, and the Naval Mutiny. All these had their contributions.

In 1945–46, following Jinnah's call for Direct Action, there was violence in Bengal and Bihar. Perhaps a few hundred or a few thousand people were killed. That unnerved the Congress (Hindu) leadership. They tamely agreed to the partition, though they had taken solemn oath to oppose it, and had won a thumping mandate from the people on that platform. The partition was ostensibly agreed so that violence could be avoided. However, what followed partition was violence on an unprecedented scale, for which there is no parallel in history. Some half a million people were killed and about 10 million displaced. That would not appear to be a great success story for *ahimsa*.

Unfortunately, things in Bharat have been over-simplified. India's independence is touted as the victory of *ahimsa*. That is used as a reason to impose *ahimsa* on the Hindus, under all circumstances and for all times. As we have brought out in the preceding paragraphs, ground facts are significantly different. The weapon of *ahimsa* devised by Gandhi was a short-term specific weapon for a specific situation. Further, *ahimsa* was a concept which only the Mahatma understood. Even his close associates were often flabbergasted by his actions in the name of *ahimsa*. The concept of *ahimsa* should have been cremated along with the bones of the Mahatma. However, we the lesser mortals, keep on parroting *ahimsa*, without understanding both its concept and context. In any case, *ahimsa* can have no space in an independent India. We are in the Kali Yuga and surrounded by people who wish no good to us, some overtly, others covertly. This, out of context, out of season and out of tune singing of the *ahimsa* tune can do us immense harm.

If they wish, saints and sages may expound on the concept of *ahimsa*; that should be done in confines of their ashrams and to senior citizens only. The youth of the country must not be exposed to the pernicious, self-defeating doctrine of *ahimsa*. It should not be propagated on TV, to which general public has access.

13
HINDU TOLERANCE
The Myth

In addition to *ahimsa*, another insignia fondly, forcedly and firmly put on the Hindu lapel is that of 'Tolerance'. It is difficult to utter the 'Hindu' word, without uttering 'tolerance' in the same breath. The Hindu is being constantly told that his religion and scriptures require him to be 'tolerant'. It is generally projected as if Hinduism has no existence independent of tolerance; a Hindu should 'walk' tolerance, he should 'talk' tolerance. During TV debates, one often hears Hindu leaders, both pseudo-secularists and 'communal', going hysterical about 'Hindu Tolerance'. The following types of statements are often heard:

- Hinduism is an all-inclusive religion, which welcomes and assimilates all types of new ideas and visions.
- In the past, Hindus have welcomed and assimilated new people, having varied faiths coming from different lands.
- Hindus are so tolerant that they have the capacity to accommodate innumerable types and all manners of gods and ideas and concepts.

When Hindu leaders say that Hindus welcomed and assimilated outsiders, they mean Parsis, Jews and a few fellow travelers. The number of these people might have been a few ten thousands. They sneaked in over unguarded coastlines, in small numbers and in small sized boats. Their arrival would have been spread over a few centuries, and would have gone, largely, unnoticed in a vast country like India.

Parsis, Jews, etc constitute less than 1% of the people who arrived in India over the centuries. Hindus opposed the remainder 99%

militarily. Unfortunately, that military opposition was not enough and those people could enter India on the strength of their sword. Muslims occupy a place of pride in this category. Hindu leaders give that military non-performance the veneer of 'Hindu Tolerance'; that is making virtue out of necessity, and is plain hypocrisy. Except for the Christians, all invaders coming to India first entered Afghanistan, and many of them settled there. By Hindu standards, Afghanistan would be another very tolerant country; we know that is not so.

The projected tolerance of Hindus, born out of bogus spirituality, is a myth. It is an artificial web woven round the Hindus by people with base instinct and baser intentions. The theory of Hindu tolerance suits the non-Hindu population of Bharat. Therefore, no one is inclined to go into any detailed study and analysis of the (non-existent) base and basis of Hindu tolerance. We attempt to do so in the next few paragraphs.

Hindu Tolerance — The Facts

In Vedic literature, there are no specific instructions requiring Hindus to be tolerant. Even if some odd *shloka* appears to recommend such a course, we have to remember that one or two *shlokas* do not make a religion. There could be ten *shlokas* for the opposite viewpoint. We have to get an overall sense of the religion. Vedic *dharma* requires a Hindu to be strong, assertive and firm, which he cannot be if he is tolerant of injustice and insults. The central message of the *Bhagwad Gita* is to resort to righteous war to fight injustice or assault on your values — no tolerance there.

We begin our study with the internals of Hinduism. Hindus are totally intolerant of a part of their own society, i.e. the Shudras. Hindus even invented lower classes called *Mllechha* and *Chandalas* to be the object of their intolerance. We have described elsewhere the inhuman treatment that Hindus meted to a part of their own society. How can such people be tolerant of others? Hindus are also intolerant of widows, whom they would like to either burn (*sati, johar*), or throw on the ghats of Benaras to live in the most inhuman conditions imaginable. Lord Vishnu showed no tolerance of an isolated act of impropriety by one Kshatriya king. In his *avatar* of Parushuram, he wiped the entire Kshatriya clans, not once, not twice, but 21 times.

The *Ramayana* and the *Mahabharta* are the barometers for Hindu behavioral standards. Let us examine some episodes in these epics to detect traces of tolerance, if any. In *Ramayana*, Laxman displays extreme intolerance in cutting off nose of a woman, Surpanakha. What was her fault? She had only made a marriage proposal to Laxman, who at that time, was without his wife. In any case, those days, rulers used to have multiple wives. That act of Laxman cannot be justified on any account, least of all on morality. During Rama Rajya, a Shudra Shambuk took to asceticism. Lord Rama cut off his head on the ground that asceticism was not permitted to the Shudras.

In the *Mahabharta*, Karan was a man of high character and caliber, who had done no personal harm to Arjuna. During the battle, Karan's chariot got stuck in mud. As per the rules of war that entitled Karan for relief; and Arjuna was inclined to give that relief. However, Lord Krishna insisted that Arjuna should show no tolerance and kill Karan instantly; Arjuna did that. There was the battle between Bhima and Duryodhana. As per rules of combat, the adversary is not to be hit below the hips; Bhima wanted to stick to that rule. Lord Krishna insisted that Bhima show no tolerance and strike Duryodhana on the thighs; Bhima did that. If the Pandavas had shown tolerance even on one such occasion, there was no way they could have won the war. One would really have to go around with a magnifying glass to look for episodes of tolerance in the two epics. A note of caution — cowardice is not to be confused with tolerance.

Now, let us examine the attitude of Hinduism to other religions. Vedic literature may have been formulated generally around 1200 BC. At that time, Hinduism was the only organized religion in the world. There may have been some early stirrings of Judaism in a far-off land; Hindus could hardly be aware of that religion. Some references in the Vedic literature to tolerating or respecting all *dharmas*, refer to viewpoints under the overall umbrella of the Hindu thought process. These do not refer to the acceptance of any different faith or religion, for the simple reason that there was no such other faith or religion in those days. Hinduism had a monopoly, at least, in this part of the world.

Hinduism's first brush with a different religion was with Buddhism in the 6th century BC. In a weaker moment, Hindus allowed Buddhist thought to grow out of their midst. However, as soon as they realized that it was developing into a new religion, Hindus took to annihilating it. The process started quite early in 2nd and 1st centuries BC during the reign of the Shunga and Kanva dynasties, which ruled in Eastern India, after the fall of the Mauryan Empire. The Gupta dynasty (320–520 AD) also worked towards that end. Hindus did not relent until they succeeded in driving Buddhism out of India. The finishing touches were imparted by Adi Shankaracharya in the 9th century AD. Presently, Buddhism prevails in a vast swathe of countries; but it is almost extinct in its country of origin, i.e. Bharat, the land of Hindus. That does not look like a great example of the much-flaunted tolerance of the Hindus. After some initial success, Jainism was also largely accommodated in the broad Hindu fold.

The Hindus' next brush was with Christianity in the 1st century AD, when Apostle Thomas is believed to have arrived in South India. Christianity is one religion, which spread very fast, after some initial resistance by the Romans. Again, the Hindus rejected Christianity; it could not establish its hold in India. It was only in the 19th century AD that there were some stirrings of Christianity in India. Even though Bharat was ruled by the Christians for some 200 years, Christianity could spread only marginally. The bulk and core of Bharat stayed Hindu.

Islam came to Bharat in early 2nd millennium AD. Now, this was a religion, which like a whirlwind, swept everything in its path, including the greatest of empires and civilizations, e.g. Persian and Byzantine. Though Hindus had to give in militarily to Islam, the Hindu religion withstood the onslaught of Islam, like a rock. In spite of all types of pressures and atrocities, Hindus refused to accept Islam; except for a small minority, Hindus did not convert.

Muslims ruled over India for 600 years, but Hindus refused to consider them as their equal, leave alone their superior. Hindus would not have any social interaction with Muslims, though they might have been forced to work for them. In this context, one can do no better than quote the Muslim historian and author Al-Beruni, who had accompanied Mahmud of Ghazni during his attacks on India. In his book India, Al-Beruni writes as follows:

"... All their (Hindus') fanaticism is directed against those who do not belong to them — against all foreigners. They call them *Mllechha*, i.e. impure, and forbid having any connection with them, be it by intermarriage or any other type of relationship; or by sitting, eating and drinking with them, because thereby, they think, they would be polluted. They consider as impure anything which touches the fire and water of a foreigner; and no household can exist without these two elements. Besides, they never desire that thing which once has been polluted should be purified and thus recovered; as under ordinary circumstances, if anybody or anything has become unclean, he or it would strive to regain the state of purity. They are not allowed to receive anybody who does not belong to them, even if he wished it, or was inclined to their religion. This too, renders any connection with them quite impossible, and constitutes the widest gulf between us and them."

People who had lived in pre-partition Bharat would recall the position as described by Al-Beruni. In 1940s and earlier, Hindus would not normally eat in Muslim houses. There used to be shrill cries of 'Hindu *pani* (water)' and 'Muslaman *pani*' on railway platforms; Hindus would not drink '*pani*' touched by Muslims. There could be no greater testimony to the myth of 'Hindu tolerance', which has been so assiduously but falsely built and spread.

It is possible that there may be a Vedic *shloka* (verse) or two, which could be construed as a reference to tolerance. That concept of tolerance could only mean something on the following lines:

- That Hindus may not be per se hostile to new ideas or concepts.
- That Hindus may examine any such concepts and ideas and may not oppose those which do not clash with the basic Hindu thoughts.
- However, Hindus must reject out of hand concepts in conflict with the Hindu value system, or harming Hindu interests.

However, after 800 years of Hindu slavery, presently, the concept of Hindu tolerance has come to acquire the following types of connotations:

- Hindus must grin and bear all types and manners of insults and humiliations. There should be no groans, and, of course, no complaints.
- If a 90 years old artist makes nude pictures of Hindu goddesses, no objections be raised; that would be Hindu fanaticism. Artist's effort should be lauded.
- Asking for a temple or two to be made at the exact spots where Hindu gods were born is a strict no-no. Making of such temples would be Hindu fundamentalism.
- If the Babri structure is accidentally brought down, the Hindu 'Lauh Purush' (Iron Man) must declare the day to be the 'saddest day' of his Life. As the Pakistan *ummah* may not be convinced with such a declaration, the *Lauh Purush* must travel all the way to Pakistan to repeat the infamy in their midst.

In grammatical terms, the opposite of tolerance is intolerance. However, on a wider canvas and at the political level, antonym of 'tolerance' is 'national interest'. You have to select one of the two; you cannot have both simultaneously. If national interests are to prevail, the flawed concept of 'Hindu tolerance' must be given a quick and decent cremation. Hindus must learn to examine every issue on merit. They need not accept what is not in national interest, just because it is being drilled into their ears that they are a 'tolerant' lot.

14
SOME HINDU SAYINGS
The Problem Area

14.1 Background

All religions have their respective scriptures; these give guidance on religious matters. These may also lay down behavioral norms for the individual, for the society as a whole, and in come cases even for the State. The State is just one-step above society. Of course, scriptural laws do not bind a secular State; though these may bind a non-secular State to some extent. Various religions claim varying degree of divine authorship for their scriptures.

The clergy, the saints and seers may follow rigidly the scriptural norms. In their sermons, they may urge upon their laity to do likewise. However, in the practical day-to-day life of a normal man, it is difficult to adhere to scriptural dictates. Often, there are deviations, sometimes insignificant and of short duration, sometimes significant and of long duration. Does the respective God punish for such deviations? There is no simple answer; the jury is out. Let us consider a few examples.

A fundamental pillar of Christianity is the 'Brotherhood of Man'. However, for centuries, Christians indulged in the African slave trade. They hunted Africans almost like wild animals, auctioned them like cattle, transported them like baggage, owned as property and generally treated them as sub-humans. That was a gross violation of their scriptural code. Did the (Christian) God punish Christians for that? It seems not. The Christians went on to enslave almost the whole world; the Islamic, the Hindu and Buddhist civilizations lay prostrate before them, for centuries. The Christians discovered an entire new continent, which became the fountainhead of slavery. That continent went on to produce the sole super power of the world.

Islam is a religion of love and compassion under the merciful Allah. However, pages of history are full of wanton death and destruction spewed by Islamic hordes on the hapless conquered people. That was against Islamic injunctions. Did Allah punish Muslims for that? It seems not. Muslims went on to conquer and establish their rule over half the world, which lasted for many centuries.

The case of the Hindu falls in a different category altogether. Hindus say that they do not have scriptural diktats, which have to be followed meticulously; they have only scriptural guidelines. If a Hindu chooses not to follow these, it does not make him less of a Hindu. A Hindu may believe in the One Supreme Brahman, or he may not. He may bow his head before gods carved in stones, or he may not. He would still be a Hindu. Hindus say that they largely live by these scriptures. They have not gone out of the way to conquer other people, or to inflict atrocities and indignities on them. Did the Hindu God reward Hindus for that? It seems not. For the better part of the last 2,000 years, Hindus have been slaves and have had to suffer all types and manner of indignities and atrocities.

Are there any reasons behind the success stories of Muslims and Christians, and the failure story of Hindus? It is a complex question without a simple answer. We can only attempt to give a tentative, half-hearted answer, in the following paragraphs.

It would appear that Christians and Muslims could distinguish between the applicability of the scriptural norms at the individual level, and at the State level. At the individual level, Muslims and Christians were very scrupulous in adhering to the scriptural norms, e.g. regular visits to the church or mosque, total faith in one God, etc. However, at the State level, they were choosy and adopted a practical approach. Brotherhood of man was alright, but slaves were an economic necessity. Love and compassion may be the scriptural norm; but fear of the Allah had to be put into the minds of the conquered people. Therefore, at the State level, Muslims and Christians would interpret the scriptures to suit the situational requirement. If that looks like hypocrisy, no surprise it is hypocrisy. But, it yields results; in fact, excellent results, as pages of history show.

Contrary to the above, Hindus could not distinguish between the individual and the State. Take the case of *Ahimsa Parmo Dharma*. It is a *shloka* (verse) which Hindus picked up from Buddhism and Jainism. If at all, it was for application at the individual level. But Ashoka prescribed it for the State, with such a degree of vehemence that he almost outlawed war. Not satisfied with that, he enjoined on his sons and great grandsons to do likewise. He put up edicts to this effect all over his kingdom. Results were quick to show; Hindus went under. They recovered for some short period (Gupta Empire), but soon relapsed into the *ahimsic* mould. When Hindus fought the Muslims, they were always vastly superior in numbers. The Hindu generals were also of good caliber. Hindus were fighting on home ground. In spite of all these favorable factors, Hindus generally always lost. One inference could be that the Hindu soldier did not go all out in war, with his heart and soul; the *ahimsa* bug kept troubling him in his sub-conscious (However, this is a very simplistic explanation, the actual situation was for too complex).

Presently, on paper, we have a strictly secular government in Bharat. However, this 'secular' government is quite liberal in applying and enforcing Hindu scriptural norms at the State level. The Hindu slogan, *Satya Mev Jayate* is the national motto. In 2007 AD, the Government of India officially adopted another Hindu slogan '*Atithi Devo Bhava*'. During TV debates, heavyweight politicians, including of the secular parties, are often heard quoting Hindu *shlokas*, during debates on major national issues. We will discuss applicability of some of these *shlokas* at the State level. We take up the following *shlokas* for discussion:

- *Satya Mev Jayate* — Truth Always Triumphs
- *Atithi Devo Bhava* — Guest is God
- *Sarv Dharam Sambhav* — Respect for all Religions
- *Vasudeva Kutumbhakam* — World is One Family
- *Om Shanti Shanti* — Om, Peace, Peace
- *Jan Jaye per Vachan na Jaye* — Give Life for a Promise

Before we discuss the individual *shlokas*, we make a few general points:

- These *shlokas* are from the *Satya/Treta Yuga*, when truth and righteousness prevailed. Now, we are in the *Kali Yuga*, when evil forces have taken over. It would be quite in order to review the applicability of these *shlokas* in the vastly changed circumstances.
- We are not concerned with the applicability of the *shlokas* at the individual level. Individuals may practice what they must and what they wish to.
- We would discuss the applicability of the *shlokas* at the State level only.

14.2 *Satya Mev Jayate*
(Truth Always Triumphs)

First things first, at school, young children must be taught to speak the truth, and nothing but the truth. At religious gatherings, saints and seers may impress on their laity to follow the path of *satya* (truth). However, having '*satya mev jayate*' (a Hindu religious *shloka*) as the motto of the secular Indian State is an entirely different matter.

At the state level, what matters are national interests; nothing else is of significance. If national interests are served by *satya* (truth), well and good; we will welcome *satya*. However, if for national interests, one has to resort to *asatya* (untruth), the state should have no hesitation in adopting *asatya*. Rather with practice, the state should make *asatya* into a fine art. The state may not only lie, it may fudge and feint and resort to ruse and cunning, as necessary. If required, the state must indulge in skull-drudgery of all type and manner.

The more respectable name for the above type of activities is statecraft. Chanakya and Machiavelli have defined that with a degree of clarity. Chanakya has clearly prescribed that at the state level, the sole purpose of every action is 'safety and profit' of the state; abstract questions of *satya* (truth), ethics and morality are not to be raised. Murder, poison, false accusations and subversion were to be freely used by the king to protect the state interests.

In Hindu mythology, even gods are often shown deviating from the path of *satya* (truth), viz:

Kreta (Satya) Yuga: The bowl of *amrit* (nectar) emerged during the *samundra-manthan* (churning of the oceans). It was pre-agreed that that *amrit* would be shared equally between the *devas* (gods) and the *asuras* (demons). However, the *devas* tricked the *asuras*, and got all the *amrit* for themselves. If that could happen in the *Satya* Yuga, what can we say about the Kali Yuga.

Treta Yuga: This Yuga had ¾ *satya* (truth) and ¼ *asatya* (untruth). *Satya* dominated the *Ramayana*, which was in the Treta Yuga. However, we do find some (very few) deviations here and there.

Dvapar Yuga: It was half and half of *satya* and *asatya*. *Mahabharta* was in the Dvapar Yuga. We find many wrong practices and deviations from *satya* (truth) in the *Mahabharta*. But for many half-truths and compromises with the 'war code', the Pandavas could not have emerged victorious.

Kali Yuga: As per scriptural norms, *asatya* dominates in the Kali Yuga. As such, the state should have no hesitation to adopt *asatya* as a cardinal principle (Remember, we are not talking of the individual).

As a national motto, *Satya Mev Jayate* exhorts the state functionaries to stick to the narrow and straight path of *satya*. That can be very unsettling for functionaries like Army Commanders and diplomats who have to be ever ready for various types and degree of untruths, camouflaged or otherwise. In war and diplomacy, morality just does not work except as a ruse. A study of history shows that whilst *satya* may have triumphed once in a while, *asatya* has never failed, or faltered; the *Mahabharta* is a shining example.

Churchill was perhaps the most prescient man of the 20th century. His views on relationship between man and truth are a follows:

'Man will, occasionally, stumble on the truth; but most of the times, he will pick himself up and continue on.'
'In war, every truth has to have an escort of lies.'

The practical position in the real (not the fantasy) world is that it is not *satya* that wins; whatever wins becomes *satya*; it is the strong who always prevail. The Hindi equivalent of this is '*Jis ki lathi, uski ki bhains*' (the strong gets everything). Almost every episode of world military history proves this hypothesis. We reiterate as follows:

- We are talking here of the practical world at the state level. We are not talking of individuals; they may resort to *satya* to their heart's content.
- We are keeping the religious realm out of our consideration.
- We must learn to distinguish between the state and the individual, as well between the official and the religious.

The moral of the story is that there is no way we can have *Satya Mev Jayate* as the state motto. *Satya* may be left to be practiced by individuals. The state must be ready for *asatya*, at all times and under all circumstances.

14.3 *Atithi Devo Bhava* (Guest is God)

Perhaps, the *shloka* (verse) is not meant to be taken literally. However, when you hear it often enough, the 'Guest–God' equation tends to get stuck in the mind.

Now, God is way beyond anything that the human mind can perceive; that is the very nature, concept and majesty of God. Especially in the case of the monotheistic religions, nothing in this world or even the next can be compared to God; such is the exalted position that God occupies. For Muslims, comparison of anyone or anything with God is most offensive and blasphemous. Hinduism as presently prevalent in Bharat, perhaps has a more flexible image of God, and hence this *shloka*. Hindus need to ponder whether in their attempt to exalt the guest, are they not demeaning God, i.e. reducing Him by a few notches.

Without doubt, this is out and out a Hindu *shloka*; no Muslim or Christian can even think along these lines. In 2007, the Ministry of Tourism, Government of India, officially adopted this slogan to encourage tourism. The government needs to pause and ponder whether the *shloka* goes with the (proclaimed) secular credentials of the government. Is the government not offending Muslim and Christian susceptibilities by adopting this slogan?

The *shloka* does not differentiate between the various types and manner of guests that may land up. In Vedic times, if a *rishi* (saint) showed up as an (uninvited) guest, he was to be welcomed. What happens when a disguised Ravan shows up at Sita's doorstep? One small wrong step by Sita under the influence of this *shloka* led to disaster.

In the modern times, a good percentage of guests (tourists) coming to India are hippies and flower people. Many of them are drug addicts, and some sex-maniacs. They indulge in their antics, including sexual escapades on Goa beaches, Manali heights, or in seedy Paharganj hotels. Then, there are people coming with pornographic and pedophilic intents. Seventy years old tottering Arab 'guests' come to pick up 15 years old nubile girls from the environments of Hyderabad. Then, an odd terrorist or two may sneak in as a guest. Are we to treat all these as some sort of a version of God/god?

Societal norms demand that a guest should meet three criteria — he should by invitation, of short/limited duration and preferably, a known person, if not directly, through common friends. Whatever may have been the position during the *Satya*/Treta Yugas, presently not many people would be comfortable with a guest, who is short even on one of these criteria. In 2004, there was the famous case of a UP thakur leader going as an uninvited guest to Sonia Gandhi's get-together; he was not welcomed. Thereafter, the famous thakur was seen going around muttering swear words and other inanities. We are in Kali Yuga, and the *Satya* Yuga norms can no longer apply.

All the above should not be taken to mean that we should be less unenthusiastic in welcoming tourists. However, comparing tourists with God? No way, never.

14.4 *Sarv Dharam Sambhav* (Equal Respect for all Religion)

Quoting this *shloka*, pseudo-secularists would have Hindus respect other religions without prescribing any condition of reciprocity, i.e. other religions may or may not respect Hinduism.

Though one meaning of the word 'dharma' is religion, it has a number of other meanings. It could mean laws of nature, righteousness, morality, sense of justice, duty, code of conduct, law and a host of similar things, as we see from the following examples:

Raj Dharma	— Duties of a Ruler
Dharma Yudh	— Righteous Struggle
Dharma Sankat	— Moral Dilemma
Kshatriya Dharma	— Code of Conduct for a Kshatriya
Dharma Yuga	— Righteous Age
Ahimsa Parmo Dharma	— *Ahimsa* is the First Duty
Coalition Dharma	— Code of conduct for a coalition

In none of the above phrases does *dharma* mean religion. If *dharma* is to mean religion, the meaning of all the above phrases would change entirely. Dharma may also mean moral behavior and moral order.

Now, this *shloka* is from the distant past, say around 1,200 BC. At that time, other than Hinduism, there was no other formal religion. Quite unknown to Hindus, there may have been some early stirrings of Judaism in a far off desert land. *Rishis* (seers) had no reason to ask Hindus to respect other religions when, as known to Hindus, there were no such other religions. The word *dharma* in this *shloka* does not mean religion. It means code of conduct and moral laws prescribed within the overall Hindu religious thought process. All those laws/codes should get equal respect, was the central theme of this *shloka*.

In the 6th century BC, the Buddhist thought process arose from amongst the Hindus. As soon as Hindus saw it developing into a separate religion, Hindus got after annihilating it. They did not rest tell Buddhism was driven out of India; that indicates total rejection of the *shloka*, if *dharma* is taken to mean religion.

Hindus have no unilateral obligation to respect other religions; that can be done only based on strict reciprocity. In fact, being the religion of majority community, Hinduism must get a degree of extra respect in Bharat.

14.5 *Vasudeva Kutumbakam*
(World is One Family)

This *shloka* indicates the fantasy world, in which Hindus like to live. The real practical world out there is very different. Let us start with a bit of history.

On going through pages of history, emerge names like Atilla the Hun, Genghis Khan, Halagu, Tamberlane and a host of others like them. These worthies spewed death and destruction, murder and mayhem on an unimaginable scale, in large parts of the world. They seemed to be ignorant of the (Hindu) concept of 'Vasudeva Kutumbakam'. In modern times, Hitler also did not acknowledge this concept. In his effort to enslave the world, he was responsible for some 40 million deaths in a time span of just five years (World War II). A little later, USA and the Soviet Union had thousands of nuclear missiles targeted at each other. Those would have been enough to destroy the entire world more than a hundred times over. The world is quite oblivious of the Hindu concept of the 'World is One Family'; what a shame (for the world)!

It would appear that Hindu Rulers had an abiding belief in 'Vasudeva Kutumbakam'. That is why Rulers like Anandpal and Prithviraj Chauhan did not prepare themselves adequately to deal with the attacks of Ghazni and Ghauri. The result was some 750 years of Hindu slavery.

In more recent times, we helped Bangladesh achieve its independence in the gushy expectation that we would get a 'thank-you' note. We were also hoping for a life-long friendship. Today, Bangladesh has too two obsessions; first, hatred of Indians in general and Hindus in particular. Second, it wants to become the hub of Islamic terrorism.

It is Al-Qaeda, which appears to have taken the concept of 'World — one family' seriously. It wants to impose Sharia Law on the entire world; clear and explicit advance warnings have been issued to non-believes like the USA, Europe and Bharat.

In the past, the Hindus' travel on the path of *Vasudeva Kutumbakam* led to disaster. Hindus are in no mood to learn, and wish to continue on the same path.

14.6 *Om Shanti Shanti*

Hindus are obsessed with the word *shanti* (peace); they keep on chanting '*Om shanti, shanti*' in and out of season, and mostly out of tune. They have been doing that for the last two millennia and the word '*shanti*' has come to be lodged in their subconscious.

In the distant past, Hindu *rishis* (seers) came up with the *shanti shloka*, for peace of the individual and his soul; it was for chanting strictly within the confines of the household. However, at some stage, the *shanti* word got spilled onto the street and got woven in the Hindu psyche. From an individual goal, it acquired contours of a national obsession. On attaining freedom, Bharat, striking a highly moralistic posture, started lecturing *shanti* (peace) to the whole world. When combined with their self-destroying doctrine of *ahimsa* (non-violence), it became a heady, but dangerous mixture.

The paradoxical ground reality is that *shanti* can be assured only through war. In the 4th century BC, we had Chandragupta Maurya. He indulged in all-out wars; that ensured *shanti*. His grandson Ashoka eschewed war and adopted *shanti*; 500 years of total *ashanti* (turmoil and mayhem) followed, i.e. barbaric invasions of India from 200 BC to 300 AD.

In the modern times, Hindus need to go a bit slow on their *shanti* obsession. Shanti be left strictly for the individual and his soul. The State must have nothing to do with the '*shanti*' word. Nations may love peace, but they must keep their 'powder dry'. This is the one lesson that Hindus have to learn, and learn fast. To achieve *shanti*, they must be ever prepared for war.

14.7 *Pran Jaye per Vachan Na Jaye* (Give Life for a Promise)

Translated, the *shloka* means 'Promise must be kept even at the peril of one's life'. The *shloka* is best analyzed in light of a popular fable relating to the Alexander–Porus battle in 326 BC.

As per this fable, a ladylove of Alexander preceded him to the court of Porus. She managed to tie a '*rakhi*' (sisterly thread) on Porus, and thus, became his *rakhi* sister (a Hindu concept). She extracted a *vachan* (promise) from Porus, that he would not harm Alexander in battle. As per this fable, during the battle, Porus got an opportunity when he could have killed Alexander (this is Hindu fantasy). However, recalling his promise to his 'rakhi sister', Porus let Alexander go. Alexander went on to defeat Porus, employing every subterfuge and trick of war.

By letting Alexander go, Porus put freedom and honor of the country on the bloc; that was high treason and treachery. However, Hindus like to project that (unpardonable) act of Porus as a sort of high point of their civilization and culture. You cannot get more muddle-headed than that. If Hindus had any sense of war and practical politics, Alexander's ladylove should have been put under house arrest, as a possible enemy agent. However, Hindus are over-burdened by a sense of false morality and bogus chivalry.

Porus forgot his duty to the country and sacrificed supreme national interest for some archaic concept of a 'rakhi sister' and some promise to her; only Hindus are capable of such silly formulations. They refused to acknowledge their mistake in the subsequent centuries; the question of learning from the same did not only arise.

Let us assume that the fable is true and Porus, indeed, did get a chance to kill Alexander. He should have killed him without a moment's hesitation. World history would have acquired a different complexion. Porus and Hindus would have been proclaimed 'world conquerors'. Porus could have gone on to occupy Persia, which was part of Alexander's Empire. Complete new possibilities of conquests would have emerged. That victory of Porus would have inspired the future generations of Hindus. Barbarian tribes would not have dared to attack Bharat. Actually, Porus could have gone to Afghanistan and decimated the barbarian tribes in their own den. However, all that was not to be. One moment of weakness based on some distorted value system changed the entire history of the Hindus for all times to come. However, not many Hindus are likely to agree with this formulation. They would like to wallow in their 'Pran Jaye Per—'.

Saddam Hussein, the ruler of Iraq had two sons-in-law, both Generals. In an off-guard moment, they happened to utter a word or two against the Pa-in-law. Realizing their mistake, both fled to Jordan. Saddam gave them his solemn promise (Hindu *vachan*) that he had forgiven them and they should return. On their return, Saddam had their brains blown off. The two daughters of Saddam were rendered widows at a young age; Saddam did not bat an eyelid. That is what practical politics is about, i.e. no sentimentally — personal or otherwise. That is why and how Saddam could govern an entirely ungovernable State like Iraq (as Bush later realized).

Young, growing up (Hindu) boys in their formative years are told of the 'chivalrous' act of Porus, with gusto and pride. The inference drawn by these boys is that supreme national interests can be sacrificed for a petty personal consideration, e.g. to keep a shady promise, or for some archaic concept, like that of a '*rakhi* sister'. Some of these boys would go to become Army Commanders and top diplomats. The concept of sticking to some silly *vachan* (promise), lodged in the subconscious, can prove disastrous at a crucial moment of war, or during diplomatic parleys.

It is high time that Hindus clean up their act and shed pernicious doctrines like '*Paran Jaye* —'. They should also get rid of any false sense of chivalry born out of bogus morality. In war, chivalry can mean only one of the following two things, as laid down in the *Bhagwad Gita* —

> *Kill and walk victorious out of war;*
> *or,*
> *Get killed, and not walk out of war at all.*

15
ARYAN MIGRATION
A Fading Theory

In an earlier Chapter, we have referred to the Aryan migration theory. As per that theory, the present day Indian Hindus, at least of North India, are not the original inhabitants of Bharat; they came to this country during 2000–1500 BC from somewhere outside, most probably Central Asia, or the Steppes of Southern Russia. The main author of this theory was a German scholar, Friedrich Max Muller, who was given the job of Professor of History at Oxford around 1840 AD, by Thomas Babington Macaulay, a British Minister and a self-confessed imperialist. The theory got support from some sundry historians, primarily Europeans and some Indians.

The theory was sprung upon the unsuspecting Hindus with an extra-ordinary degree of surprise and swiftness in the mid 19th century AD. At that time, Hindus were coming out of the trauma of 600 years of atrocious Muslim rule. They were in the process of adjusting to the relatively tolerant Christian rule, slavery being the common factor. If the Hindus were staggered by this new concept, they did not show it. Rather, sundry Hindu historians leapt up in support of this theory, without any research or basis of their own.

Now, by the most conservative estimates, Hindus have been living in this land for some 4,000 years; the actual figure could be near 6,000 years. Thus, for thousands of years, the poor souls (i.e. Hindus) were quite unaware of their own origin and did not know that they were guests in their own land. A 'white' European was needed to inform them of this simple fact, at such a late stage in life. It was not that Hindus were a set of illiterate ignorant people; rather, the other way around. The Hindus of those days had the most learned scholars of extra-ordinary caliber. Hindu sages and seers had come out with theories of breath-taking originality and sublimity, as contained in Vedic literature. They produced the first religious book of the universe, in the form of the *Rig Veda*. These sages were not only well acquainted with warp and woof of this world, but also of the next. However, this tiny detail of their foreign origin appears to have escaped their penetrating gaze. That does not seem possible.

As per the Aryan migration theory, the Aryans formulated the Vedas, not much after their arrival in Bharat (Punjab). Now, there is no reference at all in the voluminous Vedic literature to any former homeland, or to the hazardous journey, or the stops and challenges en route. It is impossible that the Aryans lost all remembrance of their former homeland in such a short period. If, indeed, there was a former homeland, it would have got copious references — both joyous and sweet, unhappy and bitter. However, there is not even a single line in Vedic literature.

An outstanding historian, Montstaurt Elphinstine writes as follows in his book, *History of India*.

'Neither in the Code of Manu, nor, I believe, in the Vedas, nor in any other book that is certainly older than the Code is there any illusion to a prior residence, or to a knowledge of more than the name of any country out of India. Even mythology goes no further than the Himalayan chain in which is placed the habitation of gods.'

To Elphinstine, it was quite incredible that the Aryans could have made the transition from the salubrious mountain climate of (say) Central Asia to the oppressive heat and dust of Bharat, and yet failed to record it.

The structure of Vedic literature and sublimity of its thought process indicates that it could be the product only of the people who had roamed the peaks and valleys of the Himalayas, for centuries. Gods are shown residing in the Himalayas, up to as much North as Kailash/Mansrovar. Most Hindu holy places are located along the Ganga, starting from its origin at Gangotri, up to the Bay of Bengal. Other holy places are scattered in the nooks and corners of this vast land from deep South to far East. This could not be the work of some freshly arrived people (staying in Punjab), who had yet to be acquainted with the earthy smell of the Gangetic plains and swaying pines of the towering Himalayas.

Hindus have a vast plethora of gods. Though they reside in the high heavens, these gods keep a close watch on the universe below. Many of them often journey to Bharat. Shiva with his consort *Shakti* lives as an ascetic on Mount Kailash. As a householder, he lives with his consort Annapoorna at Kashi. Then, there are the *Dev-Rishis* who are always journeying between the two worlds. We have Lord Krishna who had the whole cosmos in his mouth. Brahma is the four-headed god who is forever watchful in all directions. If, indeed, Hindus had an earlier homeland, some of these gods and *rishis* (seers) would have been aware of it. However, no god or *rishi* makes even a passing reference to any such eventuality.

The Aryan migration theory is based, largely, on the commonality of roots of some words in Sanskrit with that of Greek and Latin. Yes, that is an aspect which needs some explanation; but the Aryan migration theory is too easy and lazy an explanation. That is too slender a thread to wipe out the entire Hindu culture of thousands of years. An alternate explanation could be that in the ancient past, some people went out of Bharat to other lands. Of course, there is no archeological evidence for that; but there is no archeological evidence for the Aryan migration theory either.

Did Macaulay have a motive and hidden agenda in encouraging Max Muller in his Aryan migration theory? The answer may be in the affirmative. The argument runs as follows — The Christian missionaries arrived in India in good numbers in the 19th century AD. They were taken aback by the worship of innumerable Hindu gods, cast in stone; to them these were heathen practices. The Christian missionaries felt that the Hindu soul could only be saved by their conversion to Christianity;

that was conveyed to Macaulay. He concluded that towards that end, Hindus must first adopt English manners and mores, and learn the English language. The superiority of the English in every sphere should be demonstrated to the locals. The Aryan migration theory was ideally suited for that. It told the Hindus that in the ancient past, a set of fair-looking people had come to Bharat to teach a thing or two to the original darkish inhabitants of India; that is what led to the Hindu civilization. Now, again, thousands of years later, another set of fair-looking people had come to Bharat to uplift the Hindus again. The argument appeared to have succeeded. The Hindus offered only token resistance to the take-over of this land by these fair-looking people. In due course, Hindus started learning their language and imitating their manners and mores. We conclude by saying that the Aryan Migration theory has very weak/ shaky legs to stand on.

Section D
History of India

16
HINDU HISTORY
The Early Period

In the preceding pages, we have discussed the religious, social and cultural aspects of the Hindus. Religion and culture have a profound influence on the military mindset of a nation. In turn, that has a great impact on questions of war and peace, victory and defeat, freedom and slavery. The Hindu God had been very bountiful to them, and blessed Hindus with a large number of advantages, including a massive landmass and humongous number of people. The land was fertile, with plenty of rainfall and extensive network of rivers. There were vast plains ensuring easy connectivity. The country, the size of a continent, had natural defensive barriers in the form of mountains (world's highest) and mighty oceans. It was almost an impregnable citadel. With those advantages, Hindus could have gone on to become a major military power, and rule over half the world.

The somewhat reliable period of Indian history starts in the 7th century BC. At that time, Bharat was divided into 16 Janapadas, which mean territories or states. By about the early 6th century BC, the 16 Janapadas were converted into four major states, viz.:

State	Capital
Kosala	Savathi and Ayodhya
Maghad	Rajgir and Patliputra
Avanti	Ujjain
Vatsa (Kosambi)	Allahabad

In addition, there were two important tribes, i.e. Mallas and

Lichhavis. Of the above four states, Kosala and Magadh were bigger and more influential — Kosala being more powerful of the two. We may recall that Lord Rama had ruled over Kosala in the distant past. At some stage, Kosala annexed the state of Kasi (Varanasi). Thereafter, it came to be called Kosala-Kasi. King Pasendi (Presnjit) ruled Kosala; he claimed descent from the Suryavanshi clan of Lord Rama; however, his claim was disputed even in his lifetime.

Magadh was the first full-fledged monarchy, which may have started around late 7th century BC. The first few dynasties to rule Magadh were as follows (estimated approximate dates of their rule are given alongside):

Pradyotas	5 kings	620 to 567 BC
Sisunagas	2*+ 8 kings	567 to 402 BC
Nandas	1*+ 8 kings	402 to 322 BC
Mauryas	3*+ 5 kings	322 to 183 BC

*Indicates number of important kings

In the 6th century BC, we find King Bimbisara (or Bimbisuara) ruling Maghad, as the 5th king of the Sisunaga dynasty. A sister of Pasendi was married to Bimbisara. Around 500 BC, Magadh became the dominant power, though not yet the supreme power, in the Gangetic basin. From the sketchy details available, it appears that Bimbisara was either a degraded Kshatriya (i.e. of mixed blood), or of low caste. It has been expressed that the end word '*naga*' of Sisunaga indicate aboriginal origin. Brahmin records refer to the Sisunaga dynasty in rather derogatory terms. The Sisunagas ruled Maghad for some 160 years.

Bimbisara's son Ajatasatru succeeded him around 487 BC. It is believed that Ajatasatru imprisoned his father Bimbisara in his old age; he may have even starved him to death,. However, Buddhist records show Ajatasatru as a good and benign ruler. At some stage, Ajatasatru annexed Kosala, and Magadh became the dominant power of North India. Both Bimbisara and Ajatasatru were great Generals, who expanded the Maghdan Empire.

Note: We may mention that later (around 300 BC), there was another Bimbisara or Bimbisara, son of Chandragupta Maurya and the father of Ashoka.

Around 402 BC, a warrior Mahapadma Nanda murdered Kakavarna, the last ruler of the Sisunaga line; he founded the Nanda dynasty. Mahapadma is also believed to be of low caste. One version says that he belonged to the barber caste, and he became friendly with the queen. Mahapadma was a great conqueror; he expanded the Maghdan Empire well beyond its frontiers, both to the West and the South. It appears that Mahapadma was against the Kshatriyas, and tried to liquidate them (shades of Parshurama). During his rule, Brahmins were also not looked upon with favor. The Nanda dynasty ruled from 402 to 322 BC. It is generally believed that Mahapadma and his eight sons ruled during the period; Mahpadma himself might have ruled for some 50 years.

In the 6th century BC, in the west of India was the kingdom of Gandharva, around modern Peshawar; it included parts of Afghanistan. Taxila was its major city. At that time, Achaemidian Emperor Cyrus the great was ruling Persia. His armies defeated the Hindu king of Gandharva around 520 BC and annexed his territory. Cyrus' successor Darius I made further conquests and may have annexed the territories west of the river Sindhu. Some reports suggest that his influence may have extended up to Rajasthan. The region west of river Sindhu constituted the 20th satrapy (province) of the Persian Empire; it was considered very prosperous. It is also reported that some Indian contingents might have fought in the army of Xerxes, successor to Darius I.

At some stage during the late 5th century BC or early 4th century BC, Persians were driven out of Gandharva. In the mid 4th century BC, we find Hindu King Ambhi ruling in Gandharva and King Porus (possibly of the Puru tribe) ruling in West Punjab. Ambhi was the son-in-law of Porus.

In 326 BC, Alexander attacked India. His first contact was with Ambhi of Taxila, who did not oppose Alexander. Most probably, Ambhi collaborated with Alexander and Taxilan forces may have even fought on the side of the Greeks. Thereafter, Alexander engaged Porus on the banks of river Jhelum (Hydispes). Porus was defeated and taken prisoner. However, Alexander pardoned Porus and allowed him to rule. In addition, Alexander defeated, and in some cases decimated many independent tribes who were ruling in patches in various parts of Punjab. Alexander was forced to withdraw as his army refused to proceed further. Fear of the great Maghdan Empire in the East of India might also have been a factor.

Reports suggest that Greek soldiers wore heavy protective armor, including a helmet, perhaps of bronze. In all probability, the Hindu soldier had very rudimentary protective armor. The Hindu soldier is generally shown (in TV serials) wearing very elementary clothing, perhaps with some leather protection. If that be so, the Hindu army must have been at a serious military disadvantage.

17

THE MAURYAN EMPIRE

17.1 Chandragupta Maurya

After withdrawal of Alexander in 325 BC, a star rose on the Hindu horizon; history knows him by the name of Chandragupta Maurya. Greeks perhaps knew him as Sandrokottus. Chandragupta was the first imperial Hindu emperor who ruled over almost the whole of North Bharat from Patliputra to Punjab, and parts of Afghanistan. He was a nowhere man of humble origins.

Chandragupta, along with his mentor and friend Chanakya Kautilya (aka Vishnugupta), raised an army from among the common folk, some of whom were out-laws. He defeated the Greek General Seleucus Nikator in the West, and the Maghdhan Nanda ruler in the East. Seleucus yielded large areas, including parts of present Afghanistan to Chandragupta. Around 320 BC, Chandragupta also pushed aside Porus and in all probability, got him murdered. Chandragupta founded the great Mauryan Empire; gradually, it spread over the whole of North India and may be some parts of upper South. He was the first pan-Indian emperor, and his was the first pan-India empire.

We have already stated that the Sisunaga dynasty ruled the first major Hindu kingdom of Maghad, in 6th century BC; it is generally believed to be of low caste. So was the Nanda dynasty, which replaced the Sisunagas.

It is believed that Chandragupta was an illegitimate son of a Nanda king, through a low caste woman, who was either named Mauri, or belonged to the Mauri caste/tribe; hence the name Maurya. Thus, Chandragupta may also have been of low caste. It would thus appear that for the first 400 years (6th century to 2nd century BC) of the recorded

Hindu history, low caste kings were the rulers of major kingdoms in Bharat. Kshatriyas were in the background, perhaps ruling some small states. This would appear incongruous and fly in the face of the Hindu caste system, wherein the Kshatriyas were the ruling class.

The Mauryan Empire collapsed in the beginning of the 2nd century BC. Power in North West India passed to the barbarian tribes, referred to by the Hindus as *Mllechhas* — lowest of the low caste. In the East, Pushymitra Shunga, a Brahmin commander-in-chief of the Mauryas assassinated the last Mauryan ruler. He founded the Brahmin Shunga dynasty. Over time, another Brahmin dynasty, Kanva replaced the Shunga dynasty.

It is believed that in his old age, Chandragupta converted to Jainism, and perhaps starved himself to death. It has been expressed that possibly, a daughter of the Greek General Seleucus was given in marriage to the son of Chandragupta, Bimbisara when he was the crown prince.

Chandragupta Maurya was the greatest military Hindu commander of all times. He was not the son of a king, or even of a minor ruler. He had no formal military training, and raised his own army from scratch. He set up a pan-Indian empire, the very first in Hindu history. He also defeated the Greek General Seleucus, and extracted territory from him. He was the greatest Hindu General ever. However, for some strange reason, Indian history does not give him the recognition he deserves.

17.2 Chanakya's *Arthshastra*

Chanakya Kautilya (aka Vishnugupta), a Brahmin, was a contemporary of Chandragupta. He lived in Maghad from around 360 to 300 BC. Chanakya became friends with Chandragupta, when both were in a sort of exile and hiding in the forests. During the Nanda Dynasty, neither Brahmins nor Kshatriyas were looked upon with much favor. There is even some talk of extermination of Kshatriyas by the Nanda king Mahapadma Nanda. It is reported that Chanakya had a quarrel with the Maghdan Ruler.

Tradition says that Chanakya had his education at the University of Taxila in the North West of Bharat. Later, he might have been a professor at the University. Chanakya became a friend and guide of

Chandragupta and helped him set up the great Mauryan Empire. He became Chandragupta's minister and retired in the reign of Bimbisara. Chanakya wrote his famous book *Arthshastra*; which literally means 'science of material gains' (for the State, not for the individual). However, the book mainly covers 'political and military' subjects. The book was re-discovered in 1905 AD, may be after centuries of obscurity.

Arthshastra analyses all human weaknesses and dogmas; it does not condemn these, but uses the analysis in order to help secure political power. It could perhaps be stated that *Arthshastra's* approach to life and living is not based on morality, rather immorality. The *Arthshastra* prescribes, with great precision, the single-minded aim of grabbing political power. Chanakya does not hesitate in questioning religious creeds and other philosophies; he rejects many of those. Chanakya strongly recommends the use of brute power and crude realism. A few samples:

- A person should not be too honest. Straight trees are cut first. Honest people are victimized first.
- As soon as fear approaches near, attack and destroy it.
- Even if a snake is not poisonous, it should pretend to be poisonous.

Chanakya emerges as a champion of deceit, intrigue and all such attributes. There is not the least pretence to any type of selfless action or morality. The objective should be clear; means to achieve it do not matter, and need no justification. Moral posturing (a distinguishing trait of present-day Hindus) is to be avoided under all circumstances. Every action must be directed towards the safety, good and advancement of the state; abstract and complex questions of ethics, truth, morality and the like are not to be raised. Espionage and use of secret agents is strongly recommended. The king may resort to murder, poison, false accusations and subversion to ensure the safety of the state.

Arthshastra gives all the devices which the modern dictators have adopted; the elaborate net of espionage and the secret police; lies and deceit may be freely used in foreign policy. The art of treachery is analyzed in depth and recommended. The identified victim should be lulled to sleep with the help of '*maya*' (illusion). There is a deep analysis of various types of political concepts, i.e. enemy of your enemy is your friend. A study of the balance of forces and of geopolitical science is undertaken.

Chanakya repeatedly reminds rulers that they live in a wicked and treacherous world; to ignore that amounts to suicide (no 'Vasudeva Kutumbakam', please). In this jungle like world, the only rule of conduct is '*matsya–nyaya*' (law of the fish), in which the bigger fish gobbles up the smaller fish; or in other words, the law of the jungle. Only brutel strength and deceit will help a ruler to survive and expand the boundaries of his state. Facing an enemy, the ruler should use any and all of the following four options:

- *Saman* (Conciliation) — if the ruler is weak
- *Danda* (Assault) — if the ruler is strong
- *Dana* (Bribery) — if one is uncertain of the balance of power
- *Bheda* (Deceit) — in order to destroy the enemy through internal disorder.

Every form of treacherous and deceitful behavior, every ruthless manner of destruction of the enemy, everything that may succeed is allowed, and in fact, strongly recommended. In the world of politics and military, there are no ethics, no morality, no spirituality; one need not even seek divine guidance, and must rely on one's own means.

This approach of Chanakya had the sanction of *Mahabharta*, which has the following types of advice:

- The last world of wisdom is—'Not to trust anyone'.
- Might is always Right, which emerges out of Might; Right has its support in Might, as living beings have it on soil.
- The Strong are always Right; nothing is difficult for the Strong. Everything is pure that comes from the Strong.
- If you are not prepared to be cruel, and to kill men as the fisherman kills the fish, every hope of success may be abandoned.

Arthshastra says that in an atmosphere of danger and treachery, kingship is an extremely difficult task. A king should trust no one, not even his parents, children and relations; he must also guard against poison. A ruler should never sleep twice consecutively in the same room.

Kautilya is of the view that only conformity to all the complex and cynical rules laid down in the *Arthshastra*, will enable a king to become the master of the whole universe. The *Arthshastra* is the most ruthless and cynical political theory, ever propagated in any civilization.

Chanakya also recommends the following:

- Any power superior in might to another should launch into war. A state having a weak state next to it should capture it.
- Plans are given in the *Arthshastra* to conquer the world.
- There are detailed treatises as to how forces are to be arranged for battle.

17.3 Ashoka

Chandragupta Maurya, the founder of the Mauryan Empire died around 297 BC. He was succeeded try his son Bimbisara (different from the one who ruled in the 6th century BC). Nothing much is known about Bimbisara, except that he ruled for some 25 years. After the death of Bimbisara around 272 BC, there was a succession struggle. One of the many sons of Bimbisara, named Ashoka, was successful in eliminating other contenders to the throne, i.e. his half brothers. Thus, after a gap of about two years, Ashoka ascended to the throne around 270 BC.

Until the mid 19th century AD, Ashoka was known just as one of the many kings who ruled during the Mauryan Empire; there was nothing special or outstanding about him. In 1837 AD, a British national James Princep, an assay-master at the Calcutta Mint, deciphered the ancient Brahmi script. That led to understanding of the Ashokan edicts inscribed on rocks and pillars all over India, including the South. In many of these edicts, the king was referred to as 'Devenpriyum Piyadassi'. Devenpriyum means 'beloved of the gods'. However, nobody knew what 'Piyadassi' meant. At that time, a friend of James Princep was working on Buddhist scripts in Sri Lanka; he was approached for help. It was on the basis of the Sri Lankan Buddhist scripts that it could be finally established that Piyadassi was, in fact, Ashoka. Thus, it was only towards the later part of the 19th century that Ashoka came to be identified as the king who had put Buddhist inscriptions in the form of 'Rock and Pillar Edicts' all over India, including the South.

The empire of Chandragupta, though a massive one, was confined largely to the North; in all probability, it did not extend beyond the Vindhyas. When Ashokan 'Rock and Pillar Edicts' were found in the South, it was hastily concluded that Ashoka must have conquered the South. However, no hard evidence is available to support that view. Those southern conquests could have been of Bimbisara; some may have even been of Chandragupta. We have no way to reach any definite conclusion.

Compared to Chandragupta, the military exploits and conquests of Ashoka were quite modest. As a general, Chandragupta was head and shoulders above Ashoka. For setting up the first pan-India empire from scratch, Chandragupta has been called the Julius Caeser of the Hindus. However, Chandragupta does not get much recognition for that; in fact he is hardly remembered. Ashoka hogs all the limelight as a great Mauryan Emperor. That gives us a peak into the Hindu mindset about war and victory, generals and conquests.

In English, Ashoka means 'sorrow-less'. By disposing off his half-brothers, Ashoka displayed early signs of kingly ruthlessness. He is also believed to have ruled with a firm hand during the first eight years of his rule. However, no specific and authentic details are available of the military campaigns that Ashoka undertook in the South.

Around 262 BC, there was the Kalinga War. At that time, the Mauryan Empire was a pan-Indian empire extending from the North East to Afghanistan and covering parts of South India. Kalinga was a tiny state, adjoining the capital and power center of the Mauryan Empire. It would not have posed much of a military challenge to the great Mauryan Empire. However, the Kalinga conflict has been given the dimensions of a major war. A figure of some one hundred thousand deaths (a huge number those days) has been mentioned. That looks grossly exaggerated, and most unlikely; someone has just let his imagination run wild (There is no dearth of such exaggerations in history).

Though Ashoka won the Kalinga war, his response to victory was unnatural, undignified and unkingly, if not cowardly; he went into a sulk. That also puts a big question mark on Ashoka's earlier presumed conquests of the South. In an unkingly gesture, Ashoka declared that the violence and slaughter involved in the war was most undesirable. On the rebound, he started seeing merit in Buddhism, which advocated a pacific and pessimistic approach to life, with emphasis on *ahimsa* (non-violence), tolerance and similar shibboleth.

Buddhism was struggling to establish itself in India at that time. Buddhism is everything that the Vedic *dharma* is not. The Vedic approach to life is joyful, robust, pro-active and aggressive, believing in realistic militarism and violence in a good cause. Animal sacrifices were a part of Vedic *dharma*. Buddhism is passive, insular and defensive, with a mindset of over-emphasis on *ahimsa* (non-violence) and self-abnegation. If Buddhism was a bit short in its philosophy of nihilism, Jainism (also prevalent that time) filled the gaps. In its pessimistic approach to life, Jainism perhaps outshone Buddhism. Both these religions abhor violence and talk ceaselessly of *ahimsa*. From the military point of view, a combination of Buddhism and Jainism was a recipe for disaster.

Whilst Buddhism may be alright for the common person, it can have no space in a king's armor. If kings start dispensing spirituality and pushing morality, what would priests do? In his infinite wisdom, the Supreme God has prescribed separate roles for the king and the priest; one must not usurp the role of the other. However, that was not for Ashoka. He took to Buddhism with unprecedented fervor and hot passion. He fiercely advocated the Buddhist path of pacifism and pessimism. He put his Buddhist edicts on rocks and pillars all over the country.

Ashoka outlawed war and grossly neglected his army; if he did not disband most of it. He expressed that thenceforth, the army would be used primarily for parades and ceremonial occasions. Ashoka went on to instruct his descendents to follow his example. In one of his major rock/pillar edicts, he commanded as follows:

'The inscription of *dhamma** has been engraved so that any sons or great grandsons that I may have should not think of gaining any

---

* *Dhamma* is Prakrit for the Sanskrit word '*dharma*'. It can mean all or any of the following — universal law, righteousness, social or religious order, moral conduct. In Buddhism, *dhamma* means Budha's teachings.

new conquest. They should only consider the conquest of *dhamma* to be the true conquest.'

It is a tribute to the great sense of filial duty of Ashoka's sons and grandsons that they followed Ashoka's instructions, in both letter and spirit. In their relentless pursuit of *dhamma* (*dharma*), they pushed the mighty Mauryan Empire to disaster and Hinduism to its doom.

Just before the Mauryan empire was the great empire of Alexander. It had emerged out of the core of tiny Macedonia (a small Greek island), which had very small landmass and very few people. Almost contemporary with the Mauryan Empire was the great Roman empire; it had emerged out of the tiny core of the city of Rome — again very small landmass and very few people. Yet, each of these empires went on to establish their rule over half the world. In that process, the two empires had to subdue other great empires. Alexander shattered the Achaemeidin Empire of Persia; and Roman General Scipio destroyed the Carthagian Empire of Africa.

As opposed to the Greek and Roman Empires, the Ashokan Empire was based on the humongous landmass of Bharat and the massive multitude of the Hindu people. However, Ashoka could not exploit these huge advantages. There was no major, or even a medium power around India. After the conquests of Alexander in and around Persia and his subsequent withdrawal, that area was in turmoil. In his days, Ashoka was the most powerful king of the world; he was master of all that he surveyed. In such a favorable situation, it was incumbent on Ashoka to launch a campaign for all-round conquests, perhaps, starting with the outskirts of Persia. Chanakya had spelled that out clearly in his *Arthshastra*, written just before Ashoka's time. Chanakya had prescribed that the conquest of (weak) neighboring states was the sacred duty of the king. However, Ashoka showed no such inclination.

On his commitment (and likely conversion) to Buddhism, Ashoka started ceaseless talk of *ahimsa*, tolerance and similar unkingly shibboleth. Kingly duties required Ashoka to send 'soldiers with swords' to neighboring countries. Instead, Ashoka took to dispatching '*bhikshus* (monks) with begging bowls'. Ashoka put the unbounded martial spirit of the Vedic people first to drowsiness, and then to slumber. Hindus were pushed into the '*bhool-bhuleyan*' (dark alleys) of *ahimsa*; they could never get out of those. Sometimes, the Hindus would emerge for a short period (Gupta dynasty), to relapse again. Buddhism was finally driven out of Bharat by about the 10th century AD. However by then, the cult of *ahimsa* was lodged deep in the Hindu subconscious. They were to pay dearly for that, as we shall see as our narrative progresses.

We have stated earlier that Ashoka's greatness started emerging only in the late 19th century AD, after a British national deciphered the Brahmi script in the mid 19th century AD. Thus, the British had a distinct role in bringing out and emphasizing the greatness of Ashoka. That suited the imperialist design of the British of those days. It would help consolidate British rule, if Hindus could be told that *ahimsa* and pacifism was their honored tradition and main calling. It would not have suited their purpose to stress on the militarism of Chandragupta Maurya. The almost simultaneous arrival of Mahatma Gandhi, with his message of *ahimsa*, perhaps made the British task easier.

Ashoka's reputation was given a big boost, when Jawaharlal Nehru selected Ashoka as the role model for free India in 1947. He was perhaps taken in by the dubious role of the British in playing up the *ahimsic* model of Ashoka; at that time, Nehru was under the spell of Gandhian *ahimsa*. In 1947, India had emerged out of 750 years of unbroken slavery. It is a great irony that free India selected as its role model the very person (i.e. Ashoka), who was constructively responsible for pushing the Hindus into slavery in the first instance. India is still caught up in the '*chakra-vue*' (circular trap) of *ahimsa*; it must break out of it sooner, rather than later.

Ashoka was a king who mixed up religion with politics. It has been expressed that during about the last 10 years of his rule, it was difficult to tell whether Ashoka was a monk, or a king. He neglected his kingly duties to concentrate on his functions as a monk. By today's yardstick, that would be arch-communalism. Now, Jawaharlal Nehru was the high priest of secularism. It did not occur to him that he was opting for a 'communalist' as the role model for the secular India of Nehru's dreams.

18
THE BARBARIC INVASIONS

Neglect of the armed forces, first by Ashoka and then by his sons and grandsons could have only one result — slavery; that is what actually happened. The Mauryan Empire started shrinking after the death of Ashoka in 232 BC. It collapsed in 183 BC, within about 50 years of Ashoka's death. Mauryan commander-in-chief Pushymitra Shunga, a Brahmin murdered the last Mauryan King. He founded the Shunga dynasty that ruled over a shrunken eastern empire for about 120 years.

However, disaster struck in the North West of Bharat. At that time, barbarian tribes were hovering on the North West borders of India, looking for weakness to set in. Though these tribes had not read Chanakaya's *Arthshastra*, they well understood its message. Sensing weakness, they poured over the borders. History does not record many battles that the Hindu rulers might have put up against the invading tribes. Their take-over of North West India appears to have been a smooth affair. These tribes went on to rule North West India for nearly 500 years, from around 180 BC to 300 AD.

The barbarian tribes, which poured and ruled over North West India from the 2nd century BC to the 3rd century AD, could be divided under four broad groups:

Bactrian Greeks: They were called Yavanas in India. They were the first to arrive in India in the 2nd century BC. They were soldiers of Alexander's army, who had settled down in Bactria (northern Afghanistan) and it's environ. Two of the more famous Indo-Greek kings were Demetrius and Milinda (Menander); they ruled North West India during 180–140 BC. Milinda became a Buddhist. He had prolonged discussions with the Buddhist philosopher Nagasena, which were published as the Buddhist text *Milinda Panha* (Questions of Milinda). He is supposed to have conquered some territory in the Gangetic plain, and may have even gone close to Patliputra. It is believed that some 20 Indo-Greek kings, big and small ruled over North West India, perhaps over a period of some 100–120 years.

Scythians: They were called Sakas or Shakas in India. They were Central Asian tribes who were pushed out of their homeland by the Kushanas. King Maues or Moga established Saka power in Gandharva (Afghanistan) in 80 BC. His successor Azes annexed territories of the last Indo-Greek king. Azes ruled in the middle of the 1st century BC. His influence may have extended up to Mathura. He is sometimes associated with the start of the Hindu calendar, the Vikram Samvat (57 BC).

Parthians: They were called Pahlavas in Indian sources. Their coming to India is intermingled with the Sakas. Parthians were hardy people from the North Eastern Iranian plateau. King Mithradates II established the Parthian presence in India in the late 1st century BC. Tradition has it that the Christian Apostle Thomas arrived at the court of Parthian King Gondophares in the first half of the 1st century AD. From there, the Apostle Thomas traveled to South India.

Kushanas: They were called Tusaras or Tukharas in India. Kushanas were the Yueh-Chih tribe of Central Asia. King Kanishka of this tribe ruled over India from 78 to 114 AD. An alternate period of 120 to 150 AD has also been suggested. He was one of the greatest Kings ever to rule India. At its zenith, his empire may have spread from the Bay of Bengal (i.e. covering the entire North India) to Central Asia (i.e. much beyond Indian borders). Initially, his capital was at Peshawar where he is reported to have built a 700 feet Buddhist stupa, perhaps, the finest structure of those days. As his empire spread towards the East of India, he shifted his capital to Mathura. A headless statue of Kanishka has been found in the area around Mathura. Kanishka converted to Buddhism and became its committed follower. He is believed to have called a council of Buddhist Sangha to sort out their differences. Sometimes, the possible ascension of Kanishka (a Kushana) to the throne in 78 AD is associated with the commencement of the Hindu calendar system Saka, which started in that year.

The above tribes ruled over North West India for around 500 years. During certain periods, their influence might have extended to Central India, around Mathura, Ujjain, etc. These 500 years are called the Dark Ages of Hindu history; the historical records are quite patchy and lot of confusion and controversies prevail.

There were four or five waves of the invading barbarians tribes. Each of the tribes may not have come in one wave, and may have descended in batches. As the new tribe would defeat and replace the earlier tribe, there are no particular reports of the earlier tribes having gone back to their original homeland. The only partial exception could have been the Bactrian Greeks who might have gone back to Bactria (modern Afghanistan). In all probability, the various tribes like the Sakas, the Pahalvas, the Kushanas, and later in the 5th century AD, the Huns appear to have settled down in Bharat. These tribes, unlike the later Muslim invaders, had no particular religion of their own. So, they may have gradually got absorbed in the Hindu mainstream. In the initial stages, Hindus called them *mllechhas* (low untouchable). However, as these tribes were all victors, they may have been able to gradually get to the higher caste of Kshatriyas. A view has been expressed that Rajputs are descendents of these tribes. There may be some truth in this thesis as we start hearing of Rajputs only around 8th century AD onwards. There are no particular records of exploits of the Rajputs before that period.

At this stage, we may note the following interesting points in respect of the above referred foreign tribes:

- The first man of Hindu mythology, Manu, evolved his Manusmriti sometime in the very distant past; we are in no position to put any date to it. However, it has been expressed that Manusmriti may have been reduced to writing somewhere around the 2nd–1st century BC (or, even AD).
- In Chapter X, Entries 43 to 45 of the Manusmriti, there is a reference to Yavanas (Greeks), Sakas (Scythians) and Pahlavas

(Iranians), as lapsed Kshatriyas, meaning degraded or fallen Kshatriyas.

- There are also some references to the presence of foreign tribes in Bharat at a very early stage in history, rather mythology. We have all heard of the Yadavas of the *Mahabharta*; they were an on-going tribe from the very early times, even before the *Ramayana* days. The Yadavas had an off-shoot called Haihayas (see chapter 8.3). Tradition says that 25 generations before Lord Rama, there was a Yadava Haihaya king, Talajangha. It has been recorded that this king took help of foreign tribes like Yavanas (Greeks), Sakas, and Pahlavas (Iranians) to defeat the 40th generation Suryavanshi king Bahu. Later, Bahu's son Sagara won back the kingdom from Talajangha. We are in no position to offer any comments on the possible presence of foreign tribes so far back in history (rather mythology).

19
THE GUPTA EMPIRE & ITS AFTERMATH

We have recorded earlier that around 320 BC emerged a star named Chandragupta on the Indian horizon; he founded the great Mauryan Empire. Around 320 AD, emerged another star on the Indian horizon; he was also named Chandragupta I. He seized power from the barbarian tribes and established the Gupta Empire which reigned over North India for some 200 years, i.e. from 320 AD to 520 AD approximately; the last, about 50 years of their rule might have been rather weak. After Chandragupta I, the more important rulers of this dynasty were:

Samudragupta (335–375 AD): He made major conquests and is known as the Napoleon of India.

Chandragupta II (375–415 AD): He is identified with the legendary Hindu king Vikramaditya. The majority view is that the great poet and dramatist Kalidas lived during this period (though not everyone agrees). The Chinese pilgrim Fa-Hien visited India during 400–411 AD. However, strange as it may seem, Fa-Hien does not make any great references to the Gupta Empire, and its famous kings.

Skandagupta (455–467 AD): It is believed that he expelled the ferocious Huns who had established their hold on some areas of Bharat.

The Gupta period is generally considered the golden period of Hindu history in which art and culture reached its zenith. However, there is a minority view that the Gupta period has been somewhat over-rated and that it was one of the many states in Bharat at that time.

The Huns returned to India towards the end of the 5th century

AD. Mihirkula the Hun was a ferocious ruler. He carried out his depredations from Punjab to Gwalior during the period 500 to 530 AD. Yashudharman, the king of Mandsor (530–540 AD) finally defeated the Huns.

Towards the end of the 6th century AD, a medium sized Hindu dynasty was established in North Central India. King Harshavardhan of this dynasty ruled from 606 to 647 AD. His rule is well documented by the Chinese pilgrim Huan-Tsang who visited India during 630–644 AD. At some stage, Harsha shifted his capital from Thanesar to Kannauj. After the death of Harsha, his empire disintegrated, as he had no male heir.

Thereafter, the history of India again is blurred. For the next 300 years (say, 700–1000 AD), a number of small and medium sized states were established in Bharat; these included the Rajput states in Rajputana.

20
HINDU CALENDAR SYSTEMS

Presently, the Hindus have two current calendar systems, viz.:

- The Vikram Era — starting 58–57 BC
- The Saka Era — starting 78 AD

Both the calendar systems started during the dark ages of Hinduism, as stated in chapter 18. That was the time when there was no centralized Hindu rule and the barbarian tribes dominated North West India. The origin of both the calendar systems is mired in controversy.

The Vikram Era (58-57 BC)

The Vikram Era is traditionally associated with a great Hindu king named Vikramaditya. However, his historicity is not firmly established; there are various theories on the subject. The main-line history does not record any great King Vikramaditya around 58–57 BC. In fact, there does not appear to have been any centralized Hindu rule in Bharat at that time. The barbarian tribe of the Sakas (Scythians) was ruling over North West India, after defeating the Bactrian Greeks. It is generally believed that the Saka King Azes was ruling over North West India, around 58–57 BC.

During the Gupta period, there was the King Chandragupta II (375–415 AD); one of his titles was Vikramaditya. He defeated the Sakas, and drove them out of Ujjaini. If the Vikram Era is named after this Vikramaditya, it would mean a mismatch of some 460 years, i.e. the difference between 57 BC and say 400 AD. To overcome and explain this dichotomy, various versions are in vogue.

As per one version, the Vikram Era was initially called Krta Era for many centuries. In the 5th century AD, the Krta Era came to be called the Malava (or Malwa) Era. Perhaps, that signified the victory of Malavas (residents of the Malwa region bordering Gujarat) over the Sakas. Later on, during the 10th or 11th century AD, the era came to be called the Vikram Era.

A second version says that in the 4th century AD, Chandragupta Vikramaditya might have re-christened an ongoing era (starting 58–57 BC), after his name.

Another version is attributed to the Jain tradition, which was recorded in the 10th century AD. As per that version, a sister of the Jain patriarch Kalaka was abducted by a king of Ujjain named Gardhabhilla. That led to events which resulted in the defeat of Sakas by King Vikramaditya who ruled over Ujjain in 58–57 BC. The tradition says that the Vikram Era was founded by him. There are further variants of this legend.

The historian Farishta states as follows, "Bikramjit (Vikramaditya) was the king of Ujjain and was the contemporary of King Ardashir of Iran. According to some, he lived in the period of Shahpuhr. People associate an era with the date of his death."

Now, Ardashir founded the Sassanid dynasty of Iran in 226 AD. His descendents were all called Shahpuhrs. Many Shahpurhs (4 or 5 of them) ruled from 231 to 388 AD (may be with breaks). Thus, in all probability, one of the Shahpuhr was contemporary of Chandragupta Vikramaditya. So may be, reference of Farishta was to that Vikramaditya.

The historian Al-Beruni, who had accompanied Mahmud of Ghazni, came to know about the various legends of the Vikram Era of 58–57 BC. His writings appear to show that there were two Vikramadityas. The latter of the two, perhaps recorded a victory over the Sakas; he might have christened an era after his name. The earlier Vikramaditya flourished long before; may be, he had started the Vikram era. In other words, the later Vikramaditya, to celebrate his victory over the Sakas, founded the Vikram Era, which was tagged on to an earlier ongoing era.

We are sorry that we cannot make the origin of the Vikram Era clear. May be, all the above is quite confusing; that cannot be helped. There is a minority view that the Vikram Era might have been founded by the Saka king Azes who ruled over North West India around 58–57 BC. However, there are not many takers for this theory.

The Saka Era

The Saka Era started in 78 AD. Now, the Sakas or Shakas (Scythians) were a barbarian tribe who invaded and ruled over N.W. India in and around the 1st century BC. It would appear most strange (to say the least) that a Hindu calendar system came to be named after a foreign tribe.

Secondly, the Saka rule appears to have declined in the mid 1st century AD. Around that time, the Kushanas defeated the Sakas. It is generally believed that the famous Kushan, King Kanishka ascended to the throne in 78 AD and ruled till 114 AD (some alternate dates are also mentioned). A theory has been advanced to link the founding of the Saka Era with the ascension of King Kanishka, who however, was a Kushana. Perhaps, it cannot get more confusing than that.

21
MUSLIM INVASIONS

21.1 Background

Muslims gave an early knock on the doors of Bharat in 710 AD, when Muhammad Bin Qasim attacked Sind. That was the only invasion by the Arabs. Thereafter, there was a gap of 300 years, before Turko-Afghans started their invasions of Bharat. The first was Amir Sabuktudin of Ghazni; he carried out a raid around 992 AD, in which he defeated Raja Jeypal of Punjab. Sabuktudin was followed by his son Mahmud, who started his raids in 998 AD. There was another gap of around 150 years, before the second Turko-Afghan Muhammad Ghauri started his forays around 1176 AD. We would study the Muslim invasions under the following headings:

- Muhd. Bin Qasim (710 AD) — Defeat, Loot and Scoot
- Mahmud of Ghazni (998–1030) — Defeat, Loot and Scoot
- Muhammad Ghauri (1176–92) — Defeat and Set up Empire
- Timur (1398 AD) — Defeat, Loot and Scoot
- Babur (1526 AD) — Defeat and Set up Empire
- Nadir Shah (1739 AD) — Defeat, Loot and Scoot
- Ahmed Shah Abdali (1748–61) — Defeat, Loot and Scoot

The following comments are offered on the above table:

- Of the seven invaders, five were of the "Defeat, Loot and Scoot" type; they were not interested in setting up an empire in India. Out of these, Muhammad Bin Qasim, Timur and Nadir Shah made only one invasion each.
- Two invaders set up empires in India. The first, i.e. the Delhi Sultanate lasted 325 years; the second, i.e. the Mughals ruled for some 225 years.

- Except Muhammad Ghauri, all invaders were always victorious, some of them in their repeated forays; Mahmud of Ghazni17 times, Ahmed Shah Abdali, some 9 times.
- From 1176 to 1192 AD, Muhammad Ghauri undertook several invasions. He suffered two reverses; a major one in 1191 AD, which he avenged the following year, and went on to set up Muslim rule in India. There was an earlier minor reverse around 1176 AD, against a Solanki King in Gujarat.

21.2 Muhammad Bin Qasim

He was the first Muslim and the only Arab to attack Bharat around 710 AD. His thrust was towards Sind, where the Hindu king Dahar was ruling. In the engagement, the Hindu armies were having an upper hand, when fate intervened. In keeping with the tradition, Dahar was riding an elephant. An arrow stuck Dahar and he was killed. Panic set in, and the Hindu army was routed. Dahar's son Jaisima retreated, leaving Dahar's widow to continue the battle. She fought bravely, but was killed. It is reported that Jaisima later converted to Islam. Qasim did not attempt any further conquests and withdrew. However, some Muslim states were established in Sind, and may be up to the borders of Gujarat.

21.3 Mahmud of Ghazni

1000 AD marks a watershed in Hindu history. Around that year, rose on the scene the scourge of Hindus and Rajputs, i.e. the one and only Mahmud of Ghazni, a Turko-Afghan. He was the ruler of the tiny state of Ghazni tucked in a corner of East Afghanistan. He attacked India for the first time in 998 AD. King Jeypal of Punjab organized a confederate of Rajput rulers of North West India. The Rajput confederate was defeated; unable to stand the shame, Jeypal is reported to have committed suicide.

Mahmud came next in 1009 AD. Jaipal's son Anandpal was ruling Punjab, with his capital at Bhatinda. Anandpal also organized a huge confederate of almost all the rulers of North West India. These included rulers of Ujjain, Gwalior, Kannauj, Delhi and Ajmer. The battle was joined in the Kurram Valley near Peshawar. As usual, Anandpal was riding an elephant. The Hindu Army was on the winning spree.

However, at a crucial moment in the battle, Anandpal's elephant took fright and fled. Panic spread in the Hindu army, and it dispersed. The Muslim cavalry pursued it and massive slaughter followed.

That battle opened the gates of Bharat for Mahmud of Ghazni. Thereafter, he decided to make more frequent visits to India; in all he made some 17 forays. He would normally come to India in winter, when his capital was bitterly cold. After the engagement in 1009 AD, Mahmud met with very little resistance from Hindu rulers over the next 20 years. A few examples are given in the following paragraphs.

In 1018 AD, Mahmud attacked Kannauj. Its Raja Rajyapal did not offer much of a fight, and surrendered tamely. The fellow Hindu rulers were highly enraged; they attacked Rajyapal under the leadership of Raja Ganda the Chandel, and slew him. However, when it was the turn of Raja Ganda to face Mahmud, he fled on the very eve of battle.

En route to Kannauj, Mahmud first prepared to attack Baran or Bulandhahr; its Raja Haradatta tendered his submission. It is reported that some 10,000 of his men may have converted to Islam.

In 1024–25 AD, Mahmud decided to go to Somnath on the far side of the vast Rajputana desert. The Hindu Raja of Multan provided camels and other wherewithal to cross the daunting desert. It is reported that the Rajas of Bikaner and Ajmer opened their gates to Mahmud. Raja Bhima of Anhilwara (near Ahmedabad) fled on his approach. On reaching Somnath, Mahmud shattered the idol (most probably in the form of a Shiv Lingam). He is reported to have slaughtered the entire population of 50,000 Hindus, including women and children. Some pieces of the shattered idol were carried back to Ghazni to be buried in the pathway of the grand mosque there; that enabled the faithful to tread over the lingam. None of the 'brave and chivalrous' Rajput rulers dared to challenge Mahmud even on his return journey over the vast and inhospitable desert. By then, the Muslim army must have been quite weary, if not hungry and thirsty.

Here, we give a fanciful story. In pre-Islamic Arabia, there were some 360 idols in and around the Kabah in Mecca. The three most important idols were Al Lat, Al Uzza and Monat. The Prophet on his return to Mecca, destroyed all the idols; but, Monat was found missing. It was informed that Monat had been smuggled to India (Gujarat), which had flourishing trade with Arabia at that time. It has been expressed that the Prophet decried that some believer in the future should seek out the Monat and destroy it. In Gujarati language, the word 'su' is used for auspicious or good; so the word 'su' got added to Monat, which became su-monat. Over a period of time, su-monat got changed to Somnath.

Mahmud, unlike Alexander, was no emperor of an empire. At his start, he was a petty chief of the tiny state of Ghazni tucked in a corner of East Afghanistan. There were a number of other small states in Afghanistan, mostly hostile to the Sultan of Ghazni. Mahmud had not read Chanakya's *Arthshastra* (see chapter 17.2), but he understood its message. Taking courage in his hand and unsheathing has sword, he set about conquering small states of Afghanistan in his neighborhood. Fortune favors the bold. Mahmud was remarkably successful in uniting these states under his banner. Thereafter, he turned his attention towards Bharat. Initially, Mahmud met a reasonable amount of resistance from the Hindu rulers, i.e. Jeypal and his son Anandpal of Punjab. However, in his subsequent 15 or so raids, he got almost a walk-over. History does not record many cases where the Hindu rulers offered him any significant resistance from 1010 AD to around 1030 AD.

Mahmud has been called the sword of Islam. He was a fanatic and an iconoclast. He felt that Hindu idols were an affront to Islam. As per one version, the Sultan destroyed some 10,000 temples in the city of Kannauj; the figure appears to be an exaggeration. Mahmud admired the grandeur and architecture of the Krishna Janamsthan temple at Mathura; having done that, he ordered it to be burned down. His jihad was for the attainment of paradise, as well for getting the riches of Bharat.

21.4 Muhammad Ghauri (Muhammad of Ghur)

The depredations of Mahmud of Ghazni lasted around 30 years (998–1030 AD). Thereafter, Muslims gave Hindus some 150 years to learn from the lesson that Mahmud had taught them; but that was of no avail.

In 1176 AD, it was the turn of another Turko-Afghan, Muhammad Ghauri aka Muhammad of Ghur. He was the chief of the small state of Ghur in Afghanistan. His first step was to depose the descendents of Mahmud from Ghazni and capture that state. Thereafter, he launched his attacks on Sind and recorded a number of victories. However, he suffered a defeat at the hands of a Solanki Rajput ruler in Gujarat. Thereafter, he turned his attention to Punjab. In a series of attacks from 1177 to 1186 AD, Muhammad captured Indian cities like Peshawar, Sialkot, Lahore, and finally the fort of Bhatinda. That brought him in almost direct confrontation with Prithviraj Chauhan (Chahamanas), the Rajput ruler of Ajmer and Delhi.

The first confrontation between Muhammad Ghauri and Prithviraj Chauhan took place in the first Battle of Tarain (near Thanesar and Karnal) in 1191 AD. As per one account, the fate of this battle was decided by a personal contest between Muhammad Ghauri and Govinda-raja of Delhi, a deputy of Prithviraj. Govinda-raja's lance wounded Muhammad Ghauri, who lost his balance over the back of his horse. One of his warriors leapt up behind Muhammad Ghauri, and rode away from the battlefield. Seeing that, the Muslim army fled. On their return to Ghazni, soldiers of the routed Muslim army were made to tread the streets of Ghazni, munching on gram from the horses' nosebags put on them. They were also forced to hang their heads in shame. There are some versions which say that Muhammad Ghauri was captured. But he was pardoned by Prithviraj, in spite of his courtiers' advice to the contrary.

Elementary military strategy demanded that Prithviraj should have pursued the wounded Muhammad Ghauri to Lahore, Peshawar, and to Ghazni. He should have shattered the Muslim army once and for all. That was very much in the realm of possibility and within the power of Prithviraj. In that case, the history of Bharat would have evolved on an altogether different line. It would appear that Prithviraj did not pursue that (aggressive) line due to the Hindu mindset. We will study about the Hindu mindset in later chapters.

In keeping with the Muslim mindset, Muhammad Ghauri was not the one who would rest on his defeat, or for that matter victory. He came back within one year, with a reinforced army. The two armies met again at Tarain in 1192 AD. One version refers to 1,20,000 horses with Muhammad Ghauri and 3,00,000 horses with Prithviraj. These appear to be highly exaggerated figures. The historian Farishta talks of 150 royal vassals, mostly Rajput rulers, on the side of Prithviraj; the figure of 150 also appears to be an exaggeration. Prithviraj was successful in organizing a massive confederate of rulers of North West India, except Jaichand of Kannauj; Prithviraj had a running feud with Jaichand. At least, 8 to 10 major rulers were with Prithviraj. Main-line history does not support the popular perception that it was Jaichand who had invited Muhammad Ghauri. Muslim invaders needed no invitation.

The great Guhila ruler of Mewar was on the side of Prithviraj. It is possible that the Hindu Army was not under a single command. As a general, Prithviraj was in no way inferior to Muhammad Ghauri. However, in the second battle of Tarain, Ghauri outmaneuvered and out-foxed Prithviraj, and inflicted a crushing defeat on him. As per one version, Prithviraj got down from his elephant, and mounting a horse sped away. He was captured and killed. As per another version, Prithviraj was captured and taken to Ghazni. There, he was tortured, blinded and then killed. Govind-raja of Delhi, the hero of the first battle of Tarain was killed in the second battle; and so were the Guhila king of Mewar, and brother of Prithviraj. In all, some one-lakh people mostly of the Hindu army are reported to have perished in the second battle of Tarain.

A Sanskrit text is much less complimentary of Prithviraj Chauhan; it attributes victory in the first battle of Tarain to Skanda, Prithviraj's general. Skanda was not fielded in the second battle of Tarain, due to some court intrigue.

Medieval India had the institution of bards. They were court poets who were encouraged to write verses praising the exploits, both real and imaginary, of the king. As bards were on the pay roll of the king, they would often get their imagination run riot. What they wrote was more of fancy and less of history. One such bard was Chand (Chund to Col Todd); he has written his *Chandrasa*, the heroic history of Prithviraj. That includes the saga of elopement of Prithviraj with Samjukta, the daughter of Jaichand, ruler of Kannauj.

The rout of the Hindu army in the Second battle of Tarain in 1192 AD was the most decisive battle in the history of Bharat. It not only sounded the death-knell of the 'idea of Hindu India', but the Hindu India itself. It brought disaster to the whole of India; the morale of the public in general and of Rajput rulers in particular was shattered. Hindu India lay prostrate — an easy prey for the Muslim forces commanded by Qutab-ud-din Aibak, ex-slave and successor of Muhammad Ghauri. He went to subdue one state after the other, with relative ease. Aibak was successful in laying the foundation of an empire by the early 1200s AD. In 1206 AD, Muhammad Ghauri was assassinated in Lahore.

Amury De Rencourt, in his book *The Soul of India* has described the situation of that time in Bharat. An excerpt:

'Taken over politically and militarily by a completely alien civilization, the Hindus withdrew into a psychological ivory tower, leaving the substance of the world power and riches to the alien 'impure' barbarians. Nowhere can this attitude of a materially defeated civilization be studied as well as in the famous epic to the great hero of Hindu resistance, Pritviraj Chauhan who struggled fiercely against Muhammad of Ghur, and eventually found death in the battlefield of Tarrain. In this epic, the *Chandrasa*, written in Arabic Hindi, all the self-pity that began to characterize the Hindu outlook from then onwards wells up dramatically. Ceaseless lamentations over the barbarous depredations of the Muslims are intermingled with the mythological age of decline (The Kali Yuga) during which, the decadence of Hinduism (the only true civilization in the eyes of the Hindus) had become inevitable. There are complaints against Buddhism and its soul-destroying doctrine of non-violence (*ahimsa*). For centuries thereafter, generation after generation of Hindus wept as they heard the pitiful tales of the Muslim conquests and then drying their tears, for the most part went to work for their Muslim overlords.'

21.5 Tamerlane (Timur, the Lame)

Qutab-ud-din Aibak established the Delhi Sultanate in 1206 AD with the Slave Dynasty. During 1351 to 1388 AD, Firoze Shah of the Tughluq dynasty ruled over India. His death in 1388 was the signal for struggle and civil war in India. For almost ten years from 1388 to 1398 AD, there was no effective ruler in Delhi, and the imperial crown was for anybody to take.

At that time of utter anarchy in India, there was the Mongol conqueror Timur the Lame or Tamerlane, who was making the conquests of Persia and Bhagdad. He was somewhat of a new convert to Islam. In 1398, he invaded India. In his march to Delhi, he laid waste the whole of North West India. When he was told that his prisoners were *kafirs* (infidels), he ordered killing of some one lakh (hundred thousand) Hindu prisoners on the outskirts of Delhi. He reduced Delhi to rubble. It has been recorded in some texts that the death and plunder of Delhi left towers built high with the heads and bodies of the Hindus. One version says that the 'Siri' Fort of Delhi is so called, as its walls were made of skulls (*siri* in Hindi) of the murdered Hindus. Tens of thousands of Hindus were taken away as booty. Timur let loose the most wanton horrors on the plains and cities of India. He has himself recorded his misdeeds of barbarism. Timur inflicted more misery on India than any previous conqueror. After Timur's departure, complete chaos prevailed in India. For 16 long years, almost no one was in charge in Delhi.

21.6 Babur

To welcome the next Muslim invasion of Bharat, we shift our gaze to the early 16th century AD. In the far-off backwaters of Farghana near Samarkand in Uzbekistan, there was a petty tribal chief. His name was Zahir-uddin-Muhammad, otherwise known as Babur; he had a high pedigree. From his father's side, Babur was the great grandson of Timur the lame, a one-time ruler of Asia. From his mother's side, he was a distant descendent of the world conqueror Genghis Khan. Thus, Babur had Tartar/Mongol blood flowing in his veins. He was educated in the Turkish and Islamic milieu. Turki was his language and he often called himself Turkish.

At the age of 15, Babur occupied the rich city of Samarkand; but was expelled from it by his Uzbek rivals. He had to spend some two years in exile, living in absolute penury and surviving on the diet of cat and dog meat. At the age of 20 in 1504 AD, Babur occupied Kabul and spent the next 20 years establishing his rule in Afghanistan. In 1525 AD, he got an invitation to invade India, which was at that time ruled by Ibrahim Lodhi. One version says that the invitation was extended by some disgruntled elements at the Lodhi Court; there are other versions too.

Some accounts say that Babur landed with about 12,000 horsemen, plus a few ten-thousand foot soldiers. Ibrahim Lodhi had some one lakh troops. The two armies met in the First Battle of Panipat (north of Delhi) in 1526 AD. Ibrahim was totally outmaneuvered by Babur and was defeated. Strange, almost unbelievable as it may seem, this Muslim from the backwaters of Farghana, had come with the most advanced technology of those days; the gunpowder, the cannon and the matchlock. Bharat, the cradle of civilization, did not appear to have this technology — neither the Muslim Lodhis, nor the Hindu Rajputs.

Babur occupied Delhi. Babur was in no mood to return to the cold and inviting heights of Afghanistan, despite the unbearable heat of India and the pleadings of his troop. Babur set about subduing the Afghan and Rajput rulers, one by one. He defeated the redoubtable Rana Sangha aka Raja Sangram Singh of Mewar, in the battle at Khanuha in 1527 AD. Rana Sangha fought bravely, but he was wounded and withdrew. There are some references to Rana Sangha having been let down by a traitor Rajput.

Babur lived only for four years after his great victory in 1526 AD. He had to spend all this time in campaigns to overcome the various Rajput and Afghan rulers right up to Bengal. In these campaigns, his son Humayun assisted him. Babur heard that a Hindu god called Rama was born at some place called Ayodhya. In spite of the short time and his busy schedule, Babur tasked his general Mir Baqi to trace the exact spot of the birth of the Hindu god, and build a mosque on the spot. Mir Baqi duly complied and built a structure which came to be called the Babri Mosque. Babur died in 1530 AD.

21.7 Nadir Shah

The last effective Mughal King Aurangzeb died in 1707 AD. That signaled the start of the end of the Mughal Empire. In the 1730s, an ineffective Mughal king Muhammad Shah was on the Delhi throne.

At that time in Persia, there was a man of a humble family called Nadir Shah; but he was brave and courageous. In 1732 AD, Nadir Shah was successful in toppling the Shah of Iran. He ascended the Persian throne in 1736 AD.

The fabulous wealth of Bharat and weakness of the Mughal king, was a great attraction for Nadir Shah. In 1739 AD, Nadir Shah attacked India and defeated the imperial Mughal forces at Karnal. Muhammad Shah was taken prisoner. Thereafter, Nadir Shah made his headquaters in the Mughal palace. A rumor started that Nadir Shah had died; the locals attacked some Persian troops. This infuriated Nadir Shah and he gave orders for all-round general slaughter (*katl-e-aam*), loot and plunder. The carnage lasted for the whole day and thousands were slaughtered; huge quantities of wealth fell into the hands of Nadir Shah. That included the world famous diamond, the Koh-i-noor. Nadir Shah also seized the Peacock Throne. Nadir Shah left India prostrate and bleeding and went back to Persia. There, he was murdered by one of his nobles in 1747 AD.

Nadir Shah's invasion of India further hastened the process of downfall of the Mughal Empire. The Afghans from the North West and the Marathas from the South were highly encouraged in their raids.

21.8 Ahmed Shah Abdali

We have seen that in the 11th and 12th centuries AD, it were the Turko-Afghans, i.e. Ghazni and Ghauri who had trounced Hindu India. Thereafter, it was the turn of Tartar/Mongols, i.e. Timur and Babur and then a Persian, Nadir Shah. Now, it was again the turn of another Afghan. He was Ahmed Shah Abdali, aka Durrani who ascended the Afghan throne in 1747 AD. That was the time the Afghans got rid of the Persian over-lordship, after the murder of Nadir Shah. Abdali had served in the forces of Nadir Shah. As usual, the wealth and weakness of Bharat attracted the attention of the Afghans.

Abdali started his attacks in 1748 AD. In a series of raids, he conquered Kashmir and wrested most of Punjab from its Mughal Governor. Then in his fourth invasion in 1756 AD, he sacked the city of Delhi. He put the city through a nightmare of loot and murder. The (nominal) Mughal King Alamgir-II ruling at Delhi at that time, was murdered by his own courtiers. The Delhi crown lay at the feet of Ahmed Shah Abdali; but he did not evince any interest. The intense Indian heat may have been one factor. It has been reported that after his Delhi raid, Abdali carried his loot in 28,000 bullock-carts, camels and horses. His son had preceded him with his own booty in a few thousand carts and camels. The Sikhs looted some of this booty, when Abdali and his son were passing through Punjab on their return journey.

Abdali also tormented the Sikhs who were dominating Punjab those days. It is reported that he had the Golden Temple at Amritsar blown up thrice with gunpowder. The holy *sarovar* (tank) of the temple was filled with carcasses of slain cows. In 1761 AD, he carried out the 'Bada Ghalughera', in which some 20,000–25,000 Sikhs may have been slaughtered.

Abdali made some 9 forays to India during 1748–1761 AD. In his last visit in 1761 AD, he inflicted a crushing defeat on the Marathas in the Third Battle of Panipat. The Maratha Empire was split into several states. Abdali withdrew after this great victory. The net gainers of the battle at Panipat were the British, who could then gobble up the smaller Marathas states, one by one.

22
THE MUSLIM RULE

22.1 The Background

Qutub-ud-din Aibak, ex-slave and commander of Muhammad Ghauri established the Muslim rule in India in 1206 AD; he founded the Slave Dynasty. The defeat of Prithviraj Chauhan at the hands of Muhammed Ghauri in 1192 AD is a well-known fact. What is less known and amazing, is the ease with which Qutub-ud-din could overcome resistance in the rest of India. It was almost a walkover.

The continuous Muslim rule lasted some 600 years, i.e. from 1200 to 1800 AD. We would study this rule in three distinct periods:

- The Delhi Sultanate : 1200–1525 AD.
- The Mughal Period : 1526–1707 AD.
- The Mughal Meltdown : 1707–1800 AD.

The end of the Muslim (Mughal) rule could be placed in 1750 or 1800 AD, depending upon the date/period taken as the establishment of British power in India.

22.2 The Delhi Sultanate (1206 to 1526 AD)

Five dynasties ruled during the period of the Sultanate from 1206 to 1526 AD:

1. *Slave Dynasty* (1206 to 1290 AD) : In addition to Qutab-Ud-Din Aibak, two other famous rulers of this dynasty were: Illtutmish (1211 to 1236 AD) and Balban (1266 to 1288 AD). Illtutmish, displayed great diplomacy and prudence; he skillfully avoided an attack by Genghis Khan who turned away from Sindh. A daughter of Illtutmish, Razia Sultan ruled during 1236–40 AD,

ahead of her brothers. She wore male attire.

2. *Khilji Dynasty* (1290 to 1320 AD) : In 1290, Jalal-ud-din Khilji murdered the last ruler of the Slave Dynasty and founded the Khilji Dynasty. The most famous ruler of this dynasty was Alauddin Khilji (1296–1316 AD). He was one of the greatest monarchs to rule over Delhi. In 1320 AD, one Ghazi Malik murdered the last Khilji ruler; he founded the Tughlak Dynasty.

3. *Tughlak Dynasty* (1320 AD to 1412 AD) : Ghazi Malik ascended the throne as Ghiyas-ud-din Tughlak. The two famous kings of this dynasty were Muhammad Bin Tughlak (1325–51 AD), and Feroz Shah Tughlak (1351–88 AD). Muhammed Bin Tughlak was a very learned king, but cruel to the Hindus. He shifted the capital from Delhi to Daulatabad, some 1100 kms away; that was highly unpopular and was a disaster. He was somewhat eccentric, and is sometimes called Muhammad the mad. Feroz Shah was a bigoted ruler who imposed Jazia Tax on the Hindus. It was 10 years after the death of Feroz Shah that Timur invaded India in 1398 AD.

4. *Sayyid Dynasty* (1414 to 1451 AD) : For some 26 years after the death of Feroz Shah in 1388 AD, complete anarchy prevailed in India; no one was in charge in Delhi. After a while, a nowhere man called Khizar Khan captured power and founded the short-lived Sayyid Dynasty.

5. *Lodhi Dynasty* (1451 to 1526 AD) : Last of the Sayyid rulers handed over power in 1451 to a powerful Afghan Chief Bahlul Lodhi, who founded the Lodhi dynasty. Two of the famous rulers were Sikander Lodhi (1489–1517 AD) and Ibrahim Lodhi. Sikander, though born of a Hindu mother, was a fanatical Muslim. It was from Ibrahim Lodhi that the crown of India was snatched by Babur in the First Battle of Panipat in 1526 AD.

The Sultanate kings, were armed with a religion, highly militaristic and monotheist. They were particularly harsh on the Hindus. Public exercise of Hinduism was discouraged, and even banned during certain periods. Some of the rulers were most capricious and cruel; a few sample examples:

- *Qutub-ud-din Aibak:* A ferocious and cruel ruler
  — In the words of a historian, Aibak's gifts were by lakhs, and so were his slaughters of Hindus.
  — After one particular victory of Aibak, some fifty thousand men came under slavery and the plain became black as pitch with Hindus.
- *Sultan Ala-ud-din Khilji:*
  — He laid down: 'The Hindu was to be so reduced to poverty that he should be in no position to keep a horse, or to carry arms; he should not wear good clothes, or enjoy luxuries of life. The Hindu will not become submissive till he is reduced to poverty. I have, therefore, given orders that only the bare minimum be left to the Hindus of corn, milk and curds; they shall not be allowed to accumulate and hoard property'.
  — He asked his advisers to draw up 'rules and regulations for reducing the Hindus to poverty', and for depriving them of their wealth and property, which fosters rebellious conduct.
  — A Muslim historian has expressed that no Hindu was allowed to hold his head high; no sign of gold or silver or of money was seen in their homes. Physical beatings, imprisonment and chains were employed to enforce payment.
- *Sultan Muhammad Bin Tughlak:* He killed so many Hindus that, in the words of a Muslim historian, there was always a mound of dead bodies and a heap of corpses in front of his royal pavilion. The sweepers and executioners were wearied by their work of putting people to death, and dragging their bodies.
- Feroz Shah Tughlak invaded Bengal and offered rewards for every Hindu head; he is reported to have possibly paid for some 1,80,000 of those.

The long and harsh Muslim rule drove the Hindus further into their soul. It advanced the weakening of Hindu physique and their morale, which was already under pressure due to a variety of factors, viz.:

- A pessimistic religion like Buddhism
- Exhausting hot and humid climate
- An inadequate and vegetarian diet

- Lack of a spirit of nationalism

The Delhi Sultanate had many kings who were weak (some very weak) and their rule short-lived. During their rule of 325 years, out of the 35 sultans, as many as 19 were assassinated — mostly by their own courtiers. There would be periods of severe instability and weakness whenever power would pass from one dynasty to another and sometimes, within the dynasty itself. Some of the more important periods of power vacuum were:

- After the death of the Slave Dynasty King Illtutmish in 1236 AD, it took Balban 30 years to finally establish his rule in 1266 AD. Some weaklings ruled during the interim period, and uncertainty was the order of the day.
- Feroz Shah Tughlak died in 1388 AD. Following his death, complete chaos and anarchy prevailed for some 26 years. Tamberlane attacked in 1398 AD, but withdrew. It was only in 1414 AD, that a nowhere man Khizar Khan could establish the next dynasty.
- After every 19 kingly assassinations, periods of instability and sometimes chaos prevailed for varying periods.
- Rule of the Sayyid dynasty for 30 years was a period of extreme weakness. There were open rebellions and provinces were slipping away from the central authority. Sayyids had a hard time controlling these.

It would appear that during the innumerable periods of chaos and uncertainty in Delhi, any medium-sized Hindu Power with say 20,000 horsemen could have taken over Delhi. Some of the Rajput rulers were in a position to do that. However, they do not appear to have bestirred themselves and taken the journey to Delhi. They would rather wait for the Delhi sultan to establish himself, gain strength and then come after them. Only when attacked, would the Rajput rulers offer resistance, many times a courageous one. However, Rajputs were never victorious against the Muslim armies.

During the period of the Sultanates, two foreign travelers visited

India. One was Marco Polo — a Venetian, who came to South India on his way back from China towards the end of the 13th century. The second was Ibn Batuta, who reached Sind in 1333 AD and stayed in India for some 18 years. He was a native of Tangiers in Morocco. He served Muhammad Tughlak as Qazi of Delhi for 12 years. His account of travel is called *Rahela*.

22.3 : The Mughal Period (1526 to 1707 AD)

Babur founded the Mughal dynasty in 1526 AD. The first phase of Mughal dynasty, a stable one, lasted 180 years from 1526 to 1707 AD. Six kings, all mughals, ruled during this period. Their bloodline was Tatar/Mongol. The six kings were:

Babur	1526–30 AD
Humayun	1530–40 AD & 1555–1556
Akbar	1556–1605 AD
Jahangir	1605–28 AD
Shahjahan	1628–58 AD (died in 1666)
Aurangzeb	1658–1707 AD

Of the above, Akbar was the most famous and Aurangzeb the most notorious (as far as the Hindus were concerned). In his early years, Akbar displayed the true spirit of a Muslim *ghazi*. At the tender age of 16, he beheaded a Hindu, so that he could earn the title of *ghazi*. In his early years, Akbar was not averse to inflicting atrocities on the defeated Hindu population. Later on, he became a symbol of tolerance and Muslim — Hindu unity; he treated the Hindus rather well. Hindus were given important posts in his court. All discriminatory taxes were abolished. It is reported that even cow slaughter may have been banned. Some texts of those days record that it was difficult to find copies of the Koran and functioning mosques in the later years of Akbar (that may or may not be so). A few texts even say that Ramadhan (Ramzan), Haj, public prayer and *azan* (religious call) were not encouraged. Akbar founded a new religion called Din-i-Ilahi. For that, he was an object of hatred of the Muslim clergy; it is reported that the mullahs refused to bury Akbar when he died. It has also been expressed that not many Muslims followed his funeral cortege.

Aurangzeb was the very anti-thesis of the later day Akbar. He went after the Hindus from day one and re-imposed the hated Jazia (special tax on Hindus). It has been reported that he would take his meal only after a given amount of the *'jinnous'* (brahmin threads) had been removed and destroyed. In Varanasi, there is the most holy Hindu Vishwanath Temple of Shiva. Aurangzeb ordered that a mosque be built abutting this temple and grander in outlook; that was meant as an act of deliberate insult. After the mosque had been built, he was asked whether the temple could be brought down. He is reported to have expressed that the temple be allowed to stand as a reminder to future generations of Hindus, as to their station in life. The mosque still stands at Varanasi.

Both Akbar and Aurangzeb expanded their empires; Akbar even without any great inheritance. They ruled over the greatest empires in India; Aurangzeb even bigger than Akbar. True to Tartar blood at the time of each succession, there would be warfare and bloodshed; brother would kill brother, sons would revolt against the father. Aurangzeb imprisoned his father for eight long years till the latter's death.

Mughal rule was a period of stability. However, there were occasions when uncertainty, chaos and weakness prevailed; a few examples:

- In 1540 AD, Humayun was defeated by an Afghan Sher Shah Suri, and was forced to wander around in the deserts of Rajputana. It was during those wanderings that his son Akbar was born. The Suri dynasty ruled for 15 years from 1540 to 1555.
- Humayun recovered the Delhi throne in 1555. Before he could even start consolidation, he died suddenly of an accidental fall. At that time, Akbar was just 13 years old, and not even at Delhi or Agra, but in Punjab. This lad of 13, under the inspiring advice of Bairam Khan, defeated the Afghan General Hemu who had occupied Delhi and Agra, on the death of Humayun. Akbar then went on to establish the great Mughal Empire, almost out of a vacuum.
- At the time of each succession, when brother would fight brother, a lot of uncertainty and often chaos would prevail.

22.4 The Mughal Meltdown (1707–1800 AD)

The death of Aurangzeb in 1707 AD signaled the start of the end of the Mughal Empire. Normally, the process would have taken 20–30 years. However, unbelievably, it lasted some 100 years till 1800 AD.

Starting 1707 AD, Mughal kings were regularly installed and deposed. Kingmakers emerged both internally in the court, and externally in the form of Marathas, Nawab of Awadh and later, the British. Some examples;

- In a span of first 50 years (1707 to 1757), 8 kings were squeezed in. During that period, in just 7 years, five kings were changed in quick succession.
- In 1719 AD, King Farruk-Siyar's own courtiers dragged him from his harem and blinded him; he was tortured and killed. The Nizam and the Marathas had a role to play in that episode.
- In 1759, King Alamgir-II was assassinated by his own minister. This was during one of the raids of Ahmed Shah Abdali.
- Shah Alam-II ascended to the throne in 1759 AD. In 1788 AD, he was blinded by a Rohilla Afghan Chief, Nazib-ul-Daula. A later Maratha Chief Mahadji Scindia put Shah Alam II back on the throne. He went on to rule (nominally) till 1806 AD. In 1803, Shah Alam II was taken under the protection of the British, for the second time.

Notwithstanding the turmoil, some Mughal kings were able to stick to the throne for long periods; though most of the times their rule was nominal. These kings were:

Muhammad Shah	1719–48 (29 years)
Shah Alam-II	1759–1806 (47 years)
Akbar II	1806–37 (31 years)
Bahadur Shah Zafar II	1837–57 (20 years)

The last two were British pensioners and were sometimes located in Lucknow. The British had established their effective presence in Delhi by about 1800 AD. Most of the Mughal kings during the period of the Mughal melt-down were incompetent, ineffective and mostly degenerate. However, so strong was the prestige of the Mughal name that a myth of central Mughal authority and rule continued to prevail for some 100 years. During that period, there were four dominant powers in the North and West of India, i.e. the Marathas, Nawab of Awadh, the Ruler of Bengal and later, the British. Perhaps, the Sikhs could also be included in this list, but not the Rajputs who appeared to be happy with a passive role. Even these powers continued to see a mirage of the Mughal rule at the Centre. The Marathas over-ran Delhi at least twice; but, instead of establishing their own rule, they would help install a quisling (including a blind king) on the so-called Mughal throne. It sounds unbelievable, but it is true.

23
THE RAJPUTS

The Hindi world 'Rajput' means the son of a king; there are also some other interpretations. Rajputs are a cluster of clans who from about the 8th century AD, ruled over various big and small states in Rajputana, Malwa and Gujarat in North West India. Who are the Rajputs? The simple answer is 'No one knows for sure'; various theories are prevalent.

Rajputs call themselves Suryavanshi or Chandravanshi Kshatriyas, and trace their ancestry to the *Ramayana* and *Mahabharta* clans respectively. That requires some labored linkages. The physical events (wars) of the *Ramayana* and *Mahabharta* are generally believed to be pre-2000 BC, if not pre-3000 BC. The locations of the two wars were outside of Rajputana, where the Rajputs are presently located. The two epics might have been actually written in the closing centuries of the BC era. There are no specific references to Rajputs in these epics, or for that matter, in any other literature of those times, including the early Puranas.

There is also a theory that some of the Rajput clans emerged out of a *yajna* (fire sacrifice) at Mount Abu, conducted by rishi Vasishta. We have no way to know, or even guess, when that *yajna* might have been undertaken. Tradition has it that four mythical figures (warriors) emerged out of the fire (Some accounts say that there was only one mythical figure). They went on to found the Rajput clans of Paramara, Pratihara, Chahamanas (Chauhan) and Chaulukya (Solanki). The Paramaras meaning 'slayer of the enemy' were perhaps the first to emerge. Prithviraj was a Chauhan. These Rajputs are called *Agni-Vanshi*, or of *Agni-Kul* (*Agni* means fire). Following two reasons (largely mythological), have been cited for conduct of the *yajna*:

- Rishi Vasishta's wish-cow had been taken away by force (that was in the distant past). Vasishta conducted the *yajna* to get warriors to help him recover his cow.
- After the Rajputs had established their kingdoms, they asked Brahmins to undertake their 'raj-tilak' (coronation ceremony). Brahmins refused, saying that the Rajputs were not real Kshatriyas (see the following paragraphs); hence, they were not entitled to 'raj-tilak'. Rajputs then expressed that actions be taken to make them true Kshatriyas, i.e. undertake a sort of purification ceremony. The Mount Abu *yajna* may have been undertaken to that end.

A view has been expressed that Rajputs may be descendents of the foreign tribes (chapter 18) who invaded India from the 2nd century BC to the 2nd century AD, and of the later Huns who came in the 5th and 6th centuries AD. There are no specific references to these foreign tribes having gone back. In all probability, these tribes settled down in India in and around Rajputana, which was adjacent to the area of their operation. Thus, the Rajputs could be descendents of Yavanas (Bactrian Greeks), Pahlavas/Parthians (Iranians), Sakas (Scythians), Kushanas and Huns, who were all war-like people. These people might have gradually got absorbed in the Hindu fold, as unlike the later Muslims, these tribes did not have any particular religion of their own. The Hindus, perhaps, did not object to accepting them in the higher Kshatriya caste, as they were all victors. If the Hindus objected, they might have been over-ruled by the power of the sword.

Here, the story gets rather complex. In Chapter X (Entries 43 to 45) of the Manusmriti, there is a reference to the Bactrian Greeks, Scythians, and Iranians as lapsed or degraded Kshatriyas (see chapter 8.4). There are also some references to the presence of foreign tribes of Yavanas (Bactrian Greeks), Pahlavas/Parthians (Iranians) and Sakas (Scythians) in pre-historic (mythological) times. 25 generations before the *Ramayana*, there was King Talajangha of the Haihaya tribe, which was as an off-shoot of the Yadavas (see Chapter 8.3). It is recorded that

Talajangha took help of the above quoted foreign tribes to defeat the 40th generation Suryavanshi king Bahu. There are some hints that the Haihayas might have emerged out of those foreign tribes. Presence of foreign tribes on the soil of Bharat, so far back in pre-historic (mythological) times would be most baffling, to say the least. Could there be some linkages of Rajputs to the foreign tribes as referred in the Manusmriti, and those at the time of King Talajangha? We have no definite answer.

If Rajputs were on-going people from the epic periods, power should have rested with them in and around Rajputana (North West India), on the dawn of recorded history in Bharat. However, the factual position was different. Rajputs were nowhere on the scene at the advent of history, as seen from the following:

- At the dawn of recorded history in the 7th/6th century BC, the power center was situated in the East of India (in and around present Bihar), far away from Rajputana, which hardly figures as a major power center in the records of those days.
- For the first about one thousand years of recorded Indian history, the power did not rest with the Rajputs, and not even with the Kshatriyas. The main ruling dynasties were non-Kshatriyas (see chapter 16). The broad picture was as follows:

6th to 2nd centuries BC
Sisunaga, Nanda, Maurya dynasties — All Low Caste
2nd century BC to 1st century AD
Shunga and Kanva dynasties — Both Brahmin
4th and 5th centuries AD
Gupta dynasty — Vaishya

- The barbarian tribes started their invasions of North West India in the 2nd century BC, in the very backyard of the supposed habitat of the Rajputs; these lasted for about four centuries. There are no great tales of any resistance having been offered to them by the Rajputs. Rajputs should have been present in and around Rajputana at that time, if they were on-going clans from the *Mahabharta* period.

- In the 4th century AD, power from the foreign tribes was snatched not by Kshatriyas or Rajputs, but by Vaishyas — the Guptas. The Rajputs appear to be nowhere on the scene at that time. The Guptas went on to rule for some 200 years.

Even during the 6th and 7th centuries AD, there is no great talk of the Rajputs. In the early 6th century AD, Huns under Mihirkula played up hell in North West India. King Harshavardan ruled over North India in the first half of the 7th century AD (606 to 648 AD); in all probability, he was not a Rajput. It is only in the late 7th or 8th century AD that we hear of Rajput states emerging in and around Rajputana. Except for some very broad contours, very little is known about the history of India from the 8th to 10th centuries AD, i.e. until the start of the Muslim invasions. We give below names, old and new, of some of the more important Rajput states:

Mewar — Udaipur (Ex-Chittor)
Marwar — Jodhpur
Amber — Jaipur

Around 1176 AD, a Turko-Afghan Muhammad Ghauri started nibbling at Indian territories. He took over many cities in Punjab, without any response from the Rajput rulers. Then in 1191 AD, the Rajputs recorded their only victory against foreigners, when Prithviraj Chauhan defeated Ghauri. However, the celebrations were short-lived. Muhammad Ghauri returned a year later, and smashed the Hindu (Rajput) power in the 2nd Battle of Tarain in 1192 AD. To students of military history, it would appear inconceivable that one defeat in the 2nd Battle of Tarain shattered the Rajput morale for almost all times to come. After that one battle, Qutab-ud-din Aibak, ex-slave and commander of Ghauri, could run through the rest of the Rajput and other Hindu states, almost like a knife through butter.

The Muslim Sultanates ruled over Delhi for 325 years. During that period, there were innumerable occasions when utter chaos and weakness prevailed in Delhi for prolonged periods. We have given those details in chapter 22.1. One such period was for 26 long years — 10 years before the invasion of Timur in 1398 AD and 16 years after that;

no one was in charge in Delhi during that period. The Rajputs did not think of unsheathing their swords to march to Delhi to wrest control and to teach a lesson or two to those usurpers. They would rather sit in their house, give time to the Delhi Sultan to gather muscle and then come after them. When attacked, the Rajputs did fight back, sometimes most courageously. However, the result was always the same, i.e. Rajput defeat and Muslim victory.

At this stage, we may mention the two cases of most heroic fights by Rajput rulers. One was by Rana Sangha against Babur in 1527 AD, and the second in 1576 AD, by Maharana Pratap against Akbar at Haldighat. Though defeated and in spite of unfathomable difficulties, Maharana Pratap kept his struggle going for some 25 long years, and was successful in recovering most of his lands.

By the early 1700s, the Rajputs appeared to have lost the last vestige of their zest for warfare. We have recorded earlier that after the death of Aurangzeb in 1707 AD, there was a power vacuum in Delhi; for some 100 years, almost no one was in charge there. One incompetent (nominal) Mughal king would replace another. Powers from far places in India like the Marathas and Nizam would often capture Delhi, and exercise a sort of suzerainty over it. During certain periods, the Marathas had the power to remove and install kings on the (nominal) Mughal throne. Even the British threw their hat in the ring, and in mid 1700s, took the Mughal king in their protective custody. However, the next-door Rajputs would not think of exploiting the situation and marching on Delhi, with say 20,000 horsemen. Rather, they contended themselves by accepting the over-lordship of their religious brethren the Marathas, who treated the Rajputs rather shabbily. Rajputs also did not think of challenging Ahmed Shah Abdali, who played up havoc in their backyard, i.e. Punjab and Delhi, during 1748–1761 AD.

The Rajputs were conspicuous by their absence in mounting a significant struggle against the British take-over of Bharat. Most of the Rajput states yielded rather tamely to the British, who for their own purpose, decided not to incorporate these in British India. They would rather put a lowly British officer, called Political Agent in these

States; he would be the Ring Master, with the great maharajas acting as puppets.

Kshatriyas are the special class of Hindus for warfare and for the defense of the motherland and its people. Rajputs like to call themselves as the highest of high Kshatriyas. Rajasthan (old Rajputana) resounds with tales and legends of Rajput chivalry and bravery. History records many cases of exceptional Rajput courage at the personal level. There were also many good Rajput generals; they performed well during their internal conflicts.

However, the Rajputs generally gave a poor account of themselves when pitted against invading forces. Whenever the Muslim armies attacked from the North West, the Rajputs did not succeed in their resistance. They were mostly defeated on the battlefield, with the exception of 1191 AD. If we take an overall broad view of history, say from 1000 to 1800 AD, it is difficult to avoid the impression that the 'sentinels of the Hindus', the Rajputs, failed in their primary duty of guarding the honor, freedom and integrity of the motherland and its people. This is no reflection on the personal courageous qualities of the Rajputs.

Another peculiarity needs to be noted in respect of the military mindset of the Rajputs. All the major battles of Rajputs recorded by history were in the defensive mode, i.e. they generally fought when they were attacked. On the contrary, their religious brethren, the Marathas were always on the look-out for offensive action (see chapter 24). There does not appear to be many cases when the Rajputs unsheathed their swords, mounted their horses and marched against their Muslim adversaries, who were their chief tormentors. We are talking of a major offensive compaign, not some small engagements here or there. We find no reason why Prithviraj Chauhan, a great general with a massive army, did not march to Lahore, Multan, Ghazni, Ghur, or even Persia and beyond, to wave the Hindu and Bharat flag. There appeared to be a fundamental fault-line in the military thinking of the Rajputs, i.e. a lack of aggressive spirit. That proved to be their and their country's undoing.

No study of the Rajputs would be complete without referring to the practice of Rajputs marrying their women to Mughals. It was a one-way passage; Mughal princesses were not given to Rajputs. In 2008, a film titled *Jodha-Akbar* was released. The film depicts one Jodha as the daughter of Raja Bharmal of Amber, which was under a military threat. It shows the Raja meekly offering his daughter in marriage to Emperor Akbar, with a view to seek protection from him for his State; the film makes him look quite pathetic in the process. The Raja calls for an assembly of other Rajput rulers. He tells them that if a war was forced on him, it would result in widows and orphans in his state; thus he must avoid war at all cost. This must be the most ingenious and original excuse for avoiding war, coming from a warrior, that too a Rajput.

Raja Bharmal had reconciled himself that his daughter would have to convert to Islam after marriage. It was the daughter, who put the condition directly to Akbar that she should be allowed to stay a Hindu; Akbar agreed. The movie also shows the mother offering a vial of poison to Jodha, to enable her to get out of the humiliating position she was in.

All hell broke loose on the release of the film. Rajput organizations protested that the film was a distortion of history, and that it showed Rajputs in a bad light. It is by no means established that there was any queen of Akbar named Jodha, even though the common public perception appears to be so. Neither biography of Akbar (*Akbar-nama*), nor of Jahangir (*Tuzuk-e-Jahangiri*) mention any such name. There are some references here and there, to a Mughal Hindu queen, rechristened Mariam Zamani (Mariam of the Age). There are various Hindu names associated with her, the most popular being Hira Kanwari, perhaps of Amber. Some historians opine that Mariam Zamani may have been the wife of Jahangir, rather than his mother.

Rajput organizations took to protests, and even resorted to violence. Newspapers ran headlines, wrote editorials. TV debates were organized, and panelists got worked up on the issue; accusations were freely hurled. However, not one panelist was willing to refer to the actual cause of the Rajput ire. It was that giving of their women for the Mughal harems was a shameful episode in the history of the Rajputs. Rajputs would not like to be reminded of that episode in this time and age.

Before we are accused of a bias against the Rajputs, we may record that the strongest opponent of this practice was the great Rajput Hindu hero, the one and only, Maharana Pratap. He considered the practice degrading, and had to suffer for that. Pratap refused to have social relations with the Rajput rulers who gave their women to the Mughals. There is an interesting anecdote told by Col Todd, a British expert on Rajputs. It appears that a sister of Prince Mann Singh of Amber (Jaipur) was married to Akbar. Mann Singh was made commander of the Mughal forces. Once Rana Pratap invited Mann Singh for a meal, but failed to turn up himself. Pratap's son came and informed Mann Singh that his father had developed a headache. Mann Singh, mounting his horse, expressed that he would avenge that insult; that he did. Maharana Pratap had to spend long periods in wilderness and in exile; but he did not give up.

History records innumerable instances of extreme courage and bravery displayed by Rajputs. Notwithstanding that, the net contribution of these sentinels of Hindu society was to hand over this holy and ancient land and its innocent and trusting people, to invaders of all shape, color and religions, from different lands, both near and far. If all this looks like harsh criticism of the Rajputs, our apologies tendered in advance. We end the piece by saying that national interests are not served by sweeping the truth under the carpet. Sometimes or the other, we have to face it.

24
THE MARATHAS

Marathas started coming into prominence around the mid 17th century AD. Shahji Bhonsle laid the foundation of the Marathas. His son Shivaji (1627–80 AD) built the Maratha power. Shivaji was a leader of extra-ordinary caliber who was wise, dashing, clever and courageous. In addition to wielding the sword with a firm hand, he could fudge and feint and fox-play, to achieve his objective of victory. Shivaji did not suffer from a sense of false chivalry; if pressed, he would comprise to refute the agreement at the first opportunity. Shivaji was inspired by the idea of Hindu self-rule and greatly influenced by his mother Jija Bai, and his tutor Dadaji. As a patriot, general and statesman, Shivaji stands high in the history of India. He is a national hero.

Studying Maratha history, one cannot help noting the sharp difference in the approach to war between the Rajputs and Marathas; both were 'sword arms' of Hindus. Rajputs always acted in the defensive mode, i.e. they fought when attacked. On the contrary, the Marathas were generally on the lookout for offensive action. As soon as the Marathas would gather strength, they would attack other powers, both near and far off, Hindu and Muslim. The Marathas under Shivaji and later, had no hesitation in attacking settlements and outposts of the mighty Mughal Emperor Aurangzeb. Shivaji captured many of Aurangzeb's forts.

Having set up a mini empire, Shivaji crowned himself king, called Chhatrapati, in 1774. There is a view that the Marathas might have been of non-Kshatriya origin. However, efforts were launched to trace the Maratha lineage to the Sisodia Rajputs; we are not sure how successful these efforts were.

The Marathas were one power who repeatedly defeated (internal) Muslim armies — Mughals, the Nizam of Hyderabad and Bijapur. In 1737, Peshwa Baji Rao I inflicted a crushing defeat on the combined (Muslim) forces of Mughals and the Nizam, at Bhopal. The Mughal King was forced to cede major territories like Malwa to the Marathas. During many periods of the Peshwa regimes, the Marathas exercised suzerainty over Delhi, and could install or remove a Mughal king on the (nominal) Mughal throne. At one time, the Maratha Empire extended in pockets, up to Tamil Nadu (Jinji Fort) in the South, towards Orissa in the East, Gujarat and outskirts of Sindh, in the West. They also exercised over-lordship over Rajputs for prolonged periods. Other than the kings of South India, the Marathas appeared to be the only Hindu power, which had a naval fleet. However, that fleet may not have been a match for the Portuguese and English fleets, which were active around Indian shores those days.

Shivaji died in 1680 AD. His son Shambuji had gone over to the Mughals during the lifetime of Shivaji; he had spent time at the Mughal court, as a mansabdar (court noble). Shambuji succeeded Shivaji. In due course, Shambuji started attacking Mughal forts. The Mughals captured him in an ambush, and took him to the Mughal court. It is reported that he refused to embrace Islam, and heaped insults on Emperor Aurangzeb. Shambuji was tortured and his body cut limb by limb in 1688 AD. His brother Rajaram (second son of Shivaji) succeeded him. Rajaram had to spend a lot of time in the fort of Jinji in Tamil country, which was encircled by Mughal forces. Rajaram died in 1700 AD. After his death, his wife Tarabai, a woman of substance and grit, continued to rule in the name of her minor son.

Shambuji's son and Shivaji's grandson, Shahu (aka Shivaji II) had been brought up at the Mughal court, as a sort of prisoner prince, from the age of 7 to about 25 years. After the death of Aurangzeb in 1707 AD, the Mughals released Shahu to sow discord amongst the Marathas. The scheme was successful and there were hostilities between Tarabai and Shahu. Ultimately, Shahu was successful, with the help of a wily Chitpavan Brahmin, Vishwanath Balaji. Shahu died in 1748 AD.

Now, we come to an unique institution exclusive to Marathas.

Shahu appointed Vishwanath Balaji as his Chief Minister called 'Peshwa'. However, gradually over time, Peshwas assumed all powers and the post became hereditary. The actual Maratha ruler became a mere titular head. The first four Peshwas were:

Vishwanath Balaji — 1713 to 1720 AD
Baji Rao I — 1720 to 1740 AD
Balaji Baji Rao — 1740 to 1761 AD
Madhav Rao — 1761 to 1772 AD

After Shivaji, Baji Rao I was the most charismatic and dynamic leader in Maratha history. Under his leadership for 20 years, the Marathas carried out all-round raids with impunity — North, South, East and West. They reached Rajputana in 1735, Delhi in 1737, Orissa and Bengal by 1740. By his exceptional talent, Baji Rao I also established central control over various Maratha commanders, who were functioning semi-independently.

Balaji Bajirao, also known as Nana Sahib, continued the forward policy of his father. Around 1757, Maratha forces went up to Sindh, and their horses quenched their thirst at Attock, with the water of the Sindh River. He defeated the Nizam of Hyderabad in 1760, obtaining Bijapur, Aurangabad, parts of Bidar and the Daultabad fort.

Around 1758, Balaji's brother Raghunath Rao over-ran Punjab and captured Lahore. That offended the Afghan King Ahmed Shah Abdali. In 1760, Abdali landed in Delhi at the head of a large force. A huge confederate of Marathas from all over the Maratha country was sent to oppose Abdali, under the command of Sadesheo Bhao, a cousin of the Peshwa. The two forces met at the historic field of Panipat in mid January 1761. Abdali inflicted a crushing defeat on the Marathas. Around 40,000 Marathas may have been killed and some 30,000 taken prisoners; some Marathas fled to the hills of Punjab. The Peshwa Balaji Bajirao died of the shock of the defeat. The Maratha's dream of a Hindu rashtra was shattered.

Madhav Rao took over as the Peshwa after the defeat of the Marathas at Panipat. He appointed an able general Nana Phadnavis as his lieutenant. Madhav Rao and Nana Phadnavis were successful in consolidating Maratha states and power. In 1769, Madav Rao sent a force to the North to subdue the Rajputs, Rohillas (Muslims) and Jats. In 1771, (nominal) Mughal Emperor Shah Alam II was living at Allahabad under the protection of the British. The Marathas brought him to Delhi, and installed him on the (nominal) Mughal throne. Phadnavis played an important role in the affairs of the Marathas. Even after the death of Madhav Rao, Phadnavis worked with many Peshwas, until his own death in 1800 AD. Maratha power reached great heights, first under Nana Sahib, and later under Nana Phadnavis.

A new star in the form of Mahadji Scindia rose on the Maratha horizon in the 1770s. He set up a rival center of power to the Peshwas represented by Nana Phadnavis. Mahadji rose very fast in power and fame. In the 1780s, Mahadji had the (nominal) Mughal Emperor in Delhi Shah Alam II at his bidding. In 1788, a Rohilla Chief captured and blinded Shah Alam II. Mahadji got the (so-called) Emperor released and re-installed him on the throne. Mahadji died prematurely in 1794, at the pinnacle of his power.

We give below two events to show the courageous character of Shivaji:

- The Sultan of Bijapur sent an army under Afzal Khan in 1659 to subdue Shivaji. As a military stalemate was reached, Shivaji and Afzal Khan met to sort out the issue; they were alone and supposedly unarmed. Shivaji suspected Afzal Khan of treachery. As the two embraced, Shivaji pushed the steel claws that he was wearing into Afzal's stomach and killed him. The Marathas who were hiding nearby attacked the Muslim army, inflicting heavy losses.
- In 1662–63, Aurangzeb dispatched his uncle Shayista Khan to suppress Shivaji. Shayista captured Pune and made it his temporary headquarter. Shivaji, with about 50 followers entered Pune in the form of a *barat* (marriage party). They attacked the Mughal army. Shayista was injured, but he escaped by jumping out of a window. That event added greatly to Shivaji's prestige and the Maratha morale.

The Marathas' period of glory could be considered to have lasted some 100–120 years, with many ups and one major down. The Marathas defeated other Indian powers in India, including the Nizam of Hyderabad and Tipu Sultan of Mysore. From time to time, they controlled vast parts of India. Their major reverse was the defeat at Panipat in 1761. However, they could rise after that under the leadership of Madhav Rao and Nana Phadnavis. During most of the period of the 18th century, the Marathas exercised influence over the Mughal court in Delhi. There were many occasions, when Marathas could have established their rule in Delhi; however, they hesitated. Having captured Delhi, the Marathas would withdraw, after installing an ineffectual Mughal king on the (nominal) Mughal throne. Was it some sort of a Hindu mindset at work, i.e. the inability or hesitancy to deliver the final blow? The Marathas came quite close to realizing their dream of a Hindu rashtra; they missed that by a whisker, to the great misfortune of this great country.

There is a view that at some stage, the Marathas became arrogant. They maltreated Hindus of non-Maratha origin — Rajputs, Sikhs and Jats were at the receiving end. Ahmed Shah Abdali was the enemy not only of the Marathas, but also of the Rajputs and Sikhs. But for the somewhat arrogant attitude of the Marathas, the Rajputs and Sikhs could have stood shoulder to shoulder with the Marathas on the Panipat field. In that case, the history of Bharat would have evolved along different lines.

25
SOUTH INDIA

Unfortunately, a tradition has developed in India that when we talk of the history of India, it is mostly taken to mean the history of North India, more particularly of North West India. It could be due to two reasons. Firstly, all invaders with the exception of Europeans came from the North West. Secondly, the lack of adequate data on the dynasties that ruled South India. We attempt to give below a short resume of the history of South India.

The earliest dynasty to rule South India was the Satavahanas from about 250 BC; they had three streams:

- *Cholas:* On the Eastern Coast in the Kaveri Delta around Tamil Nadu
- *Cheras:* On the Malabar (Kerala) Coast
- *Pandyas:* Southern tip of the Peninsula (Madurai)

Over the years, the above dynasties fought each other for supremacy — defeating and being defeated and re-emerging. As time flowed by, their territories waxed and waned. Satavahanas lost their power by the 3rd century AD, when other Powers rose in the South. However, the Cholas re-emerged later in the 9th century AD.

We give below a bird's eye-view of the dynasties that ruled the various parts of South India till the mid 2nd millennium AD:

3rd to 5th century AD	Vakatakas	Upper Western Deccan (Maharashtra).
300–900 AD	Pallavas of Kanchipuram	Tamil Nadu (North)

550–760 AD	First Western Chalukyas of Badami	Western and Central Deccan (Maharashtra)
630–970 AD	Eastern Chalukyas	Eastern Deccan (Andhra)
760–970 AD	Rashtrakutas	Western, Central and Southern Deccan (Maharashtra plus Karnataka).
850–1270 AD	Cholas of Tanjore	TamilNadu and Madurai
975–1190 AD	Second Western Chalukyas	Western and Central Deccan
1110–1330 AD	Hoysalas	Central and South Deccan
1190–1300 AD	Yadavas	North Deccan
1200–1330 AD	Kakatiyas of Warangal	East Deccan
1330–1560 AD	Vijayanagara	Karnataka, Tamil Nadu
1340–1518 AD	Bahmanid	Mid Deccan (Western)

The above periods may not be taken too rigidly; there could be substantial variations, i.e. a view has been expressed that Pallavas perhaps ruled from 100 BC. We give below a brief account of the more important dynasties.

Western Chalukyas of Badami (550-760 AD)

The First Western Chalukyas were ruling areas around Maharashtra; to their South were the Pallava ruling areas around Tamil Nadu. Some of the more important Chalukyian Kings were:

Pulakesin I	543–566 AD
Pulakesin II	609–642 AD
Vikramaditya I	654–668 AD
Vikramaditya II	733–744 AD

Of the above, by far the most famous was Pulakesin II. He made extensive conquests. These included the Konkan Coast, Malawas of Malwa (adjoining Gujarat) and Gurjaras of South Rajasthan. Pulakesin II also challenged Harshvardhan and claimed victory. He conquered Orissa and Andhra Pradesh. Lands around Andhra Pradesh were placed in the care of his younger brother, whose descendents constituted the Eastern Chalukyas (630–970 AD). Hsuan Tsang visited India in those days and refers to him in glowing terms. Pulakesin II also overran the rich Pallava country of Tamil Nadu. But, later, the tables were turned and Pallava King Narasimha Varman I claimed to have defeated Pulakesin II many times and captured the Chalukyan capital Badami. Still later, Vikramaditya II reconquered Kanchipuram of the Pallavas three times. The last Chalukyan king Kirtivarman was overthrown in 755 AD by Danti Durga; he founded the Rashtrakuta Dynasty. Pulakesin II was the master of South India and probably the greatest king of his times.

Pallavas of Kanchipuram (300–900 AD)

The Pallavas ruled over Tamil Nadu touching the Eastern Coast and Southeast of the Chalukya territory. Some of their more important Kings were:

Mahendra Varman I	590–630 AD
Narasimha Varman I	630–660 AD
Parameshnara Varman I	670–700 AD
Narasimha Varman II	700–739 AD

The Pallavas had a running feud with the Chalukyas. They would frequently invade each other — winning sometimes, losing other times. Narasimha Varman II built the temples at Kanchi and Mamalapuram. Mahendra Varman I was one of the greatest Pallava kings.

Rashtrakutas (760–970 AD)

Danti Durga, founder of the Rashtrakuta Dynasty was an official in the northern territories of the Chalukyas, who were getting exhausted by their constant conflicts with the Pallavas. Danti Durga kept on accumulating power by nibbling territories, till he was in a position to side-track the Western Chalukyas. He expanded his territory to include much of Madhya Pradesh, parts of Gujarat and northern Maharashtra. After the death of Danti Durga (childless), his uncle Krishna I succeeded him. Krishna I decisively defeated the Chalukyas. Badami, the capital of the Chalukyas was captured; Karnataka was also added to the Rashtrakuta empire.

When Krishna I died in 773 AD, the Rashtrakutas were undisputed masters of the entire Deccan. Krishna I was succeeded by Dhruva. He secured his empire in the South by subduing the Pallavas (of Kanchi) and the Gangas. He was possibly the first Southern king to make conquests north of the Narmada. He conquered Malwa (abutting Gujarat) and headed for Kannauj. At that time, other Powers in race for hegemony of the North were the Gurjara–Pratihara from Western India and the Palas from Bengal. For almost the next two centuries, Kannauj kept on changing hands amongst these three Powers.

At the zenith of their power, the Rashtrakutas controlled a territory in the form of a rectangle placed vertically — some 1500 km long (North-South) and about 400 km wide (East–West), touching the Western Coast and running in the Center of India (Berar) on the Eastern side.

Cholas of Tanjore (850–1270 AD)

In the opening paragraphs of this chapter, we have talked of the Cholas ruling over the territory of Tamil Nadu (East Coast) from about 250 BC. They lost their primacy somewhere in the 3rd century AD, when Pallavas established their rule which continued up to 900 AD, or so.

The Cholas re-emerged and established a Chola empire when its founder Vijayalaya captured Tanjore in 846 AD. Vijayalaya's successor, Adita expanded the Chola Empire; he interfered in the Pallava succession struggle, and defeated Pandians, the southern neighbor. Thereafter, the Rashtrakutas who were ruling at the same time, reduced the power of the Cholas.

In 985 AD, Rajaraja I ascended the Chola throne. He undertook a series of campaigns, starting with the Pandians and then the Cheras (Kerala Coast). Though he subdued these powers, he could not conquer them, or establish his rule over them. Rajaraja I then attacked Sri Lanka, where he indulged in a lot of loot and plunder. He is reported to have also taken over Maldives. At that time, arose the later or Second Chalukyas in the Western Deccan, by ousting the Rashtrakutas. There were confrontations between the Cholas and the Second Chalukyas; but the Eastern Chalukyas considered the Cholas as their allies.

In 1014 AD, Rajaraja I was succeeded by Rajendra I. He constructed the Tanjore temple to celebrate the conquests of Rajaraja I. Rajendra I re-invaded Sri Lanka, and more treasures and priceless regalia were seized. Cheras and Pandians were again subdued. Then, Rajendra's General proceeded towards Kalinga (Orissa) to get the water of river Ganga.

Rajaraja I and/or Rajendra I dispatched naval expeditions to South East Asia. The exact results achieved by these expeditions are not known precisely. An inscription in the Tanjore temple refers to take-overs of some places in Burma, the Malayan peninsula, Sumatra and Nicobar islands.

Rajendra's reign lasted some 30 years (1014–1044 AD), during which he raised the Chola Empire to great glories. It has however been expressed that he could not conquer his immediate neighbors, though he did subdue them from time to time. His successors had to vacate Sri Lanka. The supremacy of the Cholas lasted till the early 13th century and the dynasty collapsed in 1270 AD. There are references to the Cholas having established and maintained contacts with China and South East Asia.

Vijayanagara (1330-1560) and Bahmanid (1340-1518)

Vijayanagara and Bahmanid, two contiguous kingdoms, were both founded in 1330–40 — the former in the deep Southern part of India (touching both coasts), and Bahmanid, North West of it. Bahmanid is the only Muslim kingdom of the South, which we are considering. Two brothers, Harihara and Bukka founded Vijayanagara. A Muslim Gangu Bahman Shah, earlier known as Hasan, founded the Bahmanid. There are legends connected with the founding of both kingdoms, which we do not cover here.

Bahmanid was constantly at war with its Hindu neighbors, especially the kings of Vijayanagara; they repeatedly threatened Vijayanagara. In addition to the difference of religion, a major bone of contention was the intervening rich area known as Raichur *doab* (area between two rivers). In spite of the continuous conflict, neither was able to destroy the other.

Deva-Raya II, the last effective ruler of the Sangama dynasty ruling Vijayanagara died in 1446 AD. There was general Narasimha who had captured the lost territory in Andhra Pradesh and Tamil Nadu. He ascended to the throne of Vijayanagara and founded the Saluva dynasty. Thereafter, there was another Narasimha, who founded the Tuluva dynasty. Babur's contemporary, the great Krishna Deva-Raya succeeded the second Narasimha (*deva*-raya means god king).

During the rule of Krishna Deva-Raya, Vijayanagara soared to spectacular heights. He recorded great all round conquests of Raichur Doab and Orissa, and claimed new territories in Andhra Pradesh. Loot and plunder poured into Vijayanagara. The Vijayanagara kingdom has been seen as the epitome of traditional Indian kingship and a spectacular finale to the Hindu empire.

The great Vijayanagara Empire suffered a major defeat at the hands of combined forces of four Muslim Powers in the upper Western Deccan at the battle of Talikot in 1565 AD. That was the end of this great Hindu empire in the South; its capital Vijayanagara was plundered and destroyed by the Muslims.

The following explanatory statements are made in respect to the Empires in South India:

1. The Vakatakas ruled in the area of Bundelkhand Vidharba. Their origin may be possibly traced to the Scythians (Sakas).
2. During the first half of the 7th century AD, i.e. around 600–650 AD, the following kings were contemporaries, ruling over different parts of Bharat.

   Harshvardhan (North India)   606 to 647 AD
   Pulakesin II (Chalukyas — Maharashtra)   609 to 642 AD
   Mahendra Varman I (Pallavas — Tamil Nadu) 590 to 630 AD

A version says that Pulakesin II may have defeated Harshvardhan, but this is not confirmed. Pulakesin II also defeated Mahendra Varman I, but was later defeated and killed by Mahendra's successor Narasimha Varman I.

3. The Tamil Nadu area was ruled in turn, by the following:

Cholas	250 BC to 250 AD
Pallavas	300 to 900 AD
Re-emerged Cholas	850 to 1270 AD

The area of rule of the Pallavas may have been bigger than the area of rule of the Cholas. However, the later Cholas under Rajaraja I and Rajendra I were more prominent.

4. Chalukyas ruled in three phases:

Western Chalukyas (First)	550 to 760 AD
Eastern Chalukyas	630 to 970 AD
Western Chalukyas (Second)	975 to 1190 AD

The Western Chalukyas ruled over Maharashtra, and the Eastern Chalukyas over Andhra.

5. Rashtrakutas (meaning Country Lords) had their capital at Ellora. Krishna I subsidized the excavations of India's foremost rock temple, one of the architectural wonders of the world, the Kailasanatha of Lord Shiva.

26
THE BRITISH CONQUEST

It is a law of nature that all civilizations must finally decay and be replaced; Muslims were no exception. By the early 1700s, Muslims were bored and fatigued with their 500 years of continuous rule. There were no new challenges emerging; the Muslim zest for warfare and domination was all but lost. The great Mughal Empire started disintegrating after the death of Aurengzeb in 1707 AD.

That was an ideal opportunity for the Hindus to re-establish their rule over this woeful land. Hinduism would have got back a modicum of respectability; but that was not to be. In the early 1700s, there were three Hindu powers, which could have undertaken the task.

- *The Sikhs:* The Sikhs had emerged as a martial race by the first half of the 18th century, after founding of the Khalsa by Guru Govind Singh around 1700 AD. But their enthusiasm for the national role was curbed by nine invasions of Ahmed Shah Abdali, during 1748–61 AD.
- *The Rajputs:* In view of their martial credentials, Rajputs had the wherewithal to take over Delhi on the decline of the Mughal power. However, for some reasons, they showed no inclination and appeared to have lost their zest for warfare.
- *The Marathas:* They were the dominant Hindu power in the 18th century AD. The Marathas repeatedly defeated Muslim powers, including the Mughals, the Nizam of Hyderabad and the Nawab of Awadh. They over-ran Delhi twice and Punjab once. They had the power to install or remove kings on the (nominal) Mughal throne. However, due to their Hindu mindset, they hesitated to deliver the final blow and did not occupy the Delhi

throne themselves. In 1761 AD, Ahmed Shah Abdali put an end to their dreams of a Hindu empire.

With Hindus unwilling and unready to take over Delhi, a new religion (Christianity) and a new race from the distant land of Europe filed their claim to rule this ancient holy Hindu land. Even the color of the skin of these new people was different; and so were their mores and manners.

The Muslim invaders had come from lands that were 2 to 3 weeks' horse-ride away, with lots of resting places and looting opportunities on the way. In an emergency, the Muslims could always hope to get back, or get reinforcements. Muhammad Ghauri did exactly that after his defeat in 1191 AD. The Europeans came from a land that was nearly 6 to 8 months' ship-ride away, with no resting places or looting opportunities on the way. We may recall that in those days, there were only sail ships (no steam ships), no Suez Canal, no established sea-lanes, and no navigational aids worth the name. The ships had to depend on favorable winds and were often blown off course. There were no telephones, not even telegraphic facilities. In an emergency, it would take 6 to 8 months to send a message home, and even longer to get any succor. If they lost in the battle, they did not have the option to back home. In short, the odds against the European adventurers were the most formidable that can be imagined. Once on the sea, the Europeans had to depend solely on their wit and grit, determination and raw courage. The sheer scale of their adventurous spirit and audacity of their enterprise is breath taking. If they finally succeeded in taking over Bharat, they had earned it.

A number of European countries, medium to small, threw their hats in the ring. Of these, the three more important ones were Portugal (a tiny country), France and England. The navies of each of these countries had orders to sink the ships of the other two. The Portuguese were the first to come, but also the first to withdraw, leaving the field to the French and the British. Besides fighting the Indians, these two countries fought each other in India, on the high seas and back home in Europe. In addition, to show of raw courage in the battlefield, these powers showed exceptional diplomatic skills, in pitting one Indian power against another. But for these exceptional qualities there no way 'so few could have overcome so many, in such a short time'.

On landing in India, the Europeans were met with the dust and heat of India. England is freezing cold and almost dustless. Another welcome for the incoming Europeans was in the form of tropical diseases like malaria, tuberculosis, cholera and the like. Contracting any of these diseases meant almost certain death, as those days, medicines were most rudimentary, and antibiotics unheard of. Also, contrary to popular belief, by the time the Europeans came, not many riches were left in India. Muslims had been looting the country for 600 years. Even the decorated doors of temples had been carted away. Ahmed Shah Abdali had cleaned the till to the bottom. We may recall that in one raid alone, Abdali is reported to have carried the loot from Delhi on some 28,000 carts, camels and horses.

Finally, the British won the race; they were no sultans, or sovereign rulers. They were a bunch of traders, who had formed a private company to trade; and not to invade. On landing in India, they found Bharat ready and almost willing for a military take-over. Perhaps for the first time in world history, a private company raised a private army, which started inflicting defeats on sovereign powers of Bharat — both Muslim and Hindu.

Around 1750 AD, a lad of 17 years, Robert Clive was sent to India as a sort of civilian clerk. His ship was blown off to Brazil and he landed in India penniless, after about one year. However, this civilian clerk underwent a metamorphosis, to emerge as a military genius. No Indian General, Muslim or Hindu, could stand before him. In 1757, in the famous Battle of Plassey, Clive inflicted a crushing defeat on the redoubtable Siraj-ul-Daula, the ruler of Bengal. As per one version, Clive had only about 3000 troops — 900 British, 2100 Indians sepoys, against some 55,000 troops of Siraj. Reportedly, Clive suffered only around 40 casualties out of which some 10 were British. Sounds unbelievable! Clive, in addition to being a military genius, was a corrupt person; he amassed massive wealth.

The British went on to record an equally brilliant victory in the Battle of Buxar in 1764 AD. There they defeated the combined forces of Bengal, Awadh and the Mughal King Shah Alam II. Thereafter, there was no looking back. They went on to inflict defeat after defeat on various Indian rulers — both Muslim and Hindus. These included the Marathas, the Sikhs, Tipu Sultan of Mysore, the Nizam of Hyderabad. The Rajputs did not care to put up much of a challenge. The British could achieve these spectacular results by their superior skills. They outmaneuvered the Indians in all fields — military, diplomatic and manipulative. By 1800, the British (Company) power was well established in India. Following the 1857 'Rising', India was taken under the direct and formal control of the British Crown.

Slavery is the greatest curse for humanity, and it needs to be avoided — always and under all circumstances. During the freedom struggle, the Indians had to suffer a lot of British atrocities in the form of loss of liberty, livelihood, limb and life. We may recall the indiscriminate hangings after the 1857 Rising, the Jallianwala Bagh massacre and other such episodes. We salute the Indian martyrs and stalwarts who suffered at the hands of the British, and pay them our respectful homage. With these caveats, we dare say that the British rule in India was not a curse, which slavery normally is. The British gave to India many things that were of long lasting value.

To start with, the British rid Bharat of the Muslim rule of 600 years. It is doubtful if the Hindus would been have ever able to do that on their own. The Marathas were a dominant power in the 18th century, with a sort of suzerainty over Delhi. However, in keeping with the defensive Hindu mindset, they did not establish themselves on the Delhi throne. Even in the 1857 Rising, a tottering old Mughal king was put on the Delhi throne.

After the death of Aurangzeb in 1707 AD, India was split into innumerable states. Their number may have been near one thousand. It was the British, who united them all and created one India; this included the so-called 600 odd states that were under the effective control of the British. The British made Tibet a buffer state between India and China.

They also tried to define Indian borders by the McMahon line and the Durand Line (now on Pakistan); that showed their vision. Then, there were the railways, which helped strengthen the idea of one India. The two areas presently suffering insurgency are the North East and Kashmir, where the railways did not reach. The present Indian judicial and legal system and the administrative structure are British gifts to India. Then, there is the English language. Many of the stalwarts of the Indian freedom struggle like Nehru and Gandhi were the product of the British educational and value systems. The formidable Indian Armed Forces are another great product of the British. It was the British who brought to light some of cultural jewels of Bharat. These included the Indus Valley Civilization, the Ajanta/Ellora caves and even the Khajuraho temples. It was due to the British and their English language that the Vedas and Upanishads could become part of the world heritage. The British gave India a sense of history, which was traditionally lacking in the Hindus.

In early 1900s, Indians started their freedom struggle under the leadership of Mahatma Gandhi, Jawaharlal Nehru and Sardar Patel. Jinnah jumped into the fray to demand his pound of flesh in the form of Pakistan. Earlier, it used to be a Muslim General, who would outmaneuver the Hindus in the battlefield. Now, a single Muslim Jinnah out-foxed a galaxy of Hindu leaders on the round table; Jinnah walked away with his Pakistan. This ancient holy land of the Hindus was partitioned, perhaps for all times to come.

Section E
Free India & its Armed Forces

27
PARTITION OF INDIA

Over the centuries, the Hindus of Bharat have gone through n number of traumatic experiences. Even for such experienced people, partition of this ancient holy land was a great trauma. It was particularly painful, as the Congress leadership had led the Hindus up the garden path. Congress won a resounding victory in the 1945 elections on the solemn promise of 'no partition' — not now, not ever, not under any circumstances. Jawaharlal Nehru had called the very idea of partition a fantastic non-sense. He used to wax eloquent as to how the advocates of partition were living in a fool's paradise; not realizing that the boot was on the other foot. Mahatma Gandhi had averred that partition could take place only over his 'dead body'. However, within the space of 1 to 2 years and even before the push could come to a shove, both gave in without much of a whimper, citing the old Hindu excuse of abhorrence of violence and bloodshed.

There was a set pattern. Jinnah would threaten violence; Congress would cower, pleading in their *ahimsic* style — please, please, do not threaten violence; you know we cannot stand the bloody thing as we are worshippers of *ahimsa*. Over time, the Congress with its pusillanimous actions, imparted Jinnah a halo and an air of invincibility, which he did not have and did not deserve. Jinnah was a master tactician. He exploited the situation and achieved the unachievable. Later, Jinnah was to boast that he got Pakistan with a clerk and a typewriter.

At the most crucial moment of modern India's history, Mahatma Gandhi was Lord Krishna to Nehru's Arjuna. Nehru was vacillating and refusing to engage in *dharma yudh* (righteous struggle) to save the honor and integrity of the motherland. Lord Krishna's discourse in the *Gita*, was precisely for such an occasion. But the Mahatma did not assume the role of the great Lord, and did not live up to his high stature. If Gandhi was opposed to partition, he did not make an issue of it. It was customary for the Mahatma to go on a 'fast unto death' on relatively minor issues. However, he chose not to do so on this most crucial issue of 'life and death' for the country, and the people who loved it.

There were some riots, especially in Bengal and Bihar. A few hundred, perhaps a few thousand, people were killed. During their freedom struggle, the tiny state of Vietnam had willingly taken casualties of 2 to 3 million. During World War II, Russia suffered casualties of some 20–30 million. By those standards, the death of a million or two people would have been no big deal for a humongous country of the size of Bharat of 400 million people. We could have easily called Jinnah's bluff. The Congress could have told Jinnah that it was not afraid of his '*gidar bhabhkis*' (hollow threats); that it was willing to confront these, and where necessary, reply in kind. But that was possible only if we were not stuck in the *bhool-bhulayas* (blind and dark alleys) of *ahimsa* (non-violence), *shanti* (peace) and *satya* (truth). In any face off between the Hindus and Muslims, casualties were unlikely to exceed a lakh (hundred thousand), or two. That would not have been a high price to pay for the honor and integrity of the motherland. Only when men die do nations survive; that is one immutable law and lesson of history. The Roman civilization was born in bloodshed, was sustained and expanded in bloodshed and even withered away in Roman blood shed by the barbarian tribes. Civilizations have a close relationship with bloodshed — *choli daman ka sath*, we would say:

> *Freedom is not for free,*
> *there is a price to pay*
> *the price is in human blood*
> *sheep's blood has no say.*

The Congress lost its nerve over a few hundred, perhaps a few thousand causalities. It tamely agreed to partition. It was a question of holding on for a year or two longer. After World War II, the English were exhausted both in body and in spirit. They were in no position to hold on to India for much longer. The new emerging super power, the

USA, was putting a lot of pressure on them to let go of India. Jinnah was a dying man and the Muslims did not have a charismatic leader to replace him.

In days gone by, it was the norm for the Kshatriyas to surrender Bharat to the Muslims in the battlefield without much of a fight. Now, it was the turn of non-Kshatriyas to do the honor at a conference table, again without much of a fight. Be that as it may; Jinnah was and continues to be the real villain of India's partition. The Congress failed because it did not stand up to him.

The Congress agreed to partition to ostensibly avoid violence and bloodshed. But, a great irony and a twist of destiny; it was unmitigated violence and death that dominated the partition scene. Some half a million people were reported to have been killed, and about 18 million displaced on both sides. That proved the adage that God is not on the side of the weak and the pusillanimous. He favors the strong, the courageous and the brave. There is a not so well known Hindu *shloka*: *Veera Bhogya Vasundhra* — The Brave will enjoy the Earth. The results of the Congress actions remind us of the Urdu couplet.

Na khuda hi mila, na visale' sanam
Na idhar ke rahe, na udhar ke.

(We could achieve neither God, nor our beloved; we won neither on this count, nor on that.)

It is fashionable to blame the British for the partition of Bharat. The British were neither for, nor against partition; they had no stake either way. They were prepared to go along with the stronger stream. The UK Cabinet Mission was in Delhi for about 3 months to evolve a plan, which would have avoided partition. They even presented such a plan, which was summarily rejected by the Congress.

Partition was the direct result of the perennial Hindu fear of bloodshed, and over-emphasis on *ahimsa*. It is often expressed during TV debates that partition was a good riddance. In a united India, Muslims would have been about 33% of the population. Apologists of partition (mostly pseudo-secularists) express that there was no way the

Hindus would have been able to adjust with (meaning control) such a large population of Muslims. Internally, they feared that the Muslim would have even started dominating them, as they had done in the past. This was the 'Hindu mindset' at full play, a hark back to the days of Ghazni and Ghauri; all negative thinking, and pusillanimous actions (in direct contradiction to the teachings of the *Bhagwad Gita*).

Let us for a moment think what was the number of Turko-Afghans, Mughals and lastly the British, who successfully controlled the humongous population of the Hindus in their own homeland. And, they did it not for 50 years or so, but 800 long years. Is there a message in that? They could do it due to their respective mindsets, positive thinking, ruthless approach, raw courage and manipulative skills.

Yes, there would have been a period of conflict, including violence; may be even a lot of it. There could have been casualties of say one or two hundred thousand. But over a period of time, say 10 years, matters would have settled down. There would have been no Kashmir problem, which is bleeding us endlessly. The two communities would have evolved a modus-vivendi over a period of time. United India would have emerged as a super power by now. Other countries of the world, including China, would have shown much more respect. There would have been no wars of 1965 and 1971; and in all probability, no 1962 either.

Partition is now a living reality. Though Pakistan is the homeland for the Muslims, almost all Muslim major religious-cum-civilizational symbols are in India, e.g. Taj Mahal, Red Fort, Qutub Minar, Jama Masjid, etc. Pakistan has hardly any Muslims symbols of this class. Rather, it has some famous Sikh and Hindu shrines, like Nankana Sahib and Katasraj Temples.

The Deoband Muslim seminary is in India, and so are the world reputed Muslim shrines like Ajmer Sharif and Nizamuddin Aulia. The cradle of Pakistan was the Aligarh University which is in India. The Muslim *tehzeeb* (culture) is the Lucknavi *tehzeeb* of India. Lahore culture is akin to Jullundur Punjabi culture, *'balle-balle, shava-shava, sohinye te kuriye'* (a Punjabi saying which cannot be translated); Urdu is primarily a language of the Muslims and it is mostly spoken in UP of India. In Pakistan, Punjabi is spoken in Punjab, Sindhi in Sindh and Pashto in NWFP — Urdu is in a limbo.

Pakistan is (justifiably) proud of the Muslim kings and emperors who ruled (Hindu) India for some 600 years. These include names like Aibak, Khilji, Akbar and Aurangzeb. All of them lie buried in India in a series of tombs in the Delhi–Agra belt. Most 'alams and adils' (authors and poets) of Muslim literature like Mirza Ghalib, Malauna Hali lie interned in India. So is Sir Syed Ahmed Khan, the chief sponsor of Pakistan. Homeland is where the bones of your ancestors are interned; that is why the homeland is also called the fatherland or motherland. It is the bones of the ancestors which beckon and inspire the youth for great sacrifices for the motherland. Genghis Khan died some 800 years ago. The Mongols are now searching for his bones to relocate these at a place of honor, so that these can inspire the youth.

The above types of issues have started troubling the people of Pakistan; they are facing identity crises, though no one in his right mind in Pakistan would admit to it. No one can say what shape that identity search would take, in say another 100 years. All possibilities are open, and no option is ruled out.

28
THE INDIAN ARMED FORCES

Amongst the most prized Indian inheritances from the British were the Indian Armed Forces. They were a formidable entity — one of the finest in the world, with the highest standards and lofty traditions. That fine shape of the armed forces had not emerged out of a genii's lamp; it was the result of the British politicians' vision. A good percentage of British politicians had themselves served in the army. Others had a father or grandfather, who had shed his blood in the service of the motherland in some far-off colony. As a child, the British politician, sitting in the lap of his grandfather, had been hearing heroic tales of their exploits.

The British were past-masters in handling military issues. Within a short time of their arrival in India, they could motivate Indians to fight for them and lay down their lives. As early as the Battle of Plassey (1757 AD), Clive had some 2100 Indian sepoys (soldiers) with only about 900 British soldiers; and they won the day for him against some 55,000 troops of Siraj-ul-Daula. The story was repeated again and again. We may have to live with the shame that it was the Indian sepoy who helped the British enslave India. That is a tribute to the super and supreme motivational and manipulative skills of the British.

Over time, the British built an intricate and artful web of military value systems in India. An aura of grandeur was built around the military career, especially the officer class. In order to give the military status and respect in the social circuit and hierarchical order, all types and manners of concessions and incentives were extended to them. Innumerable welfare schemes for the soldiers were introduced. It was incumbent on the District Magistrate to go (on horseback) from village to village, enquiring about the welfare of military families and ex-servicemen. Such gestures, big and small, won undying loyalty of the soldier, who fought fiercely for the Imperial Raj.

Starting with that excellent base, India could have gone on to become a major military power. But, that was not to be; there were several reasons for that. We examine these reasons in the succeeding paragraphs.

Pre-1947, the Indian armed forces were (rightly) seen by Indians, as a symbol of British imperialism and a tool for their oppression. The armed forces often acted as a back-up when the police bashed up the Indian freedom fighters. The infamous Jallianwala Bagh was carried out by the British army; the indiscriminate and heavy firing was ordered by a Brigadier. Unfortunately, the oppressive character of the forces continued to linger in the Indian mind, even after gaining independence. Perhaps, an overnight shift from an oppressive to a protective role was difficult.

As opposed to the British politician, a typical Indian politician did not have had even a distant link with the military. Further, the new rulers of the country suffered from an overdose of *ahimsa*, which had become a part of their mental make-up; it was lodged in their subconscious. Most of them felt apologetic about militarism. There was a visible lack of enthusiasm about the armed forces in the political class. The general (unstated) approach was that we may have the armed forces, as we may not be able to do without them. To the great misfortune of the Indian people, even Jawaharlal Nehru fell in this category. Now, Nehru was a most prescient person of his times. He had not only read, but written about the military exploits of various civilizations, including the Greek and the Roman. The Indian people had great expectations of him; but these were all belied.

The matters were not helped by the British patterned life-style of the armed forces. In 1947, most of the senior Indian officers were Sandhurst trained, brought up in pucca British style. They ate sausages and ham, with fork and knife. They wore Saville Row suits and spoke in pucca British clipped accent; most took care to speak in broken

Hindi, lest they he taken for 'brown sahibs'. Pre-independence, it was the norm in the officers' messes, to refer to the Indian politicians in rather derogatory terms. These officers did everything in their power to project an isolationist attitude. The feeling was fully reciprocated by the politicians. They felt uncomfortable in the presence of smartly turned out officers with western manners and mores, and with a lot of spit and polish. The meetings between the politicians and military officers were short and crisp, with each trying to avoid an eye contact with the other. Sometimes, they met at parties; but for a nod or two, there was not much material for conversation. Both respected each other's space and privacy.

The matters were made worse by a series of military putches in the neighborhood of India. The khadi clad politician had already sensed a lack of respect from the 'brown sahibs'; the military putches only strengthened his suspicion. Now, preservation of his 'gaddi' (chair) is an instinct most natural to a politician. He, therefore, felt the need to keep a tight leash on the military. He found a willing ally in the bureaucrat, who occupied every position of power. The bureaucrat had the same objective for his own reason; the die was cast.

The Hindu masses at large dived in their memory lane. They could hardly recall an occasion in the last 50 generations where the 'famed' Hindu armies had saved this ancient holy land and its trusting people from the marauding hordes of invaders. Notwithstanding the hot tales of valor that we read in some books, the cold fact is that the (trusting) Hindu masses were always let down by their armies. It did not matter whether these invaders came from the Khyber Pass or from Europe, after a hazardous journey on boats. The Hindu subconscious threw up only tales of loot and plunder, mayhem and murder, dishonor and rape. Their temples were not only looted but torn down; every type and manner of ignominy and humiliation was heaped on them, not for a year or a decade, but for centuries on end. With such a subconscious memory, the average Hindu did not have much of an expectation from this new military of his, now in western attire. The overall picture may not have been of despair, but was not of much hope either.

With the early deaths of the Mahatma and Sardar Patel, Nehru was the only colossus left of the triumvirate. He ruled the hearts and minds of the masses. Nehru was a visionary and had a world view of issues. He had the necessary credentials to build up a new military ethos for this primarily Hindu country, now clothed in a secular attire. But Nehru did not evince much interest in military issues; one can guess three possible reasons for that:

- Every Indian politician of that era was cast in the *ahimsic* mould and it was not easy to break out of it; Nehru was no exception. Military values do not come easily to *ahimsic* people.
- At an early stage of India's freedom, the bug of world peace bit Nehru. He thought that he could build a niche for himself in this field at the international level, and thus, become a world figure. He took to sterile issues like non-alignment, and co-founded a club in that name, where a lot of 'gas' was generated, but with no ground results. His over-commitment to issues like non-alignment left him with no time to think of China and its plans. He was to ignore some very sound advice on the subject from Sardar Patel.
- Even with the secular mask on, there is a Hindu lurking behind. Hindus have this hypocrisy of thinking that they are the repository of all morality and have a monopoly on truth, e.g. *'Satya mev jayate'* is their national motto. India (read Nehru) started lecturing the world and teaching them moral principles. Motivated by the Hindu slogan of 'Vasudeva Kutumbakam' (chapter 14.5), Nehru evolved the famous *Panchsheel* principle with Chou En Lai; it was soon shredded into small pieces, which we are now trying to pick up from here and there.

The net combined effect of the foregoing factors and general disinterest of the new rulers was that the Indian armed forces were no longer the priority that they were for the British. The forces were initially kept at a low key, and soon began to be neglected. Every possible opportunity was utilized to push the Service Chiefs down the Protocol List; so much so that now they appear even below the Attorney General. Soon, the romance went out of the Service career.

When India became free, one of the first tasks was to evolve a new structure for management of the Defense set-up. Therefore, it would be relevant to enumerate the broad principles on which the defense structure of a democratic country needs to be built:

- In a democracy, there has to be 'civil' control over military; But the word 'civil' means political (and only political) and not bureaucratic.
- War is the most complex and specialized activity that a man engages in. What makes a soldier give his life (for the country) is an issue far more complex than even understanding the nature of God. Over the ages, millions have claimed to understand the concept of God. But those who understood the motivation behind the soldier's willingness to die would be in thousands, may be only hundreds. It is not claimed that all generals understand these issues, but some of them do. We have to identify them and bring them up. In short, issues of war have to be left to the generals. They must be listened to with respect directly by the politicians, not through the via media of bureaucrats. In any set up, lot of space must to left to the generals to plan and maneuver.
- Each cog in the defense structure which has some degree of power must have an equivalent amount of responsibility and accountability; and that must be defined in very precise terms leaving no room for ambiguity, and manipulability to escape responsibility.

All the above principles were violated with impunity in evolving the defense structure of independent India. In view of the luke warm interest (and lack of capability) of the politicians, bureaucrats took on the job. Of course, the politician reminded the bureaucrat of the principle of a 'tight leash'. The Indian bureaucracy (the ICS and IAS) has some of the finest brains of the country. They set about designing a structure in which generals were pushed on to the periphery, from which they could:

- Neither participate in any meaningful way in the decision making process.

- Nor protest over being excluded.

In other words, the generals could be seen (occasionally), but were not to be heard.

Service Headquarters were given the status of attached subordinate offices. An umbrella-type all encompassing Ministry of Defense (MoD) was created and put above these subordinate offices. All powers — organizational, financial and promotional — were concentrated in the hands of the bureaucrats at the MoD. The post of Defense Secretary was created, who soon enough assumed powers of an ersatz Chief of Defense Staff. The Defense Secretary and as indeed, the Joint Secretaries can walk in and out of the office of the Defense Minister several times a day. The Service Chiefs generally get to meet the Defense Minister only on weekly meetings.

With their close proximity to the politicians, bureaucrats in the MoD have the ear of the Defense Minister. Whenever they find him in a relaxed mood, they can always whisper a thing or two in his ear. The note for the selection of a new Service Chief is initiated by a Joint Secretary, in which he could cleverly build in the necessary biases. Defense Secretary would pen the final note in which he would, of course, keep the 'pliability' factor in view. The deputies to the Service Chiefs, called the Principal Staff Officers at Service Headquarters have no chance of interacting with the Defense Minister. Views of the generals cannot be conveyed to the Defense Minister, unless and until these have been edited, chipped and chopped by the Deputy and Joint Secretaries. An iron wall in the form of bureaucracy came to be built between the military and the politicians, between the Defense Minister and the generals.

In the foregoing, we have laid major part of the blame for the present state of affairs in the Indian defense set up on the shoulders of politicians and bureaucrats. We have, largely, spared the generals. That is not fair. Generals must also carry a good share of blame for their many acts of commission and omission. On the advent of independence, as stated earlier by us, the politicians gave somewhat of a cold shoulder to the generals. Far from being alarmed from that, the generals might have felt even a bit relieved. The might have argued, 'Let us talk to the politician through the bureaucrat; he speaks our type of English'. The generals showed a singular lack of vision in not appreciating that the politician controls every lever of power. The overall blame could possibly be distributed as follows: — One third each to politician, bureaucrats and generals — In India, politician is the '*mai-baap*' (all-in-all); he should be given 50% of the blame; the remainder 50% being shared equally between the bureaucrats and the generals.

At this stage, it must be stated to the credit of the bureaucrat that he only moved in the space which the generals were reluctant to occupy. By the time the generals woke up, it was too late; the bureaucrat was well entrenched and had the ear of the politician. If a general was to express even a mild dissent with the state of affairs, he could be branded anti-national. Something on these lines appears to have happened to a Naval Chief, who was sacked most unceremoniously.

Throughout the history of independent India, generals have generally failed to put up their point of view with the required degree of clarity and emphasis. Is it possible that we have failed to produce generals of the right caliber? Irrespective of the actual position, the prevalent belief is that we did produce 'good' generals. Whatever, the generals did not assert when it was imperative to do so, sometimes even in national interest. There could be many reasons for that — the rat race for promotions being one of the important one. It is not easy to disregard the 'goodies' that come with the post. Another reason could be the intense Inter-Service rivalry. That exists in all countries, including the USA which even has a 'Joint Chief of Staff'. However in India, the rivalry exceeds all limits, and is the most distinguishing feature of all Inter-Service interactions. That rivalry is not going to go away even if we appoint a 'Chief of Defense Staff', which in any case, would not solve any of the problems presently staring 'India's Defense' in its face. It would just add another cog to the wheel, and make the issues even more complex (However, it is a bigger question, and needs a separate discussion.)

One way to put the Armed forces 'in their place', is through means of Pay Commissions. As such, a decision was taken at a very early stage not to allow a General anywhere near the outskirts of the Pay Commissions. The sixth Pay Commission submitted its report in early 2008. By an ingenious thought process, it upset the long established equivalence between the various ranks of the armed forces, vis-a-vis the para-military. Among other things, it pushed the police DGP of a state to a higher level than a Lieutenant General (non GOC-in-C). That meant that the DGP of even the smallest state (with a police force of say 7,000), and DGP (Housing) of UP, rank higher than a Corps Commander, with 60,000 troops, guarding the most sensitive part of the Western border.

When a hue and cry was raised, the government kept on deliberating over it for more than a year at various levels, i.e. Committee of IAS Secretaries (no generals permitted), Group of Ministers. Whilst some Relief Packages were announced, the question of status of the Lt Gen was further complicated. Displaying exceptional ingenuity, combined with rare depravity, the Lt Gens were further split vertically. It was decreed that only ⅓rd of the Lt Gens will be given a higher grade; the rest ⅔ must rot at the 'Low' grade. That decree has created three tiers of Lt Gens (perhaps unprecedented in the world):

Lt Gen — GOC-in-C
Lt Gen — High grade (⅓rd)
Lt Gen — Low grade (⅔rd)

In the IAS, some 80–90% become Additional Secretaries, and about 60–70% full Secretaries. In the Armed Forces, only about 10% reach the level of Lt Gen. After that agonizing and hazardous journey, the Lt Gen is informed, "Please cool your heels in the low grade, till we get time to look at you." That can shatter the most committed and the most loyal.

There is only one level each of (full) Secretary, and Additional Secretary. Why shred the rank of Lt Gen, all for a paltry few lakhs (hundred thousands) rupees a year; or is there a deeper scheme? It is these 'low grade' Lt Gens, who as Corps Commanders are at the cutting edge of the battle. In the final analysis, it is their plans and push and daring that determines the difference between 'Victory' and 'Defeat'. No sane nation will put its generals with 'a grievance on their mind', to face the enemy in the actual battlefield. That is how the psychology and nature of war works, which as a nation we do not understand. A mindset which can think of this type of mischievous scheme (split Lt Gens into 3 grades) can do anything to destroy the cohesiveness of the Armed Forces.

However, under a lot of pressure, the issue was partially resolved in January 2010 i.e. after about 2 years; the generals gained as Additional Secretaries had to be accommodated.

In the present defense structure, whilst all power rests with the Ministry of Defense, they have no accountability worth the name. The present defense structure has evolved not on the sacred principle of 'national interest', but on the demeaning principle of 'power grab' and keeping the Armed Forces 'on the leash'. The disastrous results are there to see as we shall enumerate in the following chapter.

We may spend a few minutes to understand China's view of its military. Chinese communists established their rule in China at about the same time as India got its freedom. Contrary to the Indian scene, leaders of the new China were product of the 'Long March'; all of them had seen actual field action, and watched their comrades fall on the battlefront. We may recall that the British politicians have had similar experience. Once you have seen death at close quarters, your view of military, if not of life itself, changes.

Mao Tse Dung was a military genius, who actually enunciated new doctrines of war; he took control of the Chinese State. Mao did not need any briefings from generals; rather he briefed them on the conduct of war. For the first 30 years or so, the effective power in China rested with the military. That period was enough to weave and integrate military thinking in the national psyche. Later, when civilian elements were brought in, the armed forces continued to be a player in the decision making process, if not up-front, at least in the background. That is the difference between China and India. For the Chinese, militarism is a sort of religion; for India, it is just one of the routine issues (like say minority affairs), which the IAS must deal with in their day to day routine manner. That is why the Chinese could wrap ropes around us in 1962. But Indians would not like to recognize that fundamental truth. They would like to continue to believe that we lost in 1962 because we did not have proper boots, or some such other similar silly excuse. The Chinese PLA had fought battles of the Long March either in bathroom slippers or bare-footed, in the biting and intolerable cold of China — some 50% having perished during the march itself.

The moral of the story is:

- Forces must aim, plan and endeavor to fight with adequate numbers and adequate equipment.
- However, if necessary under any circumstance, forces must be (mentally) prepared to fight (successfully), with inadequate numbers and inadequate equipment. That is what nationhood and (good) generalship is all about.

Whatever we have stated in the preceding two paragraphs does not mean that the Chinese have any great military advantage over India; they don't. In the level of performance, no soldier in the world can measure up to the Indian soldier. All that India has to do is to incorporate changes in its mindset, and get its act together at the higher levels of military and civil leadership. If that can be done, the Indian Armed Forces can take on the 'best' in the world; let there be no doubt on that account.

In 2009, Admiral Sureesh Mehta was the Chief of Naval Staff and Chairman of the 'Chiefs of Staff Committee'. On 10 August 2009, he dropped a bombshell on the unsuspecting Indian public. In a televised address, the Admiral emphatically declared that India was no match for China, and that there was no way the yawning gap between the two could be bridged. What a public statement to come from the Chairman of the 'Chiefs of Staff Committee'. Even if there was an element of truth in this, the top Defense functionary of the land should be the last person to say that in public.

If such a situation actually exists, it has to be the result of gross all-round neglect (political, bureaucratic and military) over a 20–30 year period; it could not have emerged in a year or two. The Admiral (actually the General) should have briefed the Cabinet Committee on Security in the close confines of the War Room; not a word to come out. And that should have been done within a few months of the Admiral taking-over, and not a few days (yes days) before his retirement.

The public statement has achieved merely the following:

- It tries to project China as a huge bug-bear, whom we must fear; this is an admirable self-goal. China's projected military advantage is largely a product of our 'Defeatist' mindset. Even the tiny Vietnam is not as afraid of China, as our public postures (including TV debates) make India to be. India is a huge country with humongous resources; it is no push-over. We do not have to keep harping on China; let us set our sights at a different level. What is needed is a change in mindset in the higher echelons of governance, as well as in society.
- It has dealt a mortal blow to the morale of the Indian soldier. Earlier, we had blamed the politicians and bureaucrats for not understanding the concept of 'military morale'. We have to now admit with great regret (and some shame) that admirals (generals) may be doing no better.

The 'earth-shaking' (only our perception) statement of the Admiral evoked no response from the government, or the ever watchful, ever alert media. Even the main opposition party, the BJP, who claims a monopoly on nationalism, just looked on half in agreement, half in bewilderment; there was not even a whimper of reaction from them. Perhaps, they, like the government, did not understand what the Admiral was trying to say.

If the Admiral was even partially correct, the government should have been in a 'tizzy', and latched on to this one single issue. If the Admiral was wrong, the government should have clarified the situation at the level of the Prime Minister. The PM spoke twice in Parliament on a non-issue like Balochistan (just mention of this name in an irrelevant piece of paper, called joint statement). Why the deathly silence on this 'life and death' issue for the nation? This is indicative of the 'sickening' low level that the defense issues occupy in this woeful land. But, who are we to complain? There are very wise and prescient men in charge of the nation's destiny. We should mind our own business, and just shut-up (which we dutifully do, except for making one last comment below).

Perhaps for the first time in the history of modern democratic nations, the top Defense functionary has informed the nation in advance of (almost) certain defeat (to us, the word 'no match' does not permit of any other interpretation; we use the word 'defeat' with utmost reluctance, and with a sense of horror). Lack of any political response, not even a 'twitter' or 'tweet', would appear to suggest the following:

- There is nothing new about it; the political class always knew about it.
- If 37 years after learning our lesson in 1962, we are still 'no match', why keep on pumping more and more money in Defense?

The political class appears to be of the view that we may not unnecessarily worry our head about this 'no match' issue. If things go wrong in any conflict, they can always explain things away. India has a ready-made set of excuses for defeat; those can be presented to the nation. Indian people are simple and trusting. They are neither prone to, nor known to object to what the government tells them. Remember, they lapped up every word that the government told them about the 1962 debacle, i.e. that the (shameful) defeat was really not our fault; some one else must have been responsible. Most Indians are still heard muttering in their sleep, 'Oh! We were defeated in 1962 due to lack of boots/socks, deficiency in equipment, may be also some shortage of food.' Others are convinced that we were defeated because the Chinese were just too strong. That has been our hallowed tradition from the days of Ghazni and Ghauri. The people never held the rulers accountable in the past; why would they do it now? Indians are a decent set of people; they do not believe in making unnecessary trouble; they are the accommodating type.

The politicians would add that the Indian public knows and appreciates that they (the politicians) are fully busy (both waking and sleeping hours) in attending to the issues of terrorism. Where is the time for them to attend to the Defense issues? As soon as they find some leisure, they would devote time to Defense; *Insha-Allah* (God willing) that should happen soon. Why make all this fuss and noise? We must have patience — another great Hindu virtue!

29
SOME INDIAN CONFLICTS

In the preceding chapter, we have studied the management structure of the Indian Armed Forces of independent India. Having done that, it would be relevant to judge the performance of the armed forces in the conflicts that India had to face post 1947.

During actual hostilities, it is customary for countries to project a favorable view of the war effort and of forces engaged in combat. Many a time, victories are claimed where they may have been nothing more than a stalemate. Rout of the enemy forces are reported, where there might have been just a tactical withdrawal. Often enough, tales of heroic deeds are exaggerated; sometimes, even fabricated. All this is done in the belief that it helps keep up the morale, both of the armed forces and the civilian population. It also helps stir up nationalist sentiment. This is a legitimate viewpoint.

However, it is essential that we do not become victims of our own propaganda. The 'cold and hard' facts of every conflict, including the most unpleasant, must be faced:

- Internally by the armed forces, immediately after termination of hostilities, say within one to two years.
- By the nation as a whole, after a decent interval of time, say 5 years or so.

In our study of conflicts, we will deviate from the normal practice of projecting a rosy picture of events, and sweeping the unpleasant under the carpet. We would confront the 'cold and hard' facts and bitter truths of battles. It is only by doing so that we can:

- Identify our weaknesses, and then
- Initiate measures to overcome these

We are aware that by adopting the above procedure, we would face criticism from people who would claim to be well-wishers of the country. We may even be accused of denigrating our gallant armed forces; we have no such intention. We take that risk in our firm belief that it is only by facing unpleasant facts that we can safeguard the long term interests of the armed force, and of the country. With these caveats, we will study the four major conflicts that India faced after independence:

India vs China — 1962
India vs Pakistan — 1965
India vs Pakistan — 1971
LTTE Interlude — 1987–88

Before we proceed further, we gratefully acknowledge the innumerable deeds of courage and bravery of the Indian soldier in the above and other conflicts. Nowhere in the world are there soldiers as gallant as the Indians of all races, color and religion. We also record our respectful homage to all Indian martyrs in wars and conflicts. The reader is requested to keep these caveats in view whilst going through our comments on the various conflicts in the following paragraphs.

29.1 India vs China — 1962

The first major test of the Indian armed forces came in 1962 when India was involved in a border conflict with China. In 1914, an Englishman, Sir Arthur Haney McMahon tried to define the border between India and Tibet (China) on the highest watershed principle. The effort was only partially successful, as the central Chinese government of that time did not ratify the agreement. In the late 1950s, the border dispute between India and China (who had incorporated Tibet) started simmering. Some border posts were set up by the Chinese; India considered it as incursions in Indian territory.

Around October 1962, Jawaharlal Nehru gave a public statement that he had asked the armed forces to get the offensive posts vacated. In the event, it appears that China took the initiative. Before the Indians could act, the Chinese attacked over the Eastern border. Skirmishes also occurred in the Western (Ladakh) region, where the Indian troops gave an extremely good account of themselves.

But in the East, the Indian army, for some inexplicable reason, failed to offer any credible resistance. There were unconfirmed reports of battalions and even perhaps a brigade, giving up their positions (hard facts are difficult to come by). The Chinese forces advanced with extraordinary ease. It was not the defeat, but the manner of defeat which was most humiliating. Matters were made worst by the Chinese declaring a unilateral ceasefire on 21 November 1962; the Chinese withdrew to their original positions.

The Indian nation was staggered beyond belief; no one had imagined that such a situation could develop. The great visionary Nehru himself was forced to declare that they had been living in a dream world of their own making. Nehru could not survive the shock, suffered a stroke and died in 1964.

After the great debacle, the market was awash with books mostly written by (defeated) generals and Intelligence top brass, whose failures in the first instance had resulted in the disastrous situation. Their first (in fact only) priority was to blame everyone else, except themselves. There was a liberal use of words like 'if' and 'but'. Over the centuries, Indian (read Hindu) commanders never learnt the basic lesson that 'victory' speaks for itself and does not have to rely on 'ifs' and 'buts'.

Every type and manner of imaginary and untenable excuses were trotted out for the defeat and the humiliation, e.g.:

— Lack of Intelligence
— Lack of acclimatization
— Shortage of equipment; proper (winter) clothing and boots were stressed
—Inadequacy of all types and manner of resources

The (trusting) Indian public was led to believe that the troops could not fight due to inadequacy of equipment; till date (2009), most Indians believe that that was the case. That it was not so is proven by the fact that the Indian troops fought well on the Western front, where the winter was much more severe.

At this stage, it would be relevant to record some views of Napoleon Bonaparte, the great French general. Napoleon was made a major-general at the age of 26, and given command of some 40,000 French troops, one of the most ill-equipped army of those days. Napoleon was asked to conquer Northern Italy, which France had been trying to occupy unsuccessfully, for a century or so. There were some two hundred thousand Italian and Austrian troops in Northern Italy at that time. Napoleon addressed his troops with words somewhat on the following lines — 'I know you have neither food nor clothing, nor boots; but, we are going to win in any case'. (These are not his exact words, but only convey the sense.) Napoleon went on to conquer Northern Italy with that army. The moral of the story is that generals may fight many times with adequate equipment; but sometimes they may have to do that with inadequate equipment. That is the nature of war, which must be won under all circumstances.

The actual reasons for the 1962 debacle were:

- Failure of higher direction and control at Army HQs and Ministry of Defense
- Almost total failure of generalship at the field level
- Failure of the troops to do what they are trained and expected to do, i.e. 'stand up and fight'.

The writer is aware that he would be criticized and pilloried for writing the last factor above, which would be projected as a reflection on our gallant soldiers and an effort to break their morale. The writer, however, believes that there must come a time in the history of nations when they must stand up and face 'cold and hard' facts, howsoever unpleasant these may be; that requires courage. The remedial actions can start only after such an acceptance of reality.

Understandably, following the debacle, there was turmoil in India; some generals, including the Chief of the Army Staff, were eased out. The Minister of Defense was asked to go. But care was taken not to touch any bureaucrat at the Ministry of Defense. An inquiry was undertaken;

but its report was kept under wraps. Soon everything was forgotten and things fell into the earlier easy groove.

The 1962 fiasco was a failure of fundamental and basic nature. It should have called in question everything connected with the armed forces. There was a requirement for a change in the very mindset of the armed forces and its controlling establishment. Wholesale changes of management structure, procedures and training patterns were called for. That is easier said than done; so, we are where we always were, i.e. nowhere. It would be a monumental error to think that we learnt any lesson from the 1962 debacle.

29.2 India vs Pakistan — 1965

The Indian Armed forces were next tested in 1965. Here, Pakistan took the initiative, and on 1 September 1965, launched an armored thrust in the Jammu area and made substantial advances. India responded by opening the Lahore front.

At the very outset, we would like to record that the Indiam Army fought bravely and most courageously at several fronts. It recorded many victories, out of which two important ones were:

— Capture of Haji Pir Pass in the Kashmir Valley
— Battle of Asal Uttar on the Lahore front, in which large numbers of Pakistani Patton tanks were destroyed.

However, our aim in this book is to examine the weak areas (if any), so that remedial actions could be initiated. A chain is as strong as its weakest link. Our this effort may not be, in any way, construed as a reflection on the bravery and courage of our gallant soldiers, which are exemplary and unmatched.

Somehow, the Indian public was led to, or came to believe that the Indian army would have breakfast of *paranthas* and *lassi* (Indian bread and buttermilk) at Anarkali (a famous area) in Lahore; that was not to be. The Indian army was bogged down at the Ichhogil Canal, which proved to be a formidable Pakistani defense line. That was a bit of an embarrassment, keeping in view the numerical superiority of the Indian army, i.e. a ratio of some 2.5:1. Powers like USA and USSR intervened, and a ceasefire was agreed upon on 24 September 1965. Both sides had captured territory and claimed victory. Pakistan annually celebrates 1965 as their victory. The Indian public was led to believe that it was India's victory; be that as it may.

Lt Gen Harbaksh Singh, an outstanding Indian General was Commander of the Indian troops in the Western Sector. In his book *War Dispatches*, he writes as follows about the 1965 Indo–Pak War (emphasis added):

- On the first day (i.e. 1 Sep.), one infantry battalion fell back after a largely *imaginary* Pakistan counter-offensive.
- Another (battalion) gave way and *broke line*, was reformed, and then broke again.
- By the afternoon of 7 Sep., in the 4th Mountain Division, two and half battalions of six had *abandoned their defenses*.
- One Division Commander just escaped capture.
- By 19 Sep. in one brigade 'all control at battalion and brigade level was lost and *formation ceased to be a cohesive force*'.
- There was nothing wrong with the machinery; *the men manipulating it were found wanting.*

In 1965, senior Indian politician YB Chavan was the Defense Minister. He was appointed to the post after the great 1962 debacle, to pull the Indian armed forces out of the morass in which they were found in 1962. He kept diaries of the 1965 war. Based on these diaries, his one time Secretary RD Pradhan has written a book *1965 War — The inside Story*. The book brings out some unpleasant facts. It endorses Harbaksh's view on the 'break ups' in the Fourth Mountain Division, mainly due to desertions. It goes on to say that commanding officers of two Battalions lost their mental balance. The 161 Artillery regiment (part of No. 10 Division) deserted en-masse, leaving their guns, ammunition and vehicles behind. When ordered to bring these back, they fed a largely false story.

The Ichhogil Canal fiasco needs to be described in some detail. After partition, Pakistan built the Ichhogil Canal a few kilometers away from the Indo–Pak border. It is a public belief that in 1965, a thrust to

Lahore was a priority objective of the Indian Army. That would have required the crossing of the Ichhogil Canal in large numbers, say by at least one Corps, i.e. some 60,000 troops. Pak troops in large numbers must have been waiting on the other side of the Ichhogil to give a 'hot' welcome to the Indians. However, the Indian troops got bogged down at Ichhogil. Only two companies of the 3 Jat Battalion could cross the canal on 6 Sept., i.e. say about 600 men, instead of the 60,000, or so. Crossing in such small numbers was like walking into a death trap. The two companies had to be called back in a hurry. Indian troops were not to cross the Canal again, because of confusion at 15 Division HQs, and laxity by I Jat (Battalion), as stated in Pradhan's book.

There were more worrying issues. General JN Chaudhari was the Chief of Army Staff (COAS). About him, based on Chavan's diary, Pradhan writes as follows:

- Sept 2: COAS — looked 'somewhat depressed'.
- Sept 4: COAS came at 8.30 PM — 'He was rather depressed'.
- Sept 9: COAS was 'somewhat uncertain' of himself.

Pradhan goes on to record that on 3 Sept., the COAS told Chavan that he was sorry that he was somewhat depressed yesterday (i.e. on 2 September). On 10 Sept., Chavan noticed and recorded that the COAS was more (?) sure of himself — not fully sure, only 'more' sure. Can you imagine the head of an 'army at war' being 'depressed' or 'unsure of himself'? And, what does the Defense Minister of the 'country at war' do? He just records the fact. The book does not say if Chavan discussed the issue with the Prime Minister, or even mentioned it to him. There is nothing to suggest that Chavan considered replacing the COAS, even temporarily.

That was not the end. There is another (unbelievable) story, which runs as follows. At an early stage of war, the Chief of the Army Staff (COAS) Gen Chaudhari asked Lt Gen Harbaksh Singh to withdraw the Indian forces behind the river Beas. That would have involved abandoning a major part of Punjab, including the most holy city of Amritsar to the Pakistan Army. In his book, Harbaksh describes the episode as follows;

"Last night on 7 September, the Chief of Army Staff rang me up to say that he had read XI Corps Commander's letter sent to me and his advice was to save the whole army from being cut off by Pakistan's armor push, I should put (pull) back to the line of river Beas. I was aghast at this suggestion and said that since it was a tactical order, he had to come to the front with me to give it, or else he had to issue an operational instruction, as is the custom."

Later, Harbaksh and the COAS had a heated discussion on the subject at the Ambala airfield.

Pradhan confirms the above episode by writing that Chavan 'heard' that someone at the top at Army HQ had suggested withdrawal to the Beas Line. Chavan does not appear to have made any effort to confirm what he had heard, or to check with Gen Chaudhari. What a novel way to run a war! Withdrawal to the Beas Line would have been a humiliation worse than the 1962 debacle against the Chinese. That such a course of action was suggested at the very top level would appear to suggest one or more, of the following:

- A very questionable state of mind of people in charge of the Indian war machine
- A defeatist mindset; an inability to understand the meaning of country's 'honor'
- We did not learn even an 'iota' from the 1962 debacle.

In August 1965, India for a change was in a mildly offensive mood in Kashmir. On 26–27 August 1965, India occupied a hill feature called the Haji Pir Pass across the cease-fire line, in Pakistan occupied Kashmir. Pakistani response was imminent and expected at any moment. It came in the form of a severe artillery barrage in the Chhamb–Jaurian sector of Jammu, at 0330 hrs in the early morning of 1 September 1965. India says that it was taken by surprise; no one can help those who always want to be taken by surprise. By daybreak or so, Pakistan armor thrust was in full swing. Let us examine the speed of the Indian response.

Chavan records in his own hand in his diary:

1 Sep

"By 4:45 p.m. COAS and CAS — came to me for orders for use of the Air Force in Chhamb Sector against tanks.

Had no time to consult ECC[*] or prime minister — Took decision on their advice and asked them to go ahead."

That means for about 12 hours, no air strikes were launched against a full-fledged armor attack on the country. If we had a functioning system, the air strikes could have been launched within about two hours (maximum), from the next door air base of Pathankot. As usual, Air HQs were waiting for government instructions. What instructions are required for strikes against advancing enemy armor columns inside your own territory? It appears that Air HQs and Army HQs are some sort of 'Non Government Organizations (NGOs)', always waiting for some 'government' instruction, or the other. Even when no such instructions are required, they will insist these are required.

Chavan records that he was informed of the attack at 1100 hrs, whilst on a visit to a hospital. We have no way of knowing why Chavan could not be informed, by say 0900 hrs; we were in the second half of the 20th century and had functioning telephones. Chavan took two hours to return to office by 1300 hrs, as recorded by him. Surely, the ECC could have met by 1330 hrs. Rather, at 1645 hrs, Chavan records 'no time to consult ECC'. These were office hours, and ECC members were within 2 to 5 minutes of walking distance of each other. Beyond office hours, they are within 2 to 5 minutes of driving distance. What were the members of the ECC busy in, when India was under formal attack? In war situations, minutes matter, hours can be like an eternity. Actually, all authorizations should have been given days earlier, as the Pakistani attack was expected and imminent. But that would have been against the very concept and spirit of the 'Indian Defense System'; we must get 'surprised', always and under all circumstances. We had been surprised earlier in 1962, and were to be surprised later in Kargil in 1999.

In the midst of full-scale operations, Chavan records as follows on 6 September:

---

[*]   Emergency Committee of the Cabinet

"We are not a war minded nation; and I think I am proud of that. Yet—"

That gives us a peek in the Hindu mindset on military issues. When under frontal attack and at the height of military operations, our prime thought goes out to our status of a 'non war-minded nation'. How can you fight a war with such a mindset? The one elementary lesson of history is that it is the 'war-minded' who have dominated and ruled the world; the 'non war-minded' have just served at their table. Why should a humongous country and a great civilization like India wish to be in the latter category? War and victory are mostly a play of mindsets. May we ask a rather 'cheeky' question? When Chavan talks of 'not a war-minded nation', does he include the Indian Muslims in that? Most Muslims may disagree with such a formulation. Muslims have gained a lot by being 'war-minded'; they ruled over this country for 600 years. Then, they walked away with a sizeable chunk of this holy land. It was the 'war-mindedness' of Jinnah which resulted in Pakistan. Why would they give up their calling?

Chavan was a Maratha. Amongst the Hindus, the Marathas were the most 'war-minded' people (see chapter 24). Shivaji challenged the great Mughal Emperor, Aurengzeb; his son Shambhuji heaped insults on the mighty Emperor, right in the Emperor's court. The Marathas were always on the look-out for offensive actions, and captured innumerable territories. It was only during the reign of the Marathas that Hindus lived with some dignity and pride. Why should Chavan wish to give up that great legacy? There is nothing in the Hindu scriptures which informs us that Hindus are not 'war-minded'. We will cover that subject in some detail in chapters 40 and 41.

Pradhan also records an extract from a purported secret order dated 29 August 1965 from the Pakistani President to his Chief of Army Staff, which has the following line:

'3. As a general rule Hindu morale would not stand more than a couple of hard blows delivered at the right time and the right place.'

We let the above line pass without any comment.

The above type of happenings should have called for a revolution in Indian thinking about the concept of war, and the ways to engage in it. After decades of training and indoctrination, the thought of even one company 'breaking line' is horrifying (authorized withdrawal is different). Here, whole battalions appeared to have done that. Military thinkers, strategists, tacticians, nationalists (who instill nationalism) and psychologists (to remove fear of death) should have got cracking. We had been taught a lesson in 1962, and supposed that we even learnt it; at least, that is what the (trusting) Indian public was led to believe.

Of course, the Indian troops engaged in fierce fights on many fronts, and gave a good account of themselves. However, the overall performance of the Indian Army in 1965 was well below their potential, all due to a defensive mindset and lack of an aggressive attitude and 'Killer' spirit. Nations require these latter attributes if they have to record victories in war, and live with dignity.

29.3 India vs Pakistan — 1971

16 December 1971 was the finest hour of the Indian Armed Forces. The Pakistan Army was decisively defeated. 93,000 POWs had been taken — perhaps a world record for one theater of war. Pakistan was vertically split into two. The Pakistani *ummah* (Muslim public) was stunned beyond belief; the 'unhappenable' had happened. The stories of Ghaznis, Ghauris and Abdalis taught to Pakistani school children were rudely interrupted. They would have to add a new (unpleasant) chapter to their history books. The Pakistani *ummah* would have to adjust to a new belief that even a *kafir* (infidel) army can win.

Lest the victory goes to our head, we may note the special circumstances in which the 1971 war took place. These were:

- The Pakistan army was isolated and completely cut off from its supply base in West Pakistan. The Indian Navy had effectively blockaded the oceans.
- The Pakistan army in East Pakistan was highly outnumbered, having only about 93,000 troops, with no scope of any enforcement. If required, India could have easily put say 4 lakh (hundred thousand) troops, out of its army of 10 lakh.

- East Pakistan did not have an Air Force of any consequence; India had complete dominance of the skies.
- The local Bengali population was very hostile to the Pakistan army.
- For a change, the Indian political leadership was outstanding, resolute and ruthless (a highly desirable attribute).
- The Indian army had Generals of caliber who clicked with each other.

There was no chance of losing for an army having so many factors in its favor. So, let us not draw any long-term conclusions from 1971 that the Indian Army has turned the corner and is on a winning spree. We have still to travel a lot of distance on that road, as the next episode with the LTTE shows.

The taste of the pudding of the '1971 victory' was spoiled by a controversy that erupted in 2008. There is a village called Longewala near the Indo–Pak border in the desert of Rajasthan. A very large number of Pakistani tanks were destroyed during the 1971 war in this sector. That was generally considered to have been done by the 'Hunter' aircraft of the Air Force operating from Jaisalmer, situated near Longewala. However, the Army claimed that a major tank battle took place at Longewala, and a number of 'bravery' awards were given for that engagement. Though a controversy had arisen in 1971, it soon died down without any damage to anyone's reputation. Matters rested at that for 37 long years. However, in early 2008, for some inexplicable reason, Army HQs took the Defense Minister to Longewala to demonstrate to him how the 'great' battle of Longewala was fought. A retired major-general, who (as a major) was an Air OP pilot in the Longewala sector, gave a statement generally implying that no tank battle was fought at Longewala, and it was merely rehearsed on a sand model. That was endorsed by a retired air marshal; he, as a wing commander, was commanding the Jaisalmer air base in 1971, from where the Hunters had operated and pounded the Pakistani tanks. Incidentally, a book of a retired brigadier of the Pakistan army was published at about that time. The brigadier, as a junior officer, was in the Pakistan tank force at Longewala in 1971. His version also appeared to support the version of the major general. We offer no comments on the episode.

29.4 The LTTE Interlude — 1987–88

The LTTE (Lanka Tamil Tigers Elam) problem in Sri Lanka had been simmering from the early 1980s. In July 1987, the Sri Lankan President signed an agreement with Rajiv Gandhi, Prime Minister of India. Sri Lanka agreed to accept Indian troops to help maintain peace in the North and East of Sri Lanka, and aid in the process of arms surrender by the LTTE.

India put in some four divisions of troops in Sri Lanka, starting August 1987. The general belief among the civil population in India, especially Tamil Nadu, was that Indians had gone to Sri Lanka to help their Tamil brethren. However, events took a dramatically different turn; full-scale hostilities broke out between the Indian army and the LTTE. The clashes were bloody and there were plenty of Indian casualties (as there were of the LTTE). Instead of peace, what developed was a climate of conflict and war. After a stay of about two and a half years and without achieving any of the objectives, the Indian forces withdrew; it was not a glorious exist. The mighty Indian army could not discipline a rag-tag guerilla organization called LTTE. Our comments are based mostly on newspaper reports.

29.5 The Conflicts Summarized

Let us try to summarize the performance of the Indian Armed Forces in the four major conflicts outlined in the preceding sections. We would do that by giving each conflict marks out of 10.

Conflict	Performance	Grading	Remarks
1962	Dismal	0/10	—
1965	Stalemate	5/10	In view of Indian numerical superiority, stalemate was an embarrassment.
1971	Excellent — Superb	10/10	A bit liberal marking; all factors were in favor of India.
LTTE	Inadequate	5/10	LTTE was a rag-tag force.
	Total	20/40	or 50%

In a rather arbitrary manner, we would like to describe what a particular percentage marking could mean in terms of Performance:

90% Excellent
80% Good
70% Passable
60% Deficient
50% Highly Deficient
40% Gravely Deficient

We see from the above table that the actual field performance of the Indian Armed Forces taken as an average of four conflicts has been 'Highly Deficient'. The writer is fully aware that such things are not to be expressed in public; he may invite criticism and ridicule for doing that. He may even be labeled as anti-national for lowering the morale of the armed forces, and as such of the country.

It is felt that there should be two stages of any conflict. During the actual hostilities, all types of claims, often exaggerated can be made, about the field performances and acts of bravery. However, after a decent interval of the conflict, say 2 to 5 years, actual 'cold and hard' facts of war must be faced in all their nakedness. However, that seldom happens. Authors of propaganda meant for public consumption, start believing their own words, and everything is left largely as it was.

Earlier, whilst considering the 1962 and 1965 conflicts, we have recorded cases of battalions, even perhaps a brigade 'abandoning position' or 'breaking line'. Efforts were made to project those as a consequence of the lack of equipment, or something similarly silly. People-in-the-know always knew that the reasons were different — of fundamental and basic nature. But few, perhaps none, had or has the courage to state these reasons in public in the mistaken belief that it would effect the national morale.

The writer is of the view that in the light of totality of circumstances, persons even remotely connected with the defense of this holy and ancient land should be a very worried lot. The list of such people would go much beyond the generals; it should include the following:

— Generals in uniform, present and upcoming
— Bureaucrats in the Ministry of Defense
— Other high-ranking bureaucrats including the Cabinet Secretary, Chiefs of IB & RAW, Home Secretary and Others
— Civilian experts (actual, not pseudo) in the defense field who might have acquired their expertise, by doing stints in the Ministry of Defense, or otherwise.
— Retired Generals of the intellectual type, who write books on defense and pontificate in TV studios on defense matters.
— Clutch of politicians across the political spectrum who have the caliber to grasp defense matters. Presently, there are some 8 to 10 young MPs, which may include a future prime minister. They must concentrate on defense as a special subject.
— Anybody else who has the good of this ancient holy land in his or her heart.

If India ever embarks on a project for a major upgrade of our 'Defense Set-up', it would be the most difficult task that we may have ever faced in our history. It will be a task more difficult than edging out Buddhism and bringing back Hinduism, which took us 800 years.

We have another specialty. As the real issues are too formidable to tackle, we pick small matters, of no great fundamental consequence. This can perhaps be illustrated by some examples, viz.:

• Listen to any Defense debate. An issue always stressed is the shortage of some 10,000 officers in the army. Yes, this is an issue. However, it was not shortage of officers that resulted in the 1962 debacle and the types of issues listed by us under the 1965 conflict. The causes were different and of fundamental nature, which we are not willing to confront.

• Some 15 years back, under the onslaught of the feminine brigade, the Army allowed recruitment of women. Lately, there are periodical reports of alleged sexual harassments. In 2008, an officer of the rank of Major General was court-martialed. TV debates are organized in which every 'feminine cause' pusher emphasize on the great merits of having women in the army. Now, in spite of what feminists may say, women are physically different (we are not saying inferior) from men. That is why we have different teams for hockey, cricket, etc for women. Even in a docile, largely cerebral game like golf, women compete only amongst women. And the army happens to be a little bit about physical attributes. If we want, we can reserve 100% of space in Information Technology for women, and up to 75% in Parliament. But, can we please treat the army on a different footing? The Army is about winning wars, and not about proving feminine concepts, or empowering women (which we all want to do). Any step which affects the 'chances of a win' even by 1 part in a million should not be acceptable. And in the view of the writer, 'women in the army' fall in that category.

- A new panacea now being recommended for the Indian Defense ills is the creation of the post of the Chief of Defense Staff; that is just a red herring. It will not address any of the basic problems facing the Indian Defense structure; may be, even make things worse (However, this is a larger issue and needs to be debated separately.)

Will we ever have the inclination, time, space and courage to confront the real core issues of the Indian Defense System? The simple answer appears to be 'No, Never'. But, there could be, hopefully, a more complex and favorable answer.

29.6 India's China Syndrome

Caveat: India and China are neighbors, both aspiring to become global powers. The two can go on their aspirational route without having to step on each other's toes. Nothing that is stated below may be taken to infer that the conflict between the two is unavoidable, or likely.

Presently, India suffers from the China syndrome. In TV debates on China, one sees a galaxy of retired diplomats and generals, and other busy-bodies; they are undoubtedly a set of most prescient men of India. What happens in TV studios is series of tired monologues symptomatic of a nation without any iron in its soul. The primary emphasis is to project China as some sort of a super military power. All sort of dooms-day scenarios are painted for India. That is just a manifestation of the Indian (read Hindu) 'defeatist mindset', in this case linked to the 1962 defeat.

The root of the 1962 debacle lay in the failure of the Indian army to put up a fight against the Chinese in the Eastern sector. As that was too shameful to be publicly admitted, a propaganda war was unleashed to project China as a vastly superior military machine, against which we could not have succeeded, even if we had tried. Attributes were assigned to the Chinese army, which it did not posses. The following types of deceptively disguised statements were let out:

- Chinese came in wave after wave — what could we possibly do?
- Chinese had vastly superior armaments.
- Their Generals out-foxed ours.
- We were shivering with cold.

The media, always hungry for news, picked up the theme and started playing it around, with a degree of vehemence. At the same time, (defeated) generals came out with a series of books, plugging the same line, i.e. projecting the Chinese army as a 'super human' one. The Government refused to give the authentic version. Even after about half a century, the official Henderson Brooks Inquiry report is under wraps.

The bitter truth is that the above types of statements were substantially untrue; in fact, most were patently false. There is no evidence to support the oft-repeated theory that the Chinese came in waves. It is highly unlikely that the Chinese had an overwhelming numerical superiority. For all we know, overall numbers might have been even in our favor; at least should have been. We were fighting in our own backyard.

However, the general public had no option but to believe the make-believe stories told to them by the media and the books penned by the generals. The result was that over time, the concept of the (perceived) superiority of the Chinese army got lodged in our subconscious. It was in keeping with that mindset that Admiral Sureesh Mehta, serving Chief of the Naval Staff, and Chairman of the 'Chiefs of Staff Committee' made a statement on 10 August 2009. He went on to publicly proclaim that India was no match for China and the gap was so wide that it was unbridgeable. What an admirable self-goal to be achieved by the Admiral?

Now, there is nothing new or novel about our investing our adversaries and tormentors, with super human attributes. We have been doing that from the days of the Ghaznis and Ghauris, Baburs and Abdalis. The sad and bitter truth is that we were always defeated in our conflicts with invading armies. Our response was to project those upstarts as masters of 'vastly superior military machines', against which we could not have succeeded. We had the following types of thought processes, even if we might not have made formal statements to that effect:

— Oh, Ghazni was a great Sultan; he was even a greater general; we stood no chance against him.
— Prithviraj had let Ghauri go after capturing him (as per our version). Ghauri did not reciprocate that act of kindness. How mean of him!
— Abdali was ruthless, and always winning. What could we have done?
— Clive was a civilian clerk. He had no business to turn himself into a great general, and start defeating Indian generals.

The refrain of our argument throughout history has been somewhat on the following lines:

'Defeat was really not our fault; they (our tormentors) were just too powerful.'

Bharat is suffering from a besieged psyche and psychology of victimhood. We are happy to project ourselves as victims. Let us explain by an example. In July–August 2009, the Indian Prime Minister issued a joint statement with his Pakistani counterpart. The statement contained a line that the Pakistan Prime Minister said that they had some threat perception in Balochistan, a border province of theirs. There was no reference, not even a remote one, that that threat was from India.

But all hell broke loose in India. The main Opposition Party, the BJP, walked out of Parliament, and went to complain to the President. Their contention was that Pakistan was trying to imply that that threat could be from the Indian Intelligence Agency R&AW, and that any such suggestion was preposterous. India can never do such a thing; we are a paragon of morality, and *satya* (truth). In any case, we are *ahmisic* (non-violent), and worshippers of *shanti* (peace). So, how can anybody dare imply to the contrary? We are the quiescent type; we cannot be active. We have never been the doers; only things have been done to us. We can only be victims; we are not used to any other role. That was the pith of their protestation and argument.

With that type of mindset, we are projecting China as a sort of super power. No doubt, China is a formidable military machine; but, so are we. Bharat is a humongous country, and is no push over. Being an ex Air Force man, the author would like to record his opinion that our three major weapon systems in Mirages, Mig 29, and Su 31 are the most formidable. In fact, awesome may be a more appropriate word, keeping in view our capability for mid-air refueling. These are amongst the best weapon systems in the world; and so are the men manning these machines. It is possible (but not necessary) that we need some additional numbers. The numbers will get looked after in due course; that aspect need not be over emphasized. The author also ventures to say that the Indian army is also generally well equipped; their T 90 tanks are among the best in the world, and so are the Bofor guns. Upgradation of weapon systems is a continuous process; we do not have to emphasize that in an out-of-the-way fashion, and project that as our weakness.

Let us have a look around. China has some 15 countries around its periphery. With the exception of Russia, all those countries are small; in fact, some are tiny. If those countries were to go by the Indian example, they would have no reason even to exist. Their faces would be forever bereft of a smile; but, we know that that is not the case. Those countries are living with honor and dignity. We have the example of the tiny Vietnam standing up to the formidable Chinese army.

Nothing stated above should be taken to mean that we ignore or sleep over external threats to our security. Rather, those threats should be the main focus of our national attention, which presently, these are not. But for that, we do not have to beat our chest in public over our perceived weaknesses, especially at the official (including highest Defence) levels. India is a big country, it must learn to think and act big. The Indian Armed Forces are capable of taking on any world class military machine, provided we can get our mental act together; some 90% of our problems is in our mind*. It may be relevant to quote an Urdu couplet here:

Khud hi ko ker buland itna	Raise yourself to a level,
Ki her taqdir se pehle	Above all others;
Khuda bande se khud puchhe	So, God himself will ask
Bta teri raza kya hai	Tell me, what do you wish to be?

---

* Of course, problems of the mind are almost impossible to tackle.

30
THE KARGIL CONFLICT

It appears from newspaper reports that some Pakistani *mujahideens* (irregulars) occupied the Kargil heights sometime around April 1999. Later on, they were joined by regular Pakistan troops. Their overall numbers might have been a few thousand (some estimates talk of about 1000 numbers). Again, going by newspaper reports it would appear that the Pakistanis built some bunkers on those heights. India remained blissfully unaware of these major incursions till some shepherds informed of the same; So much for Indian Intelligence.

When the factual position was learnt by the Indian side, panic set in and all hell broke loose. The Northern Army Command was in a frenzy and Army HQ in a tizzy. The Ministry of Defense did not know which way to look. The Government got busy in efforts to hide its embarrassment.

As the story goes, the army was caught with its pants half down. There was an intolerable pressure from the Army HQ to get the occupation vacated before the nation would come to know the full scale of all round negligence and incompetence, primarily of the Intelligence set up. The Ministry of Defense was a willing accomplice. Young officers were asked to mount frontal attacks on the enemy sitting at heights and behind bunkers and boulders. Crossing of the Line of Control (LoC) was banned. Naturally, the Indian casualties were intolerably high; 527 Indian martyrs to evict a few thousand (or less) intruders, mostly with light arms. That was a clear sign of the panic of the authorities. Our troops performed extremely well and many cases of exceptional courage and bravery were recorded. We pay our tributes to the martyrs.

During the conflict, regular statements were made by the Indian authorities, including Service Chiefs, that India would not cross the LoC. Hypocritical moral posturing? Why? Pray, why would we not cross the LoC when the other party has? If that results in a wider conflict, so be it. After all, the armed forces are for such occasions only. That was the Hindu defensive mindset on display (the BJP was in power). If we had crossed (perhaps on the quiet) the LoC a bit here and there, the intruders' supplies could have been intercepted and they could have been starved out. In that case, our casualties would have been much lower, perhaps just in two digits. In the din and euphoria over the Kargil 'victory', let us not forget that it was our pseudo-moral posturing and lack of aggressive spirit (i.e. no LoC crossing) that resulted in such high casualties. The Government of the day must take full responsibility for that.

The BJP government decided to convert the whole situation to its advantage by exploiting the emotions of the nation. For all their proclaimed independence, TV channels were roped in and they became willing tools. As coffin after coffin draped in the national flag arrived at Delhi airport, the scenes were shown on live TV. The burning funeral pyres in villages were also made a part of the exercise. As expected, all this display aroused national sentiment and won sympathy for the government.

Nawaz Sharif was the Prime Minister of Pakistan at that time. At some stage, he panicked and requested President Clinton for an interview, of all days on 4th July — a national holiday in America. Clinton agreed as a special case. Going by newspaper reports, Clinton showed Nawaz satellite photos of the movement of Pakistani nuclear missiles during the conflict. Nawaz was red-faced and denied any knowledge of the movement. Instead of showing any sympathy, Clinton was furious. He asked for the immediate withdrawal of the intruders. Nawaz Sharif rushed back, and soon the intruders melted away.

The BJP Government went on to proclaim a 'major' victory. Celebrations set in; jamborees were organized at which the BJP leaders went eloquent. Efforts were made to project the 'victory' as even bigger than 1971; fortunately, these were not successful. In 1971, we had taken 93,000 Pakistani prisoners and created a whole new country. During Kargil, we perhaps took no prisoners (as far as public knowledge goes); we just drove out a few thousand (or may be only about 1000) from our own border lands. In keeping with our (dubious) tradition, we fought only when attacked, and took care to fight only on our own land. Going over (even a little bit here and there, and on the quiet) to the enemy land would have dented our pseudo-pacifist image, and interfered with our Hindu military mindset. That is a carry-over of the Hindu military strategy from the days of Ghazni and Ghauri, i.e.:

- *Fight only when attacked; wait to be attacked*
- *No need to take the fight to the enemy's home*
- *Let the enemy bring the fight to your home*
- *No offensive action; be always on the defensive.*

The hypocritical policy of moral posturing and over-emphasis on *ahimsa* (non-violence) and *shanti* (peace) can never pay. The world only respects 'power' when projected in an appropriate manner, and at an appropriate time. Anyway, the public was duly taken in. In the ensuing election, the public gave the BJP and its allies a governing majority and they went on to rule for five years.

Even by the dismal Indian standards, Kargil was a major Intelligence failure. All organizations including R&AW, IB, Army Intelligence and others were found missing in action. Various types and manners of excuses were fabricated and let out to misguide the gullible and simple Indian public. Deceptively calibrated statements were leaked out from time to time. Let us consider an example.

It was expressed that it was a (long-standing) practice to withdraw troops from Kargil and other heights, during the winters (mind you, there is no such practice at Siachen where the winter is much more severe). It was never clarified how long this practice was in vogue, and more importantly, at what level was this practice sanctioned. Did this 'practice' have the concurrence of the GOC-in-C, Northern Command, and the Army Chief? It would appear from the information available that it was just a 'practice' indulged in, without any sanction at an appropriate level. It would also appear that it was also a 'practice' that having withdrawn, no patrols were to be undertaken, in spite of the mandatory requirement for armies to send such patrols. There are also a large number of Army and Air Force helicopters in the area; some more could have been positioned. It would seem that it was also a 'practice' not to undertake regular surveillance sorties by these helicopters.

The Ministry of Defense (as apart from Service Headquarters) is the single point organization for the defense of the country. Did it have a role and what was it doing? Where was the BSF? The simple answer is that the BSF was in Rajasthan where all the smuggling takes place. It was let out that it was a practice not to put the BSF on the LoC. Who sanctioned such a 'practice'? Was it with the knowledge of the PM, or the Cabinet Committee on Defense/Security?

There are a slew of questions as outlined in the preceding paragraphs; but there are no answers. A high powered Committee was set up to go into the Kargil fiasco. It successfully swept all inconvenient issues under the carpet. No one of any consequence, except a poor Brigadier was held to blame.

The Indian State (read BJP) claimed and hailed Kargil as a 'major victory'. What type and grade of 'major victory' was Kargil? Before we answer that question, we again record our respectful homage to the 527 martyrs and acknowledge the innumerable deeds of courage and bravery during the Kargil conflict.

India has the 3rd or 4th largest army of the world, with more than one million troops under arms. The army has an awesome array of the most modern weapon systems of tanks, artillery and mechanized infantry, backed by every type of conceivable support system. Then, there is the formidable Indian Air Force, again having the most advanced aircraft weapon systems — 3rd or 4th largest in the world. What were these formidable and awesome armed forces pitted against? A rag-tag force of irregulars and regulars numbering a few thousand (some estimates talk of about 1000 numbers), mostly equipped with small arms. They perhaps had nothing more lethal than medium/heavy machine guns and howitzers. They had no tanks, no heavy artillery and without a single aircraft. Their only advantage was that they were sitting at heights.

The above type of engagement would not qualify to be called a war; it was more like an intensive border conflict. By saying so, we are in a way trying to reduce the importance of the innumerable cases of raw courage and bravery displayed by the Indian soldiers during the Kargil conflict. We are just trying to prepare the country for major conflicts that it may have to face in the foreseeable future.

On the tenth anniversary of the Kargil conflict, there was a lead Editorial in the *Hindustan Times* of 25 July 2009. A sentence in that reads as follows:

"One of the controversial issues from the war was failure of intelligence assimilation and dissemination, a problem that seems to have outlived the euphoria of victory against *heavy odds* (emphasis added)."

What constitutes *heavy odds* — a few thousand (or even less) lightly armed irregulars and regulars, pitted against a million strong Army, backed by a formidable Air Force? That gives us a peek into the Indian (read Hindu) military mindset.

India is big; it must start thinking big, i.e. of successfully engaging armies a million or two strong, equipped with modern weapons. We should claim a 'major victory' only when we come up with good performance against a force of our own size. We may be faced with that type of situation not too far in the future, say 10–15 years down the line. That requires a type of mindset, altogether different from the one reflected in projecting Kargil as a 'major victory', which was a work of small minds. We end this chapter in the hope that our comments on Kargil are understood in the broader context in which we make these, i.e. to prepare the country for bigger conflicts — perhaps much bigger conflicts, which are in a different league altogether.

31
THE INDIAN ATOM BOMB

On 11 and 13 May 1998, India crashed into the exclusive and elitist club of the 'Nuclear Haves', with five nuclear blasts; the whole country was euphoric. Pakistan responded on 28 and 30 May 1998, with six explosions — a game of one-upmanship, if there ever was one.

Millions of words and hundreds of books have been written by prescient men on the Nuclear (Military) Power, i.e. what it can do, or cannot do. There are also some long-term issues to be considered with respect to China. We are in no position to express any strong views on such a complex issue. But in the short term, we have India and Pakistan, both nuclear armed, almost in eye-ball to eye-ball contact. For the time being, they may not be arch enemies, but are not exactly on friendly terms. There is more agenda to disagree, than to agree on. Kashmir is one bone of contention, terrorism is another. Following the Mumbai terror strikes on 26/11, war drums were sounded by both sides, for two to three months.

Throughout the history of the two countries, India has had a numerical military manpower superiority over Pakistan of around 2.5:1. That is a lot of excess muscle, and Pakistan realizes that. By letting Pakistan have the atom bomb, we have lost that advantage. If any time Pakistan feels that its vital interests are even marginally affected, it may not hesitate to drop the bomb. As against that, India may ponder agonizingly before doing any such thing. In a show of moral posturing, India stands committed to 'no first use of the bomb'. There were rumors that in an almost no-conflict situation of Kargil, Pakistan had moved its nuclear missiles. There are reports that in the late 1980s, Israel had offered to bomb the Pakistani nuclear facilities, operating from Indian soil. Possibly, the plan had the blessings of the USA. India did not agree to that.

There is even a more worrying aspect. In Pakistan, there is no particular established procedure for the transfer of political power; they may be — *'Jiski lathi uski bhains'*, i.e. the powerful takes all. It is no knowing when the power in Pakistan may pass into unsafe, even dangerous hands. During processes of such transfer of power, one or two Atom bombs can always get misplaced and go missing. No one, not even the Americans, have an exact account of Pakistani nuclear assets. The situation is pregnant with all types of possibilities, one more frightening than the other.

As proved by activities of an atomic scientist of Pakistan, the chances of the Pakistani bomb becoming an Islamic Bomb, and that becoming a Terrorist Bomb, are indeed high and in the realm of possibility. Two prime targets for such a bomb would be the great 'Satan' (America) and idol-worshiper and Kashmir occupier, the Hindu India.

Of course, the Indian know-alls would assure us that in any such eventuality, USA has plans to remove the Pakistani nuclear assets. They may or may not have such a plan. They may change that plan due to domestic politics, or for other reasons. They may try executing the plan, but may not succeed. There is no way we can rely on their plan. We have to have our own means.

But we do not have many practical options. A nuclear free South Asia could be one such option. But, every known or unknown Indian intellectual would bristle at the very mention of such a proposal.

32
BANGLADESH

1971 was a major military victory for India; but it turned out to be a strategic blunder. In the long term, our friend and 1971 ally Bangladesh would be more of a threat than even Pakistan; not in the military sense, but in many others ways, viz.:

- It has become the hub of anti-India activities, which is being exploited by some other countries not so friendly to India.
- It is emerging as a haven for Islamic terrorism; some of the latest suspected terrorists caught in India are from that area.
- It is a Gangotri (source) of illegal immigration into India — both Hindus and Muslims. These people have spread up to Delhi and Bombay, posing serious security threats. Governments, both Center and State, are unwilling and unable to cope with the situation, for a variety of reasons, including the vote-bank politics.

Statecraft demands that we should have foreseen in 1971 itself that a united, rather than a divided, Pakistan was more in the long term interest of India. As it happens, it is the nature of Punjabi Muslim to dominate and it is not in the nature of the restive and rebellious Bengali to be dominated. It was a situation tailor-made for perpetual conflict. Pakistan would have been forced to keep almost half of its army (dominated by Punjabi Muslims) in the Eastern wing to keep under control the (rebellious) Bengalis. Being isolated, the Eastern Pakistan army would have been an easy prey for the Indian army (as was proved in 1971). With the Pakistan army so nicely split, the military threat to India would have been drastically reduced.

The 1971 defeat of the Pakistan Muslim army by an infidel (that is how they look at us) army was the greatest ever trauma for the Pakistani *ummah* (Muslim public). Even more galling was the 93,000 Muslim prisoners taken by (Hindu) Bharat. The Pakistani *ummah* is brought up on stories of repeated defeats of the Hindu armies by the Muslim armies, and taking of Hindu prisoners by the thousands. In their history books they had never ever read of a Hindu victory (and Muslim defeat) and taking of Muslim prisoners. The whole of the Pakistani nation was in turmoil; their entire history was sort of falsified. The unhappenable had happened.

Bhutto was almost on his knees in Shimla, pleading for the release of the 93,000 POWs. It is reported that he expressed that he may not be able to land back at Lahore if he went back without the PoWs. Indira Gandhi could have played her trump card. She could have offered not only the return of the 93,000 PoWs, but the entire East Pakistan (an absurd and unthinkable proposal), in return for a full and final settlement of Kashmir. Bhutto would not have believed his good luck, or that he was actually hearing the words 'return of East Pakistan'. He would have asked his daughter Benazir (who was accompanying him) to pinch him to ensure that he was not in a dream. Bhutto would have asked for time to recover from the (pleasant) shock. He could have been asked to come back after about a week, with two drafts, viz.:

- One; for the en-bloc transfer of East Pakistan back to Pakistan. Actually, we could have just released the 93,000 PoWs, and allowed them to go back to the spot where they had surrendered.
- Second; for transfer of whole, or at least most of Kashmir to India in perpetuity.

We may recall that after the 1965 Indo–Pak conflict, Kashmir was on the back burner. In 1972, Kashmir was not as emotive an issue which it became later in early 1990s, on the rise of Islamic fundamentalism, after the victory of the Taliban in Afghanistan. Kashmir would have been too small a price to pay for East Pakistan. The intolerable blow of defeat by an infidel (as seen by Pakistan) army would have been substantially softened. The Pakistani *ummah* would have danced in the streets. It is a real possibility that a bust of Indira Gandhi could have been installed in Lahore.

India could have resumed its moral posture and claimed to the world that it was not interested in the break-up of Pakistan and that its aim was only to stop the flow of refugees. Indira Gandhi's stature in the international arena would have hit the upper circuit-breaker. President Nixon would have rung up to thank her.

But no diplomat of any standing would have agreed with the above approach and would frown at the very suggestion. They would express that was no way to conduct diplomacy. They would even pity the state of mind of a person who could make such a (outrageous) suggestion. They would have reminded us of the rich cultural heritage of Bharat and our national motto of '*satya mev jayate*'. India does not let down 'friends', only 'friends' let down India. How can we give up our cherished traditions?

We, being laymen, are in no position to disagree with our diplomats, who have oodles of experience in this line. We can only file in a request for a re-reading of Chanakya's *Arthshastra*. Friends are temporary; national interests are permanent. In any case, nothing spectacular or even substantial has ever been achieved without bold and innovative thinking, lately called 'out of the box' thinking. Reading of stereotype bureaucratic briefs has never taken any country anywhere. So, we rest our case.

33
CIVILIZATIONAL SYMBOLS
The Ayodhya Temple

Since times immemorial, nations have had symbols of their civilizations. These are mostly in the form of religious structures, i.e. churches, mosques, synagogues, temples, etc. All such holy structures have been zealously guarded by the respective people, at great cost to themselves in terms of having to shed copious blood. Presently, a small area of real estate in Jerusalem is the most disputed piece of land in the world. The reason being that it contains in close proximity the holy shrines of the two dominant religions, i.e. Christianity and Islam, and of an ancient religion, i.e. Judaism. Due to the dispute over this small piece of land, two wars have already been fought; world peace hangs by a slender thread; there are even chances of a nuclear war.

It would appear that the world's first grand religious shrine was the Jewish Solomon temple built in Jerusalem around 950 BC. In 597 BC, Babylonians captured Jerusalem. One of their first steps was to demolish the temple of Solomon. When freed, Jews rebuilt the temple. In 70 AD, the Jews rose in rebellion against the Romans, who were the occupying cum ruling power. The Romans crushed the rebellion. As a punishment to the Jews, they demolished Solomon's temple, which has since not been rebuilt.

Around 32 AD, Christians emerged out of the Jews. Jesus Christ spent his last days in Jerusalem, was crucified there and ascended to heaven. Thus, Jerusalem is most holy to Christians. They have their religious symbols there, including the Church of the Holy Sepulcher.

The Muslims believe that the Prophet Muhammad (pbuh) ascended to heaven from Jerusalem. Therefore, within 3 years of the death of the Prophet, Muslims captured Jerusalem in 635 AD. In 688 AD, they built the Dome of the Rock at the spot where once stood the temple of Solomon. Al Aqsa mosque was also built nearby.

Thus, all the three Judaic religions have their most holy religious symbols in Jerusalem. A small portion of the wall of the Solomon's rebuilt temple was left standing; Jews call it the 'Wailing Wall'. It abuts the Muslim 'Dome of the Rock'; the situation is tailor-made for conflict.

If Hindus say that they may also be allowed to have their religious symbols, all hell breaks loose. The most voluble dissenters are amongst the Hindus themselves, i.e. the pseudo-secularists. The Muslims are more restrained; they understand the nature and importance of religion — their own and as such, of others (including of the Hindus). Even when Muslims are prepared to understand and appreciate the sentiments of Hindus, the pseudo-secularists will not let them do that; the latter are more loyal than the king.

Probably, the four most holy sites of the Hindus in North India are:

- Krishna Janamsthan at Mathura
- Shiva Temple at Somnath
- Rama Janmabhoomi at Ayodhya
- Vishwanath Temple at Varanasi

Of course, there are also Badrinath, Kedarnath, Amarnath and others.

The above four temples were duly dealt with by the Muslim invaders and rulers. Mahmud of Ghazni took up the first two:

Krishna Janamsthan (Mathura)

This temple was built in honor of Lord Krishna at the site of his birth place at Mathura. It was burnt down by Mahmud of Ghazni around 1018 AD. He did this after admiring its breath-taking architecture. Later, a mosque was erected at or near the site.

Shiva Temple (Somnath)

It was looted and destroyed around 1025 AD by Mahmud of Ghazni. The Shiva idol (most probably in the form of a lingam) was smashed.

Ram Janmabhoomi (Ayodhya)

Babur dealt with the Ayodhya site. After capturing Delhi and Agra in 1526 AD, Babur heard that a Hindu god was born at a place called Ayodhya. Now, Babur survived in India only for about four years, during which he was fully occupied in subduing Afghan and Rajput overlords. During those conflicts, sometimes he came quite close to losing his entire empire. Still, Babur could find enough time and leisure to task his general Mir Baqi to trace the exact place of birth of the Hindu god (Lord Rama) and build a mosque at that place. Mir Baqi duly complied and built a mosque at the site in Ayodhya, the Babri Masjid.

Vishwanath Temple (Varanasi)

Varanasi is the holiest city of the Hindus, where part of the time Lord Shiva himself resides. It has the most holy Vishwanath Temple which was dealt with by Aurangzeb in 1670 AD or so. He built a mosque abutting and dwarfing the temple. The temple was allowed to stand as a message to the future generations of Hindus to remind them of their status in life.

Now there are some 50 Muslim countries in the world; some of these are formally declared as Islamic States. In many of these countries, the construction of temples though not banned, is not encouraged. But in Saudi Arabia, the construction of temples is totally prohibited, not only within a few kilometers around the Kabah, but the whole of Saudi Arabia. This is justified on the ground that Saudi Arabia is the custodian of the most holy Muslim mosques; nobody objects to that line of argument.

But, here in Bharat, some mosques had to be constructed at the spots (or close vicinity thereof) where Hindu gods were born, or are believed to have been born by the vast majority of Hindus. But try giving this argument to the pseudo-secularist; he would singe you with his scornful gaze, as if you were some low form of species just parachuted from the 3rd millennium BC to the 3rd millennium AD. The Hindus are reminded of their great virtues of tolerance and *ahimsa*; they are advised to wallow in their perceived glorious (largely mythical) past. Why build a temple when the mosque is already there? Why dig up these past 'dead' issues? Why not concentrate on development? Why, why?

Just after independence, renovation of the Somnath temple was undertaken. This was on the insistence of Dr Rajendra Prasad and Sardar Patel (both not typical communalists), over opposition from the high priest of secularism, Jawaharlal Nehru, the then Prime Minister. That renovation did not damage the secular fabric of India.

The Ram Janmabhoomi issue has been troubling the collective conscience of the Hindus for 100 years or so. More specifically, the following events took place at the Ram Janmabhoomi after India's Independence:

- 1948–49: Opening up of the locks of the Babri structure and permitting *pooja* (Hindu worship). The Ram *lalla* (infant) idol was installed inside the structure.
- 1986–87: Re-opening of the locks and permitting *pooja*.
- 1989: *Shilanayas* (foundation-laying) of the new temple.

The above actions took place under the patronage of the government of the day. But, most surprisingly, the governments at all the three times were Congress governments, which were, otherwise, avowedly secular.

In the early 1980s, RSS–BJP started the Ram Janmabhoomi movement to build a temple in place of the Babri structure. A top BJP leader, later christened *lauh purush* (iron man), undertook a *rath-yatra* (motorised journey) in 1989 from Somnath to Ayodhya to create an awakening amongst the Hindus on the Ram Janmabhoomi issue. His journey was prematurely terminated when the Chief Minister of Bihar, Lalu Prasad Yadav arrested the *lauh purush*; but, the yatra achieved its purpose.

A crowd of ordinary Hindus from all over India collected for *kar seva* (voluntary service) at Ayodhya on 6 December 1992. A galaxy of RSS–BJP leaders, including the *lauh purush*, was on the dais. The crowd started demolishing the Babri structure, with rather rudimentary tools. There were approving glances and nods from some of the leaders on the dais. To the great surprise of everyone, the structure was completely demolished by the evening. It was a sight no Hindu had imagined was possible. There were 'puppies and jhuppies' (kisses and hugging) amongst some of the leaders on the dais.

At that time, there was a BJP government in UP. At the Center was the Congress government under Narasimha Rao. The Central Government acted as per the script. The State government resigned by the evening of 6 December 1992. The Prime Minister went incommunicado till the next morning. The Central forces which were camping nearby were not moved to the site of the demolished structure for quite some time. That time was enough for the removal of the debris, and for a make-shift temple to come up, and the start of the symbolic pooja (Hindu worship).

By this time, the RSS–BJP leaders on the dais realized the enormity of the event. It also started dawning on them that they might be held responsible for the demolition; incoherence set in. They started giving out anguished cries of their innocence — no, no; we did not do it. We do not know who did it; we are, in no way responsible. You see, we cannot do it; we are Hindus — tolerant and ahimsic. That was the refrain of their argument.

No doubt Babur had built a mosque at Ayodhya. But by the mid-1900s, the idol of Lord Rama had been installed inside the structure. Namaz had not been said since that time; pooja (Hindu worship) had been permitted from time to time. Thus for all practical purposes, from the mid 1900s onwards, the Babri structure was no longer a proper mosque.

Now, there is nothing new or novel about demolitions and desecrations of mandirs, gurudwaras, mosques and churches. That has been going on for centuries. We are all familiar with demolitions of mandirs; it was a routine activity for most of the Muslim invaders and rulers — Ahmed Shah Abdali had the Sikh Golden Temple (holier than the holiest) at Amritsar blown up thrice with gun powder.

Muslims and Christians have been demolishing and desecrating each other's religious places for centuries; a few examples:

Cathedral of Saint John, Damascus

Byzantine Christian Emperor Theodosius, at the end of the 4th century AD, built a grand church in Damascus, Syria; it was named Cathedral of Saint John, the Baptist (a forerunner of Christ). Tradition has it that below the church lay the head of John the Baptist, in a silver casket. Damascus was captured by the Muslims in 636 AD. It would appear that for sometimes, both Christians and Muslims prayed at the shrine, perhaps in side by side enclosures. The church was demolished during 706 to 715 AD, under orders of the Caliph Al Walid of the Umayyad dynasty; tradition says that the Caliph himself led the demolition party. A mosque was built in its place; it is called the Umayyad or Great Mosque. It is one of the oldest, and perhaps the 4th holiest mosque of Islam. Possibly, the head of Saint John still lies underneath. In 2001, Pope John Paul II visited the place, primarily for the presumed relic of Saint John.

Santa Sophia, Istanbul

Sometime during 532 to 537 AD, the Eastern Roman Emperor Justinian built a grand church at Constantinople (Istanbul); it was called Santa Sophia. For 900 years, it was the largest Christian cathedral in the world. In 1453 AD, Ottoman Turks captured the city. The church was converted into a mosque, under orders of Sultan Mehmed II. Christian features were removed; mosaic was plastered over. Islamic features were introduced; mehrab, etc were added. It came to be called Hagia Sophia. In 1935, the mosque was converted into a museum by the (secular) Republic of Turkey. In 2001, Pope John Paul II visited the building. It is reported that he used the opportunity to pray to the Christian God in this erstwhile church, later mosque, and presently museum.

Christian Crusades

During the 11th to 13th centuries AD, European Christians undertook seven crusades, to rid Christian shrines in Jerusalem from Muslim control. Hundreds of thousands of Christians marched over thousands of miles of most hostile territory, to give battle to the Muslims. In the first Crusade, Christians captured Jerusalem in 1099 AD. It is reported that there was such massacre of Muslims and Jews that in some streets of Jerusalem, blood flowed up to the level of the reins of the horses (not to be taken literally). The Crusades lasted some 150 years.

Miscellaneous

- Christians expelled the Muslims from Spain in the 16th century AD. Following that, a large numbers of mosques were demolished in Spain.
- The Muslim shrine, the Dome of the Rock in Jerusalem stands at the spot where once stood the Jewish Temple of Solomon, which was most holy to the Jews.
- During 2007–08, there were newspaper reports that mosques were demolished by the State in Islamabad, the capital of the Islamic State of Pakistan. The Red Mosque (Lal Masjid) was raided by security forces.

The Babri Structure

What is unique and novel and disturbing about the demolition of the Babri structure, is that a 'deemed' mosque has been touched for the first time in Bharat. What is even more disturbing is that that has been done by common Hindu masses, who are otherwise, supposed to be tolerant and *ahimsic*. That is what set the cat among the pseudo-secular pigeons. For them, it was Armageddon and Apocalypse rolled into one; no bigger tragedy had overtaken Bharat in its history.

We may recall that the Hindu *lauh purush* (Iron man) had undertaken a *rath-yatra* (motorized journey) in support of the movement to build the temple at the Ram Janmabhoomi site. Of necessity, that would have involved the demolition of the Babri structure. But, when the structure actually came down, you got nothing but frowns and disapproval from the *lauh purush*. He went on to proclaim the day (of demolition) to be the 'saddest day' of his life. A few years down the line, the *lauh purush* traveled all the way to Pakistan, to repeat the infamy of the 'saddest day' to the *ummah* there. He perhaps reasoned that the Pakistani *ummah* had better claim to the Babur legacy than the Indian Muslims.

Let us revert to the RSS–BJP leaders on the dais on 6 December 1992, when the Babri structure was demolished; they refused to accept any responsibility for the demolition. Let us consider an alternate scenario. Those leaders could have walked to the nearest magistrate and claimed exclusive and full responsibility for the demolition. They could have asked to be sentenced for the same. They might have been sent to jail for a week, a month, or a year. The whole country would have been electrified. The BJP could have gone on to get a ⅔ majority in the Parliament, which could have been used to build the temple. When you dither at a crucial moment of history, you get consigned to its dustbin. History is not in the habit of giving a second chance.

In the event, in a subsequent election, people brought the BJP within sniffing distance of a majority; it formed a coalition government with some opportunistic small parties. The election was won by the BJP on three basic issues, of which 'building the temple' was the only one, which had some practical chance (howsoever small) of being achieved. During part of the period, the BJP was in power both at the Center and in UP; a situation never likely to arise again. Good opportunities have this nasty habit to never visit again.

Having got to power, the BJP forgot everything about the mandir; they even forgot that it was the main item on their menu. The party thought it would be enough to fool the people to issue a vague statement meaning nothing, every year or two. A big hunt was launched to look for a good lame excuse (to be fed to the gullible Hindu masses), for not making the mandir. They hit the jackpot with the word 'coalition-dharma'; it looked like a good excuse. The BJP was in a coalition; how on earth could anyone (in his right senses) expect them even to think of the mandir? Now, the Hindu masses are nothing, if not simple and trusting. They thought that the word 'coalition-dharma' was a sort of a new branch of Vedic dharma. They took time to realize how they were being short-changed, and taken for a ride. However, they did realize. In the next election, against all expectations and predictions, the masses threw the BJP into the dustbin of history.

The net result is that the Ayodhya issue is where it was put by the common Hindu masses on 6 December 1992. The BJP has not been able to move it an inch, or a brick forward.

34
THE HINDU BOUNTY

From the very beginning of civilization, the Hindu God, variously called *Brahman, Parmatma, Ishwar*, had been very kind and bountiful to the Hindus. He blessed them with everything that a civilization could aspire for.

As a starter, God made Hindus as the very first civilization to emerge on this planet. Almost from day one, he gave them a well-formulated religion and philosophy — a two in one package. We may recall that both the Greek and Roman civilizations in their hey days were without the base of a formal religion, though the Greeks excelled in philosophy. The Hindu religion grew up in the fertile land of Bharat, blessed with the monsoon rains and awash with great river systems. Compared to that, other major religions, i.e. Christianity, Islam and Judaism came up in the arid, sandy deserts of West Asia, with very little forest cover, few rivers, inadequate vegetation and scanty rain. Bharat is a huge (single) landmass protected almost on all sides either by the mighty Himalayas, or by massive oceans. Bharat has a strategic location astride trade routes and dominating sea-lanes.

The great Lord made Hindus humongous in numbers and homogeneous in character, being bound by one sublime religion. He made Hindus believe in the soul being the real thing, everlasting and indestructible. That should have made death meaningless for the Hindus, 'death' being merely a change of dress and a change of address. With these types of advantages, Hindus could have gone on to become a great power, including a major military power. They could have dominated the world in every sense of the word. However, things did not take that route and Hindus could not leave their footprints on the sands of time.

At this stage, we may note one uniquely outstanding feature of Hinduism, i.e. refusal of the bulk of Hindus to convert to Islam, even after they were militarily overcome. The onslaught of Muslims had started in the 7th century AD from Arabia, in the West towards Europe; and in the East towards India and beyond. Muslims made massive strides within 50 to 100 years of the Prophet's death in 632 AD. The whole of West Asia came under their sway. Mighty empires like the Byzantine and the Persian could not stand before the Muslim onslaught. These civilizations were not only defeated militarily, but their entire populations were converted to Islam — some by consent, the rest by coercion.

The march of Islam was halted in the West by the King of Franks, Charles Martel, 'the Hammer'. He defeated the Islamic hordes in the Battle of Tours/Piotiers (80 km South of Paris) in 732 AD. It has been expressed that but for that battle, the whole of Europe would have been converted to Islam, with dire consequences for Christianity and other religions.

In the East, Bharat could be subdued militarily without any great difficulty. But, the Hindus of Bharat refused to be converted to Islam. In spite of coercion over hundreds of years, only a small body of Hindus could be converted; most of these were from the lower strata of the society. The main body of the Hindus stood firm as a sort of challenge to Islam. Islam had to leap frog to Indonesia and Malaysia, leaving India largely intact as a Hindu entity.

The Muslims ruled over India for 600 years. However, the Hindus refused to consider them as their equal, leave alone their superior. Though business dealings were common, Hindus refused to have social interactions with the Muslims. Hindus stayed scrupulously away from Muslim festivals and other rituals. A (devout) Hindu would not accept a glass of water from Muslim hands; the question of eating in a Muslim house did not arise. This should not be taken to mean that in 1940s, upper class Hindu and Muslim families of Lahore were not eating chicken *biryani* (a rice dish) in each other's house. Senior citizens of pre-partition days, would recall shrill cries of 'Hindu *pani* (water)' and

'Musalmaan *pani*' on railway platforms. The two communities were living almost in perfect isolation. Though the Hindu yielded militarily, they could not be cowed down religiously. The reasons lie buried deep in the psychology of Hindu religion; we may be under-qualified to analyze the same. That can be done only by great Hindu scholars; therefore, we leave it at that.

The Modern Hindu

The Hindus were under the jackboot of slavery for 750 long years, from 1200 to 1947 AD. The period of slavery under the Muslims was difficult and harsh. In a comparative sense, British rule was somewhat benign. A slavery period of even 50 years is enough to drain a nation of its identity and pride; the havoc of 750 years of slavery cannot even be fathomed.

On emerging out of this prolonged period of slavery, Hindus are in search for their lost identity and pride as a nation; it is not an easy search. The task has been made all the more difficult by the presence of the dominant minority, i.e. the Muslims, who walked away with a chunk of a most prestigious part of this ancient holy land (where the Vedas were evolved). The Hindus also have had to cope with two competing ideologies, i.e. secularism vs communalism.

Now, there are any number of countries, especially the Western democracies, which are secular countries. Britain is a declared secular state. But the Queen of England is also the Head of the Church of England, i.e. equivalent to the Pope for this church. One of her official titles is 'Defender of the Faith' — a truly religious symbol. Nobody finds that incongruous or incompatible with Britain's status as a secular country. In its hey days when Britannia ruled the waves, the United Kingdom had no hesitation in flaunting its Christian identity. It exported thousands of missionaries to its colonies to show the true path to the heathens. However, it must be stated to the credit of the Christian Britain that they did not use force for conversion purposes.

In the 1950s, Britain's power and prestige started declining, as colony after colony slipped out of its hand. Britain's commitment to secularism increased in almost direct proportion to decrease in its power and prestige. To put its secular credentials on display, Britain started opening its shores to all types and manners of immigrants of many faiths; that was to prove very costly to them in the long run. Britain is already feeling the heat with periodic terrorist strikes by people brought up in the British community itself. In 2007, the BBC showed a conversation with two well-heeled mullahs of Britain. They expressed the view that peace would return to UK, only when Sharia Law was made applicable there. The anchor expressed as to how that could be possible; the two mullahs answered in unison — because all land belongs to the Allah. As reported in the Press in 2008, British Intelligence was monitoring some 20 active Terror Cells in the UK. Is that the wage to be paid for going overtly secular? Britain is now looking all-round in bewilderment, as whom to blame for its predicament, and what to do next.

Most of the Muslim countries are not required to display any great signs of separation between the state and religion. In many of their cases, religion is the state. Everything is Islamic — the state, the nation, the individual, the law (Sharia), and almost everything else. Some of these countries have the word 'Islamic' as a part of their nomenclature. That does not affect anybody's health; nobody cares, no one protests, not a feather is ruffled.

In India, the 'secular' word is put in usage mainly to deny the (intrinsic) Hindu identity of Bharat. In fact, the pseudo-secularists may even object to the use of the word 'Hindu identity'. As per them, Hindus never had an identity, do not have one now, and in fact, have no need of an identity. If, per chance, Hindus stumble on their (lost) identity, pseudo-secularists would lose both their past-time as well as their profession. 'How will we run our shop?' is their woeful whimper.

If a sort of Hindu identity comes to be associated with Bharat, there should be and will be no difficulty in other communities living with full honor in the country. It is a propaganda of the pseudo-secularists that that will not be the case. There is a ruling of the American Supreme Court that whilst the US government is secular, the US nation is Christian — that has not affected anybody's health.

In this country, if Hindus talk of their religious symbols, it is strictly anti-secular. Expulsion of hundreds of thousands of Kashmiri pandits is projected as a sign of vibrancy of Indian secularism. It is a non-event for the media, which like to serve a monthly fare of the Gujarat riots (most condemnable and horrific as they were). India is the only country in the world where the majority community can be bashed up, just for being that, i.e. the majority community. It would appear that Hindus are being asked to carry the major part of the burden of world's secularism.

The modern Hindu has two broad options before him. One view is that the Hindu is a secular, tolerant, *ahimsic*, peace-loving type of person with equal (even extra) respect for other religions. As per this view, the Hindu may only be vaguely aware of his own religion, i.e. if he is not actually confused about it. It is also often stressed that by his very nature, a Hindu cannot be a fundamentalist and would not take a strong position on any issue; he is always ready for a compromise. A *shloka* (verse) or two are always quoted to remind the Hindu of his docile, goody-goody nature, e.g. *Ahimsa Parmo Dharma, Vasudeva Kutumbakam.* This view is held by a group led by the Congress, with the Communists being a leading member. The group is generally allergic to use of the word 'Hindu', and sees red on its very mention. Their opponents dub this group as pseudo-secular.

The second group is led by the RSS–BJP, who are generally considered to have a 'Hindu Agenda' (whatever that may imply). This group is for the restoration of Hindu self-respect and pride; however, they have not the least idea how to go about it. If they talk on the subject in the public space, they get badgered and mauled by the pseudo-secularists. The onslaught is generally led by the media mughals, who are all high priests of secularism. The opponents of the RSS–BJP group dub them as 'communalist'. Half the effort of this group is wasted in trying to shed the appellation 'communal'. Their leaders are seen visibly squirming on TV screens when the 'communal' word is repeatedly thrown at them.

As uttering the 'Hindu' word is a strict no-no in this country, the RSS–BJP had to invent the ruse of 'Hindutva'. It is possible that some higher echelons of Hindu leaders understand the meaning of the word

'Hindutva'. However, the common Hindu folk (including the author and all his acquaintances) do not have the foggiest idea as to what Hindutva is all about (in spite of some Supreme Court ruling on the subject). The author's view is that the total number of people in India who may have a clue about 'Hindutva' may be less than one thousand. The BJP got a major drubbing in the May 2009 Lok Sabha polls. Following that, a new lexicon of 'hard Hindutva' and 'soft Hindutva' has emerged — this is meant to further confuse the public.

The RSS–BJP group is for building a temple at Ayodhya. But, when in power, they lost all interest in that project. Instead of devising some ways to go around the obstacles, they dispensed all their energy in inventing excuses as to why the Ayodhya temple cannot be and should not be built. The BJP was in power for six long years; progress equal in length to even six bricks could not be made in building the temple. When a group of Hindu unknowns did the work for them on 6 December 1992, the BJP–RSS refused to recognize their contribution, and vehemently denied any association. The RSS–BJP would shed copious crocodile tears for the Kashmiri pandits, but would not lift their little finger to help them when in power for six years. Rather, they would get busy in organizing 'Iftar' (Muslim fast breaking) parties, which even the high priest of secularism, Nehru, never attempted.

The two groups are, however, not watertight. They would cross into each other's territory as per convenience, and for vote-bank politics. The Ayodhya temple is on the agenda of the RSS–BJP. However till date (2009), all forward actions on this highly contentious issue have been taken by the Congress, e.g. opening of locks, undertaking *shilaniyas* (foundation laying), etc. As per convenience, the RSS–BJP group would join the 'secular and tolerant' symphony of the Congress, even though most of their notes would be discordant. In short, Hindus have almost no choice before them. On one side is the hard rock; on the other side, the ditch.

35
SOME WAR ISSUES

35.1 Role of Generals

As our subject relates to the military issues of Hindus, let us try to understand the nature of war, which is integral to human nature. Kipling says:

Four things greater than all things are
Women and horses, power and war

In the past, horses played a key role in war; and power emerged out of war. In early human history, wars were often fought over women, e.g. Helen, Cleopatra.

Man felt the need of war at very early stage of his development. He had to resort to war to get his share of food and thereafter, to protect it. Later, food got expanded to women, shelter and then land (country).

History tells us that civilizations and empires are made and unmade by war. And wars are made and won by generals. Thus, civilizations, wars and generals have a very close and intimate relationship. History is largely a collection of exploits of generals who won wars and established empires. Generals have always played a key and central role in the affairs of man. Civilizations that did not learn this basis tenet paid dearly for the same by losing their liberty; Hindus fall in that category. Wars are generally fought in two modes:

- *Offensive mode:* To impose your will on others, and to capture their territory.
- *Defensive mode:* To protect your land and its people .

History tells us that successful generals always operated in the offensive mode. There are hardly any tales of great generals making history while operating in the defensive mode, except in the short term, or for tactical reasons. History also tells us that the chances of success in the offensive mode are 8 or 9 out of 10; these being only 1 or 2 out of 10 in the defensive mode. In most major battles, generals on the offensive emerged victorious. Alexander recorded victory after victory as he was always on the offensive. So was the case with the Roman generals and later, with the Islamic generals.

War has been at the center of affairs of man from the beginning of pre-history. Hindu mythology begins with the war between *devas* (gods) and *asuras* (demons). The Hindu civilization resolves around the two great epics of the *Ramayana* and the *Mahabharta*, which are essentially tales of war and conflict. The Hindu celestial song, the *Bhagwad Gita* was delivered right on the battlefield when armies were lined up for battle; excerpts from the *Gita*:

1. 'Arjuna … now if you will not wage such a righteous war, then abandoning your duty, and losing your reputation, you will incur sin.' BG: 2.33.

2. 'Either slain in battle, you will attain heaven; or gaining victory, you will enjoy sovereignty of Earth. Therefore, arise, Arjuna, determined to fight.' BG: 2.37.

The message is crisp and clear — Engage in war; kill, or be killed. This establishes the centrality of war for the Hindus from the very beginning. In fact, the great Lord Krishna has called 'not engaging in righteous war', a sin.

Even at the international level, history starts with wars. History records a war in 1400 BC between the Hittites and Mittanis in Mesopotamia. Around 1200 BC, there was the 10 year Troy war. At about the same time, there were wars between Israelites and other West Asian Powers like Assyrians. Thereafter, there are wars and wars, with only patches of peace, largely to recover and recuperate. There are wars between the Persians and the Greeks, by the Greeks (Alexander) against half the world (4th century BC), between Carthage and Rome (3rd century BC), between Rome and one quarter of the world (2nd–1st century BC). Then, the Huns invade Europe including Russia, and lay it waste. Mongols (Genghis Khan and Tamberlane) play up havoc with half the world. Thereafter, emerge the Muslims, who through wars, established their rule over major parts of the world. Finally, there were the Christians led by the British; their empire spread over ¾ of the world — all through series of wars.

At this stage, we may note a few hard facts about war:

- War is not a slugfest between ageing aunts in the backyard of a house. It is a clash of titans whose outcome determines the fate of nations and civilizations, i.e. between freedom or slavery, and life or death.
- There is no scope for sentimentality, morality, truthfulness, forgiveness, and similar cant in the battlefield, nor in the events leading to the battle, or following it.
- Employment of subterfuge, trickery, fraud, chicanery and similar tricks are not only permissible, but also mandatory in war.
- Ruthlessness is the name of the game called war. One small moment of weakness or indecisiveness can spell disaster.
- In war, victory is not everything; it is the ONLY thing.
- The only barometer of chivalry in war is Victory. There is no chivalry in the saying — 'we fought bravely, but lost'.
- Defeat is a disgrace — always and under all circumstances.
- In the *Gita*, Lord Krishna stresses on the imperative of engaging in (righteous) war — to kill, or be killed. The option of surrender or retreat is not mentioned, not even in passing.
- Rules of war can be and may be bent and violated to achieve victory. Lord Krishna did that with telling effect during the *Mahabharta* war.

Having talked of war, we must talk of Generals who won wars for the glory of their nation. Pages of history are awash with names of Generals, who through victories in war, left their names on the sands of time. These include:

Alexander	— Greek
Hannibal	— Carthage

Scipio, Julius and Augustus Caesar — Rome
Attila — the Hun
Genghis Khan and Tamberlane — Mongol/Tartar
Saladin, Mahmud of Ghazni, Babur — Muslim Generals
Napoleon Bonaparte — French

Note: We are excluding generals of recent vintage from the above list. Even otherwise, the list is by no means exhaustive.

Names of Hindu generals do not appear in the above list as no Hindu general burst forth on the world scene as the conqueror of 1/16 or 1/32, or even 1/64 part of the world. Conquest of at least some parts of the world outside your own territory is a necessary qualification for inclusion in the list of 'World Generals'. Hindu generals fall in the following two broad categories:

- *Pre-1000 AD:* There were some generals who united most of North Bharat under one flag. The two outstanding names in this league are Chandragupta Maurya and Chandragupta I (of the Gupta Dynasty). However, none of them considered it necessary to wave the Hindu flag beyond the land borders of Bharat. It would appear that they had the means to do so; only the will was lacking; a sort of Hindu mindset.

- *Post-1000 AD:* These were the days of Rajput generals. They remained, largely, satisfied with their small or medium sized states. Leave alone Bharat, no serious attempt appears to have been made to bring even the whole of Rajputana under one flag; of course, by the sword. If that had been done, the history of Bharat would have developed along different and pleasant lines. The question of their bursting on the world scene perhaps never arose. These generals never went on the offensive against foreigners, and fought only when attacked; that would disqualify them from the list of 'world class' generals.

We now come back to our list of 'world class' generals; those generals had a number of common features, viz.:

- All of them ventured out of their lands to subdue other civilizations and people; sometimes because they needed their lands; but many times, for the shear thrill of subjugation. They did that as a challenge, perhaps even as a sport.

- Before venturing beyond their borders, the generals took care to unite their own people, under one command and one flag. They did this sometimes through persuasion, but mostly by the use of their sword. The two names needing special mention in this respect are Attila the Hun and Genghis Khan. They raised mighty war machines out of wandering disparate tribes, who were forever at each other's throat.

- Once they had united their own people, the generals were for ever on the look-out for offensive action. Many a time, they indulged in that for the sheer thrill of it. They never allowed the adversary to gain the initiative and seldom fought in the defensive mode.

- They were generally always victorious, and in fact recorded victory after victory. They established large empires over which they ruled, with a degree of ruthlessness.

- They often fought against overwhelmingly superior numbers and still won on the strength of their ability, superior tactics and raw courage. They always took the battle to the enemy's house, and seldom permitted the enemy to bring war to their house.

- They were all masters of strategy and tactics and regularly indulged in treachery to achieve victory which was their only objective.

- Always and without fail, they personally led the troops sitting on horseback, on the actual battlefield. They were fearless and daring.

- They were charismatic personalities who could inspire their troops to martyrdom by use of a few well-chosen words at a well-chosen moment. With that, they often got their troops to snatch victory out of the jaws of defeat.

- They were masters of diplomacy who would strike opportunistic alliances, to be broken at the first convenient opportunity. They were not committed to self-defeating concepts like morality, truth and similar cant. They might strike a temporary moralistic posture, but only to gain a tactical advantage.

- Most of them were not above indulging in plunder and loot, and raining mayhem, murder and rape on the conquered people. That would put everlasting fear in the hearts of the conquered people.
- They never heard of words like non-violence and peace. These were just not in their dictionary.

Even after thousands of years, the names of the above generals are household names. They are referred to with respect and awe in the pages of history.

Freedom is a cherished ideal and essential feature of a civilization. Survival of freedom requires war, or at least the threat of war. War is a dirty and cumbersome business; it is not a game to be played by the faint-hearted. War requires nerves of steel and a titanium-lined stomach. It is generals of vision and caliber who win wars. Such generals are produced by societies which cherish and honor their generals, not only temporarily in wartime, but also in peace time. Generals have to be groomed over a period of time in an atmosphere conducive to their growth. That is one lesson of history the Hindus did not learn in the past; and are refusing to learn ever now. When the 'chips are down and the balloon is up' (army talk for war), it is only a General or two of caliber and vision, who would save the honor of the country. Neither puffed-up politicians, nor bloated bureaucracy would be of much help at that crucial moment. We have to plan to produce such generals over the long term; they cannot be produced on the eve of war.

The aim of war and of generals has to be victory; always and under all circumstances. We would like to say:

In war, Victory is not everything
it is the ONLY thing.

In his address to the US Congress on 19 April 1951, General Douglas McArthur expressed about victory as follows: 'Once war is forced upon us, there is no other alternative than to apply every available means to bring it to a swift end. War's very objective is victory — not prolonged indecision. In war, indeed, there is no substitute for victory'

Later in 1962, while addressing the military cadets at West Point, USA, McArthur expressed as follows:

'—your mission remains fixed, determined, inviolable — it is to win our wars. Everything else in your carrier is but a corollary to this vital dedication — you are the ones who are trained to fight; yours is the profession of arms.'

Note: The above extracts are taken from the book *No Substitute for Victory* by Theodore Kini and Donna Kini (Pearson Education).

Defeat is a disgrace; always and under all circumstances. A simple victory, even a small one, is better than one thousand excuses (even valid ones) for defeat. It is better to be the foot soldier of a victorious army, than being the general of a defeated one. As someone has very aptly said:

Peace without Power is fiction
Piety without Strength is a dream
*Dharma** without *Artha*** is moonshine.

35.2 Shedding of Blood (For the Nation)

Bloodshed is an integral part of war. In fact, war is but a tale of bloodshed, and a trail of blood. There is a famous saying, little known and less appreciated in India, that says:

Civilizations that aspire for greatness must first learn to shed blood — of their enemy and of their own.

The study of history shows that no civilization, not one, became great without shedding blood — a lot of blood. Alexander on his journey from Greece to India was on a blood shedding spree — the blood of Egyptians, Persians, Babylonians, Bactrians, Hindus, of other smaller nations, and of tribes en route. After he had shed a lot of Egyptian blood, Alexander was recognized as God in Egypt, and worshipped as such. Of course, in the process, lot of Greek blood also had to be spilled. He did not hesitate to do that. It was due to the massive shedding of blood, that Alexander could set up his world empire.

---

*   *Dharma* — Righteousness.
**   *Artha* — Power, both political & economic.

Roman generals were second to none in shedding blood, including Roman blood. Julius Caesar defeated his fellow Roman General Pompey in a bloody battle. The Roman Empire expanded almost in direct proportion to the amount of blood shed. What are Attila the Hun and Genghis Khan famous for — their unsatisfied thirst for bloodshed. They shook up major civilizations, and set up their own empires, through means of sheer bloodshed. The Islamic generals excelled in shedding blood during battle. Some of them had no hesitation in shedding blood of the conquered people. Of course, shedding their own blood was martyrdom for the Muslims; it ensured direct entry into heaven and they had no hesitation in that.

During the 12th and 13th century AD, Christian crusaders were on a blood shedding spree of Muslims and of Jews. Of course, in the process a lot of Christian blood was also shed. In 1099 AD, Christian crusaders captured Jerusalem. It is reported that so much Muslim and Jewish blood was spilled that blood flowed up to the horses' reins (not to be taken literally) in some streets of Jerusalem. Later on, the Muslims paid back in kind. Nearer home in 1398 AD, we had Tamberlane invading India; on learning that Hindus were *kafirs* (infidels), he ordered some one lakh (hundred thousand) of his captives to be slaughtered during the approaches to Delhi.

Bloodshed has been a part of Hindu mythology and folk-lore. Goddess Kali is always bloodthirsty; her tongue is shown smeared with blood. In her *chinna-mastak roop* (severed-head form), Kali is shown drinking blood from her own severed head. When Lord Krishna exhorted Arjuna to battle, it was actually a call for shedding the blood of his kin, and if necessary, his own. Bhima drank the blood of Dushasna after killing him and Draupadi washed her hair in his blood.

Parshurama slaughtered the entire Kshatriya clans; not once, not twice, but a full 21 times. What a blood bath that must have been! Animal sacrifice was part and parcel of the rituals in the Vedic period; these are presently prevalent in a small measure in Hindu Nepal and certain hill areas. An odd human sacrifice (howsoever abhorrent) was not ruled out; that was perhaps meant to accustomize Hindu youth to bloodshed. However, things underwent a sea change after the advent of

Buddhism and Jainism; the cult of *ahimsa* took over. Blood shedding became a big no-no; even the sight of blood was frowned upon. As even animal blood was not to be shed, vegetarianism became the order of the day.

Coming to modern times, free and proud nations have no hesitation in shedding blood, including their own, to defend their liberty and honor. The tiny Vietnam suffered casualties of some 2 to 3 million in their freedom struggle, first against the French, and then the Americans. During World War II, Russia lost some 20 to 30 million people in their bid to stand up to the Nazi war machine. That is what nationalism is made of. A less proud people would have thrown up the gauntlet after casualties of say 1 million — but not the Russians; it was their grit and determination that came to their rescue. Both Vietnam and Russia emerged victorious. Other Allied Powers also took millions of casualties.

In 1945, Truman did not hesitate to kill around two hundred thousand Japanese in a few minutes, by dropping atom bombs on Japan. Though the Indian casualties during the freedom struggle were modest (say a few thousand), some half a million people were killed during the actual partition process. After World War II, Stalin sent hundreds of thousands of his own people to death, and to Siberia, to perpetuate his tyrannical rule; but Russia emerged as a Super Power. During 2006–08, Iraq suffered an average daily blood bath of say 30–40 persons for no particular purpose, and towards no specific aim.

It can be concluded from the above that bloodshed is a part and parcel of the march of civilization. Sometimes, it has to be indulged in as a part of policy; other times, it cannot be avoided and has to be lived with.

What is the current position in Bharat? A terror strike takes place and say, 10 people are killed; 20 TV channels land up, pitch their tents and start shouting and gesticulating as if Armageddon is already here. The same thing is repeated again and again on every channel for hours, sometimes for days on end. Terrorists are watching the whole *tamasha* (scene) on their TV screens; they cannot suppress their glee and guffaws;

some 2 or 3 terrorists have put the mighty Indian State into a tizzy. Their aim is achieved; they were looking for nothing further than wide TV coverage.

In TV debates on terrorism, two sentences most often heard with a resigned tone and a look of despair, are:

— Human life has no value
— Innocent lives are being lost

Of course, human life has value. But, it is the human life that has to be sacrificed so that liberty (and everything connected with that) can survive. Sacrifice of sheep life is of no value in such a situation. As we have said earlier:

> *Freedom is not for free,*
> *there is a price to pay*
> *The price is in human blood*
> *Sheep's blood has no say*

Of course, innocent lives should not be lost. But, in any scheme of things, it is the innocent who suffer. Towards the end of World War II, Allied bombings flattened all German cities and most of the countryside. Almost 100% of the dead (in millions) were the innocent; the real culprits were safe in their underground bunkers. Some half a million people killed during India's partition process were all innocent. That is the nature of practical life. So, let us not get ourselves overworked to frenzy over every terror strike; these have to be taken in their stride and treated as a price (in blood) that we have to pay for our liberty. The great Lord has given us the gift of life as a package. We have to learn and adapt ourselves to take the bad things along with the good.

The above should not be taken to mean that we stop planning against terror strikes. Rather, we may redouble our efforts in that direction, but on the quiet and without beating our chest in public. As a start, our Intelligence set-up needs a major upgrade. Today, we do not have Intelligence worth ten paise out of a rupee. That requires some very tough decisions.

36
INDIA GATE

It is customary for nations to build structures to honor their martyrs; sometimes these are called the 'Tomb of the Unknown Soldier'. A common feature of such structures is that these are gigantic in size and awe-inspiring in nature, reflecting the glory and grandeur of the armed forces. These are also a barometer of the pride that the nation has in its forces. Visiting dignitaries are taken to these places to pay their respects, and to lay a wreath. Some of the grandest structures in honor of the martyrs are in USA and Russia. We give below two quotes from the Washington War Memorials:

- A soldier never dies till he is forgotten by the nation.
- All gave their some; some gave their all.

In 1918, the British won the First World War. In this war, the Indian Army fought on the side of the British. At that time, India was a slave country. The British considered that even the slave soldiers who laid down their lives for the British Empire needed to be honored. Therefore, they built the India Gate, on which the names of all Indian soldiers killed in the First World War are engraved. By any standards, India Gate is the most awe-inspiring structure of New Delhi, both when it was built, as well as now in 2009. There is not a structure in New Delhi that comes anywhere near it.

When India became free, the question of making a monument to the Armed Forces of independent India came up. Some of the best brains of the Indian government were put on the job of devising a suitable scheme in line with the honor and dignity of the Indian Armed Force. The prescient men put on the job could only think of an (approximately) 6'x4'x2' *chabutra* (platform) right in the shadow of the British made India Gate, for the slave Indian Army. It has been grandly named *Amar Jawan Jyoti*. No jawan, dead or alive, can be proud of such lifeless structures, which can inspire none. Visiting dignitaries are taken to this *chabutra* (platform). No dignitary would think that the tiny *chabutra* could be representing the grand Indian Armed Forces. They consider that they are saluting India Gate (a symbol of the slave army).

It is believed that the Chief of the Army Staff has a design of a structure to honor the Armed Forces in his brief case. He has been doing rounds of the various ministries with that brief case for the last ten years or so. He gets nothing beyond sympathy, and some tea. India has to realize that it is not for the soldier to remind the nation that the 'soldier' needs to be honored. That is the duty of everyone else, i.e. other than the soldier. However, that is like asking for the moon in this woeful land; there is no way the Indian mind can conceptualize that.

37
SELF-RESPECT

Like the individual, countries, nations and civilizations have their self-respect. The term is a bit of a wrong nomenclature; what is meant by 'self-respect' is that others should respect you. Let us examine the case of India in respect of 'self-respect', with the help of a few actual episodes:

Episode 1

In 2006, an Australian cricket team visited India. At that time, a senior Indian politician, Sharad Pawar was the Chief of the Indian Cricket Association, called the BCCI. He was also a senior minister at the Center. The Australian team won the series. The Trophy giving ceremony was presided over by Sharad Pawar. Normally, players vie with each other to get themselves photographed next to, or near the chief guest. However, an Australian player took the opportunity to physically push Sharad Pawar out of the photo frame, with the Australian captain watching gleefully. As if that was not enough, the Australian captain, with a bent index finger, signaled to Sharad Pawar to get moving, pick up the trophy and hand it over to him. The whole scenario was far too offensive and humiliating to be described in words; it was watched 'live' on TV by millions of Indians and others. These TV clips continue to be frequently shown on TV, even as late as in 2008. Now, offensive and boorish behavior, especially towards South Asians countries, is the USP of the Australian Cricket team; they always seem to enjoy it and get away with it.

But, what was the reaction of Sharad Pawar; he dismissed the whole episode of being of no great consequence. The whole Indian nation was aghast and there was a massive public uproar. Following that, a (mild) protest was lodged with the Australian Cricket board, who belatedly, expressed a sort of semi-regret.

Now, let us try visualize a different scenario, i.e. we imagine that India is a self-respecting country. It would have known that verbal messages for the Australian cricket team have no effect, except perhaps to motivate them to more boorish behavior. India would have arrested the Player and the Captain on some trumped-up charge (say drunken brawl) the Indian police needs no training for that. The two could have been kept overnight in a rat infested jail and given a lice-laced blanket — two normal things for Indian jails. The duo and Australia would have realized what third world countries are all about and it is best not to mess with them.

Next morning, hell would have broken loose all over the world, especially in the Sports press. Indians would have been dubbed 'barbaric'. There would have been calls to expel India from the cricket circuit: but, no one could have dared to exercise that option as all the money in cricket comes from the (crazy) Indians. Some top Indian top cricket official would have come up with profuse apologies and expressed how the whole thing was a big mistake and would never be repeated. As a sign of our sincerity, the concerned police official could have been suspended (for the day). Things would have reverted to the normal in a week or so. However, the Australians would have realized that India had come of age, and was not to be pushed around. It is quite possible that other South Asian cricketing countries would have sent us 'thank-you' notes on the quiet.

We may reiterate that the above is a totally imaginary scenario; there is no question of it ever being put into practice. The masters of Indian culture would never permit such a thing; it is against everything that India stands for — we are used to such humiliations.

Episode 2

In the 1990s, there was a coalition Government in India, under the leadership of the BJP. A very high profile socialist politician was the Indian Defense Minister. In the course of his official duties, he visited the USA. As per newspaper reports, the Indian Defense Minister was asked to strip at the airport, in spite of protests from the Indian Ambassador, or his representative. If India lodged any protest, it is unlikely that that reached the desk of any one important in the American establishment.

If any such incident would have taken place in India with the American Secretary of Defense, it is almost certain that the 7th Fleet would have paid a visit to Mumbai.

Episode 3

In early 2008, the Chinese Foreign Ministry summoned the Indian Ambassador in Beijing, at 2.30 at night to hand over some silly message of no consequence. Moreover, the Ambassador happened to be a woman. Is that the time to call Ambassador of a major country like India? The intention of the Chinese in doing that was clear and need not be spelled out here.

The question is not why the Chinese called at that unearthly hour. The question is why the Indian Ambassador went; however, no one will ask that question. Are our diplomats not trained to notice when the country is being insulted, deliberately or otherwise? Of course, our protocol experts would quote chapter and verse as to why the Ambassador had to go; nation's honor only comes later, if at all.

Comments

Do the above episodes convey a message? Of course they do. Do the Indian antennas receive it? Perhaps not. Indians even do not understand when they are being insulted. Does it call for a change in the situation and approach? Of course yes. Will it happen? Perhaps not. We continue on our chartered path; country's honor can look after itself.

Section F
Hindu Military Reverses
An Analysis

38
HINDU MILITARY REVERSES
Synopsis

Starting with the invasion of Alexander in 326 BC to 1947 AD makes a round figure of 2,300 years. Out of this, India was under the shackles of slavery for about 1,300 years (some 60% of the time), split into two broad periods:

The Early Period: 500 years — 180 BC to 320 AD
Second Period: 800 years — 1200 to 1947 AD + another 50 years

The Early Period: 500 years

The early period of 500 years from 180 BC to 320 AD involved invasions and rule by barbarian tribes, i.e. Bactrian Greeks (Yavanas), Pahlavas/Parthians (Iranians), Sakas (Scythians), and Kushanas.

Before and after the barbarian tribes, there were the following additional invasions cum short rule:

— Alexander 326 BC: He defeated Porus, but withdrew.
— Huns: Two sets of invasions/short rule
  First :   440 to 460 AD — They were expelled.
  Second: 500 to 530 AD — Mihirkula *deva*stated North West India

Second Period: 800 years

Muslim Rule: 600 years — 1200 to 1750 AD + 50 years
English Rule: 200 years — 1750 to 1947 AD

Muslim Invasions

In all, there were seven major Muslim invaders. One of them undertook 17 invasions, and another 9, making a grand total of 31 invasions. Five of the invaders were not interested in establishing their empire in India; their main aim was to wave the Islamic flag around, and carry away the riches of Bharat. The remaining two established empires in India.

- **Invade, Defeat, 'Loot and Scoot' Types:**
    Muhammad Bin Qasim (Arab): 710 AD
    Mahmud of Ghazni*: 998 to 1030 AD — 17 invasions
    Tamberlane (Mongol/Tartar): 1398 AD
    Nadir Shah (Iranian): 1739 AD
    Ahmed Shah Abdali*: 1748 to 1761 AD — 9 invasions
    * Turko-Afghan or Afghan

- **Invade, Defeat, 'Establish Empire' Types:**
    Muhammad Ghauri (Turko-Afghan) : 1176 to 1200 AD
    — Established Delhi Sultanates — lasted 325 years
    Babur (Uzbek) : 1526 AD — Established the Mughal Empire — lasted 225 years till 1750 AD

On most of the above occasions, the collapse of the Hindu military machine was catastrophic and almost total; in some cases, it was sudden, and shattering. There is no precedent in world history of a major civilization going under militarily, in such a facile manner. During the first phase of slavery, the barbarian tribes snatched power from the mighty Mauryan Empire, around 180 BC. The Huns attacked during the rule of the mighty Gupta Empire, and snatched power from it around 500 AD.

The second phase of continuous Hindu slavery of 750 years (1200 to 1947 AD) exceeds the world record by a few centuries. During that period, Hindus had no say in the running of their own motherland and the only Hindu country in the world. During major parts of the Muslim rule, especially the Delhi Sultanates, Hindus were subjected to various types of atrocities and indignities. Whilst the atrocities might have been periodic, the indignities and humiliations were more or less continuous. In the earlier years, demolitions of Hindu temples were not by the dozens, but by the hundreds, Holiest of the holy temples like

Somnath (Gujarat) and Krishan Janamsthan at Mathura, were specially targeted. Some 40 years of relatively benign rule by Akbar was one of the few exceptions.

What was the response of Hindus to the take over of their country by foreigners? It would appear that there was no great degree of anger and indignation; no major upheaval on a national scale; no prolonged military fight back. Yes, there were three major battles fought in 1009, 1191 and 1192 AD. Hindus could have given some 50 such battles; but no way.

- After that one battle of 1009 AD, Mahmud of Ghazni roamed the plains and deserts of India for 20 years, as if he was the master of all he surveyed. None of the 'brave' Rajput rulers dared challenge him, even when he crossed the daunting Rajputana desert to go to Somnath.
- In 1191 AD, Hindus recorded their only military victory against a foreign invader in the North West; but the dream lasted all of one year.
- After the battle of 1192 AD, Qutub-ud-din Aibak could run through the rulers of North India, like a knife through butter. No major battles to oppose him are recorded.

The over-all feeling one gets is that Hindus did not think that slavery was anything extra-ordinary, leave alone a catastrophe or evil; they appeared to have accepted it as the will of God, and a fait-accompli. It also seems that Hindus thought that slavery might have been a natural thing to happen, which need not upset their (intrinsic) calm and equanimity. Neither should that take the Hindus away from their (soul and body destroying) beliefs in concepts like *ahimsa* (non-violence), *satya* (truth), *shanti* (peace), tolerance, universal love and similar shibboleth. It should also not interfere in their unremitting search for a safe and better 'next-life'. As per the Hindu thought process, the present world may be all 'maya' (illusion). Mastery and slavery were just some concepts in the mind; these were no great issues for discord or dispute, leave alone for bloodshed and war. In their search for the hereafter, the present was of no great consequence for the Hindus.

The present day Hindus do not like to spend even 750 seconds to ponder over 750 years of shame and ignominy. If the issue is ever raised, there are two types of instant, crisp and dismissive responses, viz.:

- The pseudo-secularists call it an outright communal thinking and pity the state of mind of the person raising such an unnecessary and avoidable issue. 'How can a right thinking person talk of Hindu issues in this secular country? He should be ashamed of himself,' is their stock phrase.
- The 'communal' types call the raising of such issues a mischievous effort to defame the Hindus. For them, people who could defeat Dashanan (Ravana) do not have to prove their military prowess. As per them, Hindus were always 'brave' and continue to be so now; so, where is the problem? Anybody who raises such questions must be out of his mind.

From the totality of circumstance, it would be difficult to avoid the impression that Hindus might have been even comfortable with their 'slavery'. Over hundreds of years, they made no serious attempt to throw off the yoke of slavery. The whole responsibility to defend this land was on other people's shoulders — first, the barbarian tribes, then Muslims, and finally the British. Let these foreign people scratch their head and shed their blood; that was a major worry off the Hindu mind. Babur attacked India in 1526 AD. The Muslim Ruler Ibrahim Lodhi opposed him; two Muslim armies fought for control of this Hindu holy land; primarily, Muslim blood was shed. Hindus were busy in their main occupation of *pooja* and *path* (prayer and worship) to improve their after-life, which, as per them, was the main aim of human birth and existence. The present life was far too surreal and transitory to be much worried about.

The modern Hindu mind does not like to be reminded of defeats and periods of slavery, and wants to dismiss the whole issue, by making two quick points in a matter of fact way:

- Oh, the Hindu Rulers were defeated because in those days, India was divided into small states;
- And, those states were quarrelling with each other.

If there were any other causes for defeats, the modern Hindu would not like his mind to be cluttered with such causes; that would be too much of a hassle. Hindus argue that they have stumbled on convenient excuses and justifications for their 'slavery', and the matter should be allowed to rest there. The question of veracity of these excuses was immaterial and need not be pursued unnecessarily. What matters is that the Hindus believe in these excuses, and that is enough to settle the issue. The modern Hindu has to go ahead with the task of earning his daily bread, which is quite an onerous task. These matters of victory and defeat, mastery and slavery are all in the mind, and rather inconsequential; these can be left for later, and hopefully better times.

Our first task is to examine the veracity of the above two reasons, very conveniently propagated by the modern Hindu. Actually, both the reasons have no validity, and fall flat on a little bit of examination, viz.:

- First: When the invasions of the barbarian tribes started around 180 BC, the great Mauryan Empire was ruling India, i.e. the country was not split into small states. It was the concept of *ahimsa* (non-violence) of Ashoka and the military weakness of his descendents that proved to be the undoing of the country.
- Second: At the time of invasions of the Huns, first around 440 AD and then 500 AD, the mighty Gupta Empire was ruling India. That empire also collapsed due to the incompetence of later Gupta kings.
- Third: We concede that at the time of Muslim invasions, first in 1000 AD and then in 1176 AD, there was no central rule in India. But, the so-called 'small' Hindu states were not all that small. Most of the states were medium to medium-large; Prithviraj Chauhan fell in the medium-large category, as he controlled the big states of Ajmer and Delhi.
- Fourth: Each of the so-called 'small' states of India was individually much bigger than Ghazni or Ghur, from which the invaders originally came. Some of these Hindu states were many times the size of Ghazni or Ghur. We may recall that India was humongous in both size and population, and even its 'small' was bigger than the 'big' of others.
- Fifth: In 1761 AD, Ahmed Shah Abdali defeated a vast confederate of Marathas, who were a formidable medium to

large sized power of India, at that time. Earlier, Abdali had played up havoc with the Sikhs in Punjab, who were the sword-arm of the Hindus.

The two crucial battles, which changed the course of Hindu history, and sealed the fate of Hindus, were:

1009 AD: Mahmud of Ghazni vs Anandpal, Ruler of Punjab
1192 AD: Muhammad Ghauri vs Prithviraj Chauhan

On both the above occasions, the Rajput rulers of North West India showed exceptional unity; most of them formed part of the mighty Rajput confederates, which opposed the Muslim armies (we have covered this issue in detail earlier). It is fashionable to quote the name of Raja Jaichand, the ruler of Kannauj (an estranged relation and father-in-law of Prithviraj), as an example of the discord amongst the Rajput Rulers. Actually, Jaichand was one of the very few rulers (perhaps the only one), who did not come to the aid of Prithviraj in the battle of 1192 AD. The famous historian Farishta refers to some 150 rulers, big and small, being on the side of Prithviraj Chauhan; the figure of 150 appears to be an exaggeration. What is certain is that some 8 to 10 major rulers of North West India stood shoulder to shoulder with Prithviraj Chauhan in the battle of 1192 AD. The position was no different in the earlier battle of 1009 AD; some 10 rulers of North West India were with Anandpal in that battle. Thus, the excuse of India being split into 'small states' being the cause of Hindu defeats has no basis.

Even if India was split in the so-called small states, the problem was frightfully easy to overcome. All famous generals had faced similar problems. Alexander was initially from the tiny island of Macedonia, adjoining Greece. Caesar was from the initially small state of the city of Rome. Genghis Khan, Atilla the Hun, and Babur were not even rulers of states; they were just chiefs of small disparate tribes. Ghazni was a tiny (real tiny) state tucked in the eastern corner of Afghanistan; so was Ghur in another corner. What did rulers of those small states/tribal chiefs do? They unsheathed their sword, mounted their horse, and asked the neighboring tribes or states to fall in line. Anybody not agreeing was put to the sword. Thus, from a small or even 'no-state', those generals graduated to medium states and gradually to large states in a matter of few years. That was all due to their offensive mindset and raw courage. The Hindu generals failed to follow that route.

Babur, a petty tribal chief from the backwaters of an unknown place called Farghana, captured Samarkand at the age of 15 or so. Twice, his adversaries dislodged him from there; but he bounced back both times. By the age of 20, he had captured Afghanistan. After consolidating there, he started eyeing the fabulous Hindustan, and finally became its master. If Babur had the same mindset as the Hindu rulers, he would have gone to some unmarked grave, at best as a petty tribal chief, and not as the shahenshah (king of kings) of Hindustan. The stories of Genghis Khan and Atilla the Hun are even more fascinating; these would be unbelievable, if these were not true. But for the aggressive mindset of Alexander, he would have been consigned to an unknown grave in Macedonia, without the world having even heard of him. These are just a few examples of what positive and aggressive mindsets can achieve.

Prithviraj Chauhan could have done something similar. If he felt that his state was not big enough, he should have unsheathed his sword and taken the neighboring (Rajput) states under his flag. Chanakya in his *Arthshastra*, strongly recommends that line of action; capture of (weak) neighboring states has been prescribed as a sacred duty of the king. Prithviraj did nothing of that sort. From 1176 AD onwards, Muhammad Ghauri started nibbling at territories in Punjab, the very backyard of Prithviraj. By 1186, he had captured major Indian cities like Peshawar, Multan, Lahore and the fort of Bhatinda. Prithviraj did not challenge Muhammad Ghauri till it was very late.

Now, we take up for discussion, the second excuse for Hindu defeats that 'the small Hindu states were always quarrelling with each other'. In those days, there were small states not only in India, but all over the world. Those were the days of chivalry and states (small or big) were always quarrelling with each other. There was nothing much else to do (there was no TV). Therefore, even if the Hindu states were quarrelling, that was nothing abnormal; they were just following the world practice of those day.

39
FACTORS AFFECTING OUTCOME OF WAR

It is clear from the narrative in the preceding chapter that the two most commonly cited causes of Hindu defeats, conjured by the modern Hindu mind, have no validity. Then, what were the causes for Hindu defeats? The simple and short answer is 'Military non-performance of the Hindus'. To examine the reasons of that non-performance, we have first to understand the nature of war, i.e. the factors that determine the outcome of war.

'War' is a three-letter dirty word, involving death and destruction, murder and mayhem, and everything unpleasant and unpalatable. However, it is war, which determines the fates of nations, and their pecking order in the comity of nations. Civilizations rose to their glory and grandeur on the shoulders of war; that was the case with all major civilizations, e.g. Greek, Roman, Christian and Islamic. War has dominated the human affairs right from the dawn of history, which is essentially a chronicle of wars. Those civilizations who could not understand the centrality of war in human affairs fell by the wayside; unfortunately, the Hindu civilization falls in this category. The one unimpeachable lesson of history is that maintenance of the delicate balance of civilizations requires War, or the 'Threat of War'; that is the only language the world at large understands. We have covered the issue of war in detail in chapter 35.

It is not that the Hindus could never appreciate the importance of war. Actually, one of the very first persons to understand the centrality of war in human affairs, was a Hindu; his name was Chanakya Kautilya. As early as the 4th century BC, Chanakya told everything that needed to be known about war; and he did that in a very blunt and forceful language. We have covered Chanakya's *Arthshastra* in chapter 17.2. If Hindus had paid even part heed to Chanakya's concepts, they could have gone on to dominate the world. However, the Hindus lost the script and their way very early in their history.

War is central to the issue under discussion by us. Like a computer, war has two aspects, i.e. Software and Hardware, viz.:

- *Software Factors, or 'Mind' Factors*
  Strategy and Generalship — Tactics
  Troops — Skill Levels
  Hindu Mindset (see chapters 40 and 41)
- *The Hardware Factors, or 'Muscle' Factors*
  Number of Troops
  Weapons Technology and Quantities
  Battle Venue — Distance from home base
  War Animals — Horse vs Elephant

For victory in war, it is essential that both the Hardware and Software elements are present in reasonably adequate quantity and even more importantly, in quality. No war can be fought in the absence of either of these elements. However, there can be endless arguments as to which of the above element is more important. All that we can say is that history records many cases in which armies even severely deficient in Hardware but adequate in Software, recorded victory after victory. There are not many cases in which armies short on Software could record victory, even if overflowing with Hardware. The Software factor could also be called morale, though only in a limited sense.

The primary cause of Hindus going under was their comprehensive defeats by the Muslims in the following two battles:

1009 AD; Mahmud of Ghazni vs Anandpal
1192 AD; Muhammad Ghauri vs Prithviraj Chauhan.

The first battle showed how vulnerable Hindus were. The second battle demonstrated how easy it was to subdue them for the long term. In the following paragraphs, we examine the Software and Hardware factors of the Hindu and Muslim armies in the context of the above two battles.

Software Factors

Skill Levels: In terms of skill levels of troops, the Muslim armies appeared to have an advantage; they had an edge at least in one respect, i.e. they were better horsemen. The Muslim armies had special columns of 'mounted archers', who could fire arrows with precision, whilst at full gallop; Hindu armies had no answer for that.

Generalship: It is generally believed that Muslim generals were of higher caliber. Especially, Mahmud of Ghazni is counted amongst the best generals of the world. Amongst the Hindus, Prithviraj Chauhan was a general of very high caliber. But, it would be difficult to avoid the conclusion that Prithviraj Chauhan was out-maneuvered and out-foxed by Muhammad Ghauri in 1192 AD.

Hindu Mindset: This is the subject at the center of our study. It is a rather complex issue and is discussed in detail in chapters 40 and 41.

Hardware Factors

Troop numbers: As stated earlier, almost all the Rajput Rulers of North West India contributed their troops for both the battles. That would lead us to conclude that Hindus armies must have had a significant numerical superiority; they could have easily had twice or thrice the numbers of troops when compared to the Muslim armies.

Weapons Technology: At the start of the second millennium AD, Hindu civilization was a very advanced one. There is nothing to suggest that in weapons technology, Hindus could have been in any way less advanced than the Muslims could. As such, weapons of both sides could be considered to be of comparable class.

Battle Venue: Both the above battles, as all other Hindu–Muslim battles, were fought in the very backyard of the Hindus. Hindu armies had an enormous advantage in terms of Supply and Support Systems. Muslims were fighting far away from their home base and were thus at a major disadvantage.

War Animals — Horse vs Elephant

Before the invention of the machine, the horse was the most important weapon of war. It had been used in that role from times immemorial; almost from the beginning of human history. Every famous general has ridden it; and he has ridden nothing else. One common feature of all victorious generals was their unshakable faith in the horse as a weapon of war. The most famous horse of history was Busephelus belonging to the most famous general of history, Alexander the Great.

The horse, the rider, and the sword, constitute the first example of what in modern military terminology is called the 'Weapon System'. The sword is held (firmly) in the hand of the rider, who is in (firm) touch with the body of the horse, through his legs. Through that physical touch, the horse can read the mind of his rider, including his state of confidence or panic. Thus, the horse can anticipate commands of his master and give a real-time response. This single factor of close and instant interaction of the 'horse, the rider and the sword', resulted in achieving innumerable victories.

The horse is one of the most intelligent animals. It has all the qualities essential for the battlefield, i.e. speed, stamina, agility, flexibility, easy maneuverability and endurance. The greatest factor in favor of the horse is the positive control that the rider has through means of the reins; that ensures total control and instant response, an indispensable requirement in a war situation. In addition, there is the great sense of loyalty of the horse to his master. History records many instances where the horse saved his master from hopeless situations, and in some cases from certain death. In the Indian context, we have the case of the horse named Chetak, who saved his master Rana Pratap from a very tricky situation.

Hindu armies, though using horses relied excessively on elephants. Hindu commanders used elephants in two ways:

- In large numbers to form offensive phalanxes in the battle line-up.
- As an individual mount of the commander, especially the commander-in-chief, or the king.

In addition to the Hindus, two other countries known to have used elephants in war in a major way were Iran and Carthage of North Africa. Hannibal (3rd century BC) of Carthage used elephants when attacking Europe. However, Hannibal did not ride the elephant himself; he always rode the horse.

For the Muslim commanders, it was horses all the way. In fact, all commanders all over the world used only the horse; these included Alexander, Caesar, Genghis Khan and all others. They might have heard of the elephant, but they knew its limitations as a weapon of war; it was never to be used as a mount for the commander.

Now, as a weapon of war, the elephant lacks almost all the qualities of the horse, which we have detailed earlier. All those attributes are either absent, or highly deficient in the elephant. Its most serious deficiency is the lack of any positive control; the elephant has nothing resembling the rein of the horse. Whatever limited control is there is not by the commander/rider, but through the means of a third party, called *mahout* in Hindi.

The commander/rider stands isolated in a sort of metal container (called howda in Hindi) on the top of the elephant. The commander has no interaction with the elephant and thus, lacks any type of rapport with it. The commander's orders to the animal have to be given through the *mahout*. In the clutter and din of the battle, the *mahout* may not hear the order or misunderstand it, and sometimes even pretend not to hear. In some exceptional circumstances, the *mahout* may be bribed by the enemy to let down the commander at a crucial moment of battle. The *mahout* may try to get the order implemented by the elephant, who may not respond due to the lack of any positive control. There were many instances where elephants went out of control and acted as they wished.

The elephant has no speed, is cumbersome in movement and sluggish in response — all self-destructive attributes in a war situation. It has lot of mass and strength; that often proved delusive in actual war situations. The elephant has no loyalty to the commander/rider and hardly even recognizes him. So, it cannot be expected to save its rider from tricky situations, as horses have reported to have done.

Now in an actual war situation, the commander cum king stands/ sits perched high on the back of the elephant. That might generate a feeling amongst the troops that the commander is isolated from them, and sitting rather safely on a high perch; in other words, not sharing the risks with them. Further, if a mishap were to occur to the commander, it would be almost instantly noticed by the troops; that may demoralize them and result in chaos. That actually happened in at least two cases of crucial battles, as will be seen in later paragraphs.

There is still another serious disadvantage of the elephant. In the earlier days, the sword was the primary and most important weapon of war; warrior's reputation was known by his skill and mastery of the sword. It is the only weapon useable in close combat situations, which invariably determines the final outcome of war. When the general is sitting on the top of an elephant, he is in no position to use the sword. That is a major drawback and liability.

In spite of the known and proven innumerable shortcomings and disadvantages of the elephant, the Hindu commanders, for some inexplicable reason, continued to use the elephant as their major weapon of war. Towards that end, we list below four major battles that Hindu Commanders fought against the invading forces:

Year	Invader (on horse)	Hindu Defender (on elephant)
326 BC	Alexander	Porus
710 AD	Muhd. bin Qasim	Dahar
1009 AD	Mahmud of Ghazni	Anandpal
1192 AD	Muhammad Ghauri	Prithviraj Chauhan

It is noticed from the above that.

- All four invaders were on horseback, and all four emerged victorious.

- All the four defending Hindu Commanders were on elephant, and all four were defeated. Of course, there were also other reasons for the Hindu defeats — riding the elephant was just one of those.

History records the following episodes in respect to the above campaigns:

- *326 BC:* Porus employed phalanxes of elephants as a defensive cum offensive shield. Under heavy onslaught of arrows from the Greek cavalry, the elephants panicked and started backing off, stampeding their own forces. In addition, Porus was himself riding an elephant.
- *710 AD:* There was a stage in the battle when Hindu forces had an upper hand. Dahar was on his elephant and was consequently rather exposed. An arrow hit Dahar and he was killed; panic spread in the Hindu forces and they retreated.
- *1009 AD:* At a crucial stage in the battle, when the Hindus were winning, Anandpal's elephant got panicky and bolted from the field. That was a signal for panic and the Hindu forces dispersed, pursued by the Muslim cavalry.

After the Muslims had been ruling India for 300 years, there was the last Sultanate king Ibrahim Lodhi. He also employed phalanxes of elephants in his confrontation with the fresh Muslim invader Babur. Now, Babur had come with the gunpowder cannons. The din caused by the firings of cannons caused mayhem amongst the elephants and they stampeded their own troops.

40
HINDU MILITARY MINDSET

History tells us that certain races and civilizations were traditionally war-like and aggressive. The Greek, Roman, Christian and Islamic civilizations fall in this category, though at different points of times in history. Then, there were the tribal civilizations like the Mongols/ Tartars, Huns and Goths, who spread murder and mayhem all over the world. They established major empires based on their drive for dominance, backed by raw courage.

On the other hand, there was the Hindu civilization, which, came to be (wrongly) considered by the Hindus themselves as non-offensive, docile, tolerant and the accommodating type. Even today, most Hindus like to project themselves as such. Hindu generals for some strange reason never ventured out of their land borders, even when they had the means and the duty to do so. The difference between Hindu civilization and the rest is too stark to be missed by anyone. In this respect, Hindu civilization stands almost alone in its (dubious) splendor, to its great disadvantage and at great cost to itself and its people. Over the ages, Hindus acquired a number of mindsets, some under the influence of Buddhism; these were to prove their undoing.

It is fashionable for Hindu apologists to project 'non-invasion of foreign lands' as a great virtue and an outstanding characteristic of the great Hindu culture and civilization. 'We never invaded anyone in our 3000 or 5000 years of glorious history' croak the Hindu leaders. One hears of this argument ad nauseam during TV debates and seminars. 'How unique, how wonderful is our record of non-invasions,' say Hindu leaders, with a touch of pride and twinkle in their eyes. Though we may note in passing that the great lord Rama did invade foreign lands.

for want of
assimilation

empire - building
cultural onslaught

Actually, this viewpoint is nothing but a reflection of the muddled Hindu thinking on military issues. This reasoning shows lack of the fundamental military principle, i.e. Offence is the best form of Defense; you have to get to your adversary before he can get to you. If you are on the offensive, you have already won half the battle; if you are on the defensive, you start with a great disadvantage. This is how the psychology of war works, and has always worked throughout the ages. Almost every war in history establishes the truth of this dictum. In the vast majority of battles, the party on the offensive won. Muslims won almost every battle as they were always on the offensive, and carried the battle to other people's land. The great strategist Chanakya, in his *Arthshastra* (chapter 17.2), has laid great emphasis on 'offensive actions'. The capture of (weak) neighboring states has been prescribed as the sacred duty of the king.

Let us examine the international practice on the subject of 'offensive action'. In chapter 35.1, we have listed some of the famous generals of history. In studying their careers, we find many common features. These generals spent most of their time on horseback, with their sword drawn, ready to strike. Each dawn, they would throw a challenge to a new land and a new country — 'Submit, or else', was their clear and crisp message. They took no quarter and gave none. They marched from land to land, planting their flag and moving on. They went to the enemy's lair, and seldom allowed him to come to theirs. The military initiative always lay with them. That is why history recognizes and respects these generals. Alexander became the world conqueror, as each week he had a fresh land to conquer. Genghis Khan and Timur became the terror that they were, as they identified their next victim as soon as they had decimated one. Atilla the Hun is famous, because he was always on the offensive. None of the famous generals allowed themselves to be on the defensive.

There is the case of the Punic wars; these were fought in 3rd century BC between Rome and Carthage, a country in North Africa. Around 216 BC, General Hannibal of Carthage landed in Europe for some offensive actions. He captured Spain and surrounding territories, and started attacking the mighty Roman Empire. Over a period of some ten years, Hannibal inflicted a series of defeats on the formidable Roman legions. In one engagement alone, he is reported to have killed some 50,000 Roman soldiers, a humongous number those days. Having been bored with his victories, Hannibal went back to Carthage. Now, Roman General Scipio attacked Carthage, and defeated Hannibal decisively in his own den, in the famous battle of Zama in 202 BC. Moral of the story is that Hannibal was winning as long as he was on the offensive; he lost when he adopted defensive posture.

We have earlier quoted the little known Sanskrit *shloka* (verse) — *'Veera Bhoga Vasundhra'*, i.e. Brave will enjoy the Earth. Those generals were the perfect example of that. They lived for the day, and enjoyed fully the fruit of their labor in the present life, without worrying about the next. That is what human spirit is all about, that is the human DNA. Human spirit is not about first non-performing, and then cooking up reasons for defeat.

Let us now explore Hindu scriptures on their attitude towards offensive actions. We start with the holiest of the holy, the one and only *Bhagwad Gita*; its central message is somewhat on the following lines:

Tolerate no injustice — never, under no circumstances
Engage in (righteous) war
Kill, or be Killed (nothing in between)
Kill, and enjoy the fruits of earth
If killed (in battle), enjoy the fruits of heaven

When in an offensive action, you are the winner all the way; if you win, you enjoy the fruits of earth; if you lose and are killed in battle (which you have to be, if you lose), you enjoy the fruits of heaven. This is the message of the great Lord Krishna himself. No other religion is more clear and unambiguous on the imperatives of offensive action. It was after getting this message and direction that Arjuna went on the offensive and emerged victorious.

We now take the Hindu's second most holy book, i.e. the *Rig Veda*. The *Rig* lists the innumerable military exploits of Indra, all in the offensive mode. He is shown repeatedly attacking various demons, defeating and

intolerable

killing them; those include the invincible demon Vrtra, for which Indra took help from Vishnu. The Hindu scriptures are full of tales of gods being on the offensive against the demons, and demons against the gods. In many of those offensive actions, the great lords Vishnu and Shiva play crucial roles.

Then, there are the two great epics, the *Ramayana* and the *Mahabharta*; in these lies the very soul of the Hindus. These epics are, in essence, the glorification of war and offensive actions. Rama in *Ramayana* and the Pandavas (assisted by Krishna) in the *Mahabharta*, were on the offensive, and both emerged victorious, even against heavy odds. The great goddess Kali is depicted always on the offensive against various types of demons, including the most formidable ones. She shows no hesitation even in drinking their blood. What can be a greater endorsement of 'offensive actions' in Hindu scriptures?

In view of the foregoing, it is difficult to see from where the Hindus have picked up their theory of and obsession with 'no offensive action'. It has no sanction in Hindu scriptures; Hindu culture and tradition revolves around 'offensive action'. We can keep on quoting example after example from Hindu scriptures, which stress on offensive actions and other imperatives of war. The great Lord Rama himself had gone on the offensive against a foreign land and returned victorious with full honors. Ravan had abducted just one lady (Sita); Lord Rama lost no time in crossing the seas, and dispatching that abductor to hell. The Ghaznis and Ghauris routinely abducted hundreds and thousands of Hindu women with a view to dishonor them. Some Rajput clans trace their ancestry to a Suryavanshis of Lord Rama. But, the Rajput 'Suryavanshi' blood was not stirred by these wanton acts of the Ghaznis and Ghauris. Not one Rajput ruler followed the example of their ancestor, the great Lord himself, and went to Afghanistan to teach a lesson or two to those upstarts and molesters.

There are other Rajput clans who trace their ancestry to the Chandravanshis of the *Mahabharta*. They could have taken inspiration from their ancestor, Lord Krishna. In the *Bhagwad Gita*, the great Lord has laid down the duty of a Kshatriya to get after the unjust, the tyrant and the usurper; 'Not doing that attracts sin' is the word of the great lord himself. Even that clarion call failed to stir and inspire the Chandravanshi blood.

1000 onwards

There is no way Hindus can justify their non-invasion of the Afghanistan of the early centuries of 2nd millennium AD. Such an invasion was imperative for a variety of reasons, including saving the honor of their women; a most sublime cause, if there is ever one. Even if one Rajput ruler had taken that route, there would have been no need to conduct the abhorrent practice of '*jahaur*' — mass self-immolation of women on funeral pyres.

Muhammad Ghauri had started nibbling at Indian territories in 1176 AD. By 1186 AD or so, he had incorporated large parts of Punjab up to Bhatinda in his empire. What was the response of Prithviraj Chauhan? Nil; he sat tight in his state. That was the Hindu mindset at work. The fundamental and cardinal military principle is to be always on the lookout for a 'potential' threat, and to eliminate it immediately, sometimes even before an actual threat. Chanakya's *Arthshastra* spells it out explicitly; every successful (foreign) general followed it literally. It was incumbent on Prithviraj to take on Muhammad Ghauri as soon as he saw him taking over Ghazni, and building a power base there. The Vedic *dharma* required that line of action. There is no way Prithviraj should have allowed Ghauri to take over Indian cities like Peshawar, Multan and Lahore, leave alone the next-door citadel of Bhatinda. Problems do not go away by pretending that these do not exist, or by shutting one's eyes to reality.

In 1191 AD, Prithviraj had the good fortune to defeat Muhammad Ghauri. However, Prithviraj allowed a wounded Ghauri to get away. If Prithviraj was unable to kill or capture Muhammad Ghauri on the battlefield, it was incumbent on him to pursue the Turko-Afghan to his lair in Lahore or Multan or Peshawar. Actually, Prithviraj should have traveled all the way to Ghazni and Ghur, and smashed Muhammad Ghauri's power at its very base; from the military point of view, that was the only option. That course would have prevented future upstarts to cast their lustful eyes on this holy land. Hindu history would have developed along entirely different lines.

The Hindu mind, mostly under the influence of Buddhism and Jainism, came to suffer from a series of pre-conceived concepts and notions, viz.:

- The Hindu mind failed to comprehend the centrality of war in the affairs of men, and in determining the fate of nations and civilizations. They failed to develop a military culture of offensive action (an international norm), essential for the survival as a free and proud nation.
- At some stage, Hindus started suffering from an overdose of *ahimsa* (non-violence), *shanti* (peace) and *satya* (truth) — the '*ass*' syndrome. This was partly under Buddhist influence, especially the *ahimsa* part.
- Under the influence of Brahmin priests, Hindu rulers got overtly obsessed with their next-life. If they could assure their after-life, they did not perhaps mind going-under in the present life.
- Hindus also suffered from frequent bouts of bogus morality and phony spirituality. Hindus must understand that they need not and do not have any greater share of these two attributes, than any other people on this earth do.

The above type of mindsets unnerved people from the tasks of life, especially those relating to war; that sapped the national will to fight. To date, Hindus keep on justifying the non-performance of Hindu rulers on the pretext of some non-existent tradition of 'No offensive action, please; we are Hindus'; in that one sentence, and mindset, lies the cause of Hindu downfall.

The Achilles Heel of the Hindu armies appeared to lie in the arena of pre-determined mindsets and the lack of motivation; in their inability or unwillingness to 'stand up and fight'; and to fight until victory or death, as dictated by the great Lord Krishna himself in the *Bhagwad Gita*.

41
FACTORS AFFECTING HINDU MINDSET

In the preceding chapter, we have referred to the concept of the Hindu mindset and discussed its broad contours. In this chapter, we discuss the factors which may have contributed to molding that mindset. We do that under the following headings:

- Religious Influences
- The Caste Factor
- Miscellaneous Factors

Religious Influences

For analyzing the issue in depth, we start at the very beginning of Hindu history. The Vedic people were of martial nature who loved their freedom and self-esteem above everything else. By any standard, they were a set of very proud people, who would give back in equal measure, if ever threatened or insulted. They would also fight for justice. In fact, that is the central message of the Hindu holy book, the *Bhagwad Gita*. The book also enjoins on the Hindus to engage in righteous war — to kill, or be killed. As per the *Gita*, death has no meaning for the Hindu, as the soul is indestructible and everlasting.

After recording victory over Ravan, the great Lord Rama expressed that he had attacked Lanka to avenge his clan, the Raghu-Kul, which Ravan had insulted by abducting Sita. In the *Mahabharta*, the great Lord Krishna exhorts Arjuna to pick up arms, to fight against the injustice done by his cousins, the Kauravas. War was a way of life for those early Hindu people. Their two epics, the *Ramayana* and the *Mahabharta*, are essentially tales of war.

The Vedic people were chugging along nicely with their joyous spirit and martial character. To their great misfortune, in the 6th century BC, two new religions, i.e. Buddhism and Jainism emerged on the Indian scene. These religions turned the Vedic value system on its very head; they were the very anti-thesis of Vedic religion. Vedic people lived their life to the full, enjoying the good things of life provided by the almighty God. Buddhism declared that life was sorrow and misery; and to get rid of that sorrow and misery, the individual must give up 'desire'. Buddhism preaches disengagement from normal day-to-day issues of life and living, and recommends concentration on otherworldly matters. The religion has a negative view of life, and pessimistic approach to it. In the Vedic system, animal sacrifice was the standard way of pleasing the gods, and was a routine (religious) activity. Buddhism abhorred that practice.

Now, it is all right to tell a bunch of elderly, or monks in a monastery, to live without 'desire'. However, imagine whole populations living that way. Desire provides the very reason and fuel for living. Who will carry out the daily chores of life, many of which are unpleasant and hard to implement? Who will defend the country against the invaders? We live in a wicked and wild world (yes, even in those days). For every one good man, there may be two to three evil-intentioned men out there in the real practical world. You may have good intentions; others may be full of evil ones.

That was not the end of the story. Buddhism invented and advocated the body and soul-destroying doctrine of *Ahimsa* (non-violence). The original Sanskrit word is *himsa* (violence); the word *ahimsa* is derived from it by adding an 'a'. That suggests that the word (and therefore the concept) '*ahimsa*' was a later invention. There does not appear to be any equivalent word for *ahimsa* in Persian, Arabic, Urdu and in most other languages of the world. Even in the English language, non-violence has been derived by adding 'non' to 'violence'. That would lead to the conclusion that not many, if any, civilizations believed in the concept of *ahimsa*; most may not have even a passing acquaintance with it. *Ahimsa* appears to be a typical Indian innovation, imposed by Buddhism; it is foreign to Hinduism. To have an island of *ahimsa* in an ocean of *himsa* is the surest and safest route to disaster. You cannot start believing in *ahimsa*, when others around you do not believe in that. (We make these comments with due apologies to Mahatma Gandhi, perhaps the greatest man of the 20th century).

In the 3rd century BC, Buddhism with its body and soul destroying doctrines, was struggling to establish itself in India, but finding the going tough. At that crucial moment, again to the great misfortune of the Hindus, there appeared on the scene Emperor Ashoka; he has been given the appellation 'Great' by later day modern Hindus. Ashoka lost his kingly nerve after the battle of Kalinga, and adopted the creed of *ahimsa*; he took to Buddhism as a fish takes to water. In later life, Ashoka acted less as a king, and more like a like a Buddhist monk.

Now, *ahimsa* may have a niche in an individual's personal life; it can have no space in the scheme of things of a king. Ashoka could not see this crucial difference. Ashoka, not only adopted *ahimsa* for himself, but left the following specific written instructions for his heirs, in the form of Rock edicts:

> 'The inscription of *dhamma** has been engraved so that any sons or great grandsons that I may have should not think of gaining any new conquest. They should only consider the conquest of *dhamma* to be true conquest.'

As it happened, the sons and grandsons of Ashoka followed the above instructions both in letter and spirit — disaster was the natural consequence. At that time, barbarian tribes were hovering around the borders of India, waiting for weakness to set in. They made full use of that opportunity and poured across the borders. The invading tribes appear to have met with very little resistance, due to the prevalence of the concept of *ahimsa*. These tribes went to rule major parts of North West India for some 500 years.

After the great push given by Ashoka, Buddhism caught the imagination of the Hindus and spread very fast, pushing Hinduism to the sidelines. Even Jainism did not lag behind, and prospered along with Buddhism. In such a milieu, Hindus had no chance and they

---

*   *Dhamma* is Pali for Sanskrit *dharma*, meaning righteousness

got thoroughly confused. Gradually, Buddhism started grafting itself on Hinduism, and the distinction between the two started lessening. Hindus took to identifying Buddhism-like-traits in Hinduism, and emphasizing these in an out of proportion way. That is why not many modern day Hindus are able to say whether *ahimsa* is a Buddhist/Jain or Hindu concept.   Even Christ started to sport - - - -

Buddhism prospered and prevailed in India for about 800 years, overshadowing Hinduism for prolonged periods (may be in patches). The separating lines between the two became rather blurred; it was not always possible to distinguish between the traits of the two religions. Matters got even more confused when around the 5th century AD, Hindus incorporated Gautam Budha as the ninth *avatar* (incarnation) of Vishnu. When Hindus realized the true pessimistic nature of Buddhism, they set about the task of driving Buddhism out of India. But that proved to be an uphill task in spite of royal patronage, first of the Brahmin Shunga and Kanva kings (2nd–1st century BC), and later of the Gupta Kings (4th–5th AD). It took Hindus some 800 years to get rid of Buddhism. However, many of its traits stayed put, some grafted to Hinduism, and quite indistinguishable from it. The writer would like to distinguish between Hinduism and Buddhism by making, with due apologies to Buddhism, a controversial statement, which goes as follows:

'Whatever is joyous and aggressive is from Hinduism; whatever is pessimistic and defensive has its origin in Buddhism'.

Therefore, Hindus should have no hesitation in discarding the traits coming in the later category.

Largely under the influence of Buddhism, Hindus forgot the following basic principle of life and living:

'Sword is the best, perhaps the only, guarantor of peace'.

In the days of yore, a famous general had famously said, "With sword, I bring peace to the world." In other words:

- If the king has a sharp sword, peace and prosperity prevail; even *ahimsa* can get a niche.

- If the king lets his sword get rusted, murder, mayhem and slavery follow; *ahimsa* gets wiped out.
- The strongest believer in peace must have the sharpest sword. Nehru was a follower of the first part, but he forgot the second part, with all-round disastrous results.

In the preceding paragraphs, we have laid perhaps unfairly, a lot of blame for the Hindu misfortunes, at the door of Buddhism. However, it must be admitted that Hinduism itself had a major contribution to its own downfall.

As a start, Hindus allowed the great and vibrant Vedic *dharma* to be overtaken by nihilistic and pessimistic religions like Buddhism and Jainism. These religions advocate detachment from day to day issues of life and living, and giving up desire itself. How could any person think of walking away from the glories and splendor of the Vedic and Upanishadic thought processes, and willingly walk into the uncertainties and dark alleys of Buddhism. Hindus got themselves entrapped in a *chakravue* (trap) of 'false *dharma*' or a deluded sense of *dharma*. They abandoned the sublimity of the Hindu religion, without even understanding the intricacies of Buddhism. The start of the Hindus' period of miseries is directly linked to the conversion of Ashoka to Buddhism in the 3rd century BC. For some 800 years, Hinduism in its own land was gasping for breath under the onslaught of Buddhism.

Hinduism believes in rebirth and there is a lot of emphasis on improving the quality of life in the next birth; the final aim is to attain *moksha* — deliverance from *samsara*, the cycle of birth and death. Brahmins exploited the fear of uncertainty of the next life; they used it as a means to have influence on the ruling classes. The Brahmin priest called *raj guru* (court priest), has been an integral part and an adornment of the courts of Hindu rulers from times immemorial, i.e. the *Ramayana* days. Under their influence, Hindu rulers became more worried about the next life, rather than concentrate on the imperatives of the present one.

In their worry about the next life, the Hindu rulers got busy in purifying their soul; in that pursuit, the rulers neglected organizing their forces for the defense of the country. The invaders exploited that situation to the hilt, and the Hindu rulers went under, rather easily. The invaders were ruthless, to say the least; they took to murder, mayhem and above all, all-round loot. The result was the impoverishment of the Hindus. So impoverished, the Hindu rulers started taking refuge in supernatural consolations. They convinced themselves that 'mastery and slavery' were rather superficial delusions. In so brief a life, the freedom of the nation was hardly worth shedding human blood. The Hindus were greatly assisted in going in that direction by the Hindu concept of the world being *maya* (illusion).

There are *shloka*s (couplets) in the Hindu scriptures relating to the word '*shanti*' (peace), e.g. *Om shanti shanti* (Om, peace, peace). Such *shloka*s were for peace of the individual and his soul, to be used strictly within the confines of the household. However, at some stage, these got spilled on to the streets; *shanti* became a national obsession, instead of being an individual concern. In addition, Hindus willingly boarded the (Buddhist) bandwagon of *ahimsa* (non-violence). Hindus also have this bug of *satya* (truth). Even these days, it is not uncommon to hear Hindus, as if in a trance, chanting '*ahimsa, shanti, satya*'; we will call it the '*ass*' (an acronym) syndrome. These words may be all right at an individual level, for use in the confines of the household. But there is in no way that these words may be applied to a nation as a whole. That can only be a recipe for military disaster; and that is what befell the Hindus.

Hindus also suffer from the indigestion of 'tolerance'. At some stage Hindus started to believe that they were a set of tolerant people; that is not true. Nor is tolerance a virtue that needs to be acquired. This misconception about tolerance added to the woes of the Hindus. We have already covered this subject in chapter 13 and would not labor on it any further.

When the Hindus were finally successful in driving out Buddhism, they for some strange reason did not revert to the glorious *Vedic dharma*. They brought in a new version of Hinduism, which both in essence and detail, is substantially different from the Vedic *dharma*; we have called that the *Puranic dharma*. It was perhaps not even pure Hinduism; it may have had (indiscernible) coatings of Buddhism. The *Puranic dharma* introduced a whole series of new gods and goddesses, different from the one Supreme God of the Upanishads, and from the gods of the *Rig Veda*.

The net result of all that could only have been a whole lot of confusion; and that is what appears to have followed. A set of people religiously confused cannot be militarily precise and strong; that proved to be the undoing of the Hindus. In spite of what the pseudo-secularists may say, there is a close link between religion and military value system of a society. Here, we cannot but point out to the stark simplicity and precision of Godhood of Christianity and Islam — the two successful and dominating religions of the world.

Hinduism is a mythology based religion; actually, it revolves around mythology. The *Ramayana* and the *Mahabharta* wars may have actually been fought, though no verifiable historical data is available in support of the same. Pre-1947, Hindu mythology was in the form of books and folk-tales; therefore, their spread and exposure to the general public was somewhat limited. However, with the arrival of TV, things have undergone a drastic change. The innumerable 24x7 channels have to fill their time slots. The result is that the channels have taken to Hindu mythology with a vengeance; it provides them with an inexhaustible source of stories, most exciting and entertaining, with direct appeal to the Hindu mind, if not the Indian mind.

Presently (2009 AD), there is a surfeit of serials based on Hindu mythology; these go on almost round the clock. The serials generally revolve around some god or the other. Many of these serials involve war scenes. Arrows are shown turning into snakes, spitting fires or causing tornadoes, floods and general mayhem; there is almost nothing that arrows cannot achieve. Arrows get magical powers when launched after uttering some *mantra* (spell); success gets guaranteed. Warriors are shown getting the most destructive weapons from gods, on their mere asking, even whilst the battle is going on. A great warrior like Arjuna actually traveled all the way to heaven to get *divya-astras* (heavenly weapons).

Now, the above stories are all right if read in books, as part of mythology. But to show such war scenes day in and day out on TV screens is a different matter. The raw, impressionable youthful mind, especially in the villages, may get all the wrong messages. He may start believing that war is an easy process. All that victory requires is a correct type of *mantra*, or keeping some god or the other in good humor; that god will supply the appropriate weapons. Over a period of a few years, such like mindsets have a habit of getting lodged in the subconscious. People with this type of mindset may get associated with the armed forces, with disastrous long-term consequences. These people may forget that war is a very dirty game, which requires:

— Total commitment of mind and body, both at the individual and nation level
— Hard training, and harder effort in the battlefield
— Shedding of blood, lots and lots of it
— Good generals, with aggressive and risk-taking mindset

We may recall that for some 1000 years starting 700 AD, Islam dominated the world. Now, Islam is a 100% history based religion, with no mythology whatsoever. Similarly, the second dominant religion, i.e. Christianity is also largely free of mythology.

Does all the above have a message for the Hindus? It would appear so. Whilst Hindus may continue with their mythology, it should not be spread and propagated beyond a point on TV channels. However, as the TV channels get out of other options, they would increasingly resort to more and more of Hindu mythology. They are likely to get encouragement from certain sections of Hindu society in the (mistaken) belief that that would help the spread of Hinduism. Even if it does that, it may be a type of Hinduism (all mythology based) which may not be welcome from the military angle in the present day scenario.

Hindus are unwilling to acknowledge their dismal military performance over the last 1000 years. The question of facing or rectifying their shortcomings in this field does not only arise. Whenever such a question is raised, Hindu ideologues dive headlong into their mythological past.

In 2008, author was present at a seminar to discuss terrorism and military issues. A top ideologue of the RSS was the main speaker. He shifted his timeframe to the *Ramayana* mythology (5000 or so years age). He expressed that the people who could defeat Dashanan (meaning Ravana), can have no problem dealing with any threat, military or otherwise, that may arise now or in the future; he went on and on with that theme. For that worthy, the Hindu history of the last 2,300 years was just a speck. He could not care less if a few ten-thousand horsemen (barbarian tribes) from Central Asia smashed the mighty Mauryan Empire, as early as 180 BC. He was just not worried that Ghaznis and Ghauris, Baburs and Abdalis, Clives and Dupleixs and innumerable others inflicted defeat after defeat on Hindu armies, and generally played up merry hell with the Hindu masses. The earlier ones abducted their women by the thousands; that was not reason enough for the Suryavanshi blood to come to the boil. For about 1,000 years after 1000 AD, the victors of Dashanan (Ravana), the Suryavanshis, were nowhere on the offensive scene, or on the winning side, or even seen defending the honor of their women (burning them up in mass funeral pyres does not amount to defending their honor). Hindus are often seen relying on some fanciful or exaggerated tale or two of victories told in folklores by bards. These bards were in the employment of the ruler and their main aim was praising the ruler. What they wrote was very little history; facts hardly mattered to them.

The Caste Factor

The caste system in Hindu society has been a ground reality from times immemorial. That is a fact that cannot be denied, though efforts for such denials are made from time to time. Caste is a factor exclusive to Hindus; no other civilization is known to have practiced it.

There is a lot of debate and divergence of opinion on the effect of caste on the military value system of the Hindus. The conventional view is that Hindus had the specialized warrior class called Kshatriyas; that should have helped Hindus to put up a good performance in warfare. The contrary view is that the caste system was a big drag on the military value system of the Hindus. The jury is still out and no consensus has been reached. There are two broad views of the caste system:

- The conventional view, i.e. what the castes are supposed to do theoretically.
- The practical view, i.e. how the castes actually performed in practice down the ages.

We study the above two views one by one.

The Conventional View of Castes

We have listed earlier the four castes, i.e. Brahmin. Kshatriya Vaishya, Shudra. We are not clear as to what approximate percentage of society was formed by the respective castes. In an arbitrary way and only for ease of discussion, we assume the following approximate percentages of the various castes:

Brahmins — Priestly Class — 20% 10 ∿ 5
Kshatriyas — Warriors — 20%   15 ∿ 20
Vaishayas — Traders/Agriculturists — 30% 25
Shudras — Service Class — 30% 40 50

Brahmins

Conventionally, Brahmins are the Hindu priests for proper conduct of religious rites, including animal sacrifices during the *Rig Vedic* times. Their main function was to study scriptures and give guidance on religious matters. However, in Hindu mythology cum history, there are innumerable instances where Brahmins engaged in warfare and even commanded troops; we will study this aspect a bit later.

Kshatriyas

They were the warrior class for the defense of the motherland and its people. It could be presumed that they were full of the spirit of nationalism and patriotism, and ever ready to lay down their life for the country. The common assumption is that Kshatriyas must have fought most of the Hindu wars, and were natural rulers and kings. That may not have been so in actual practice, as we shall see a few paragraphs later.

Vaishyas

Traditionally, all over the world and in all societies, it is the agriculturist classes, which provide the bulk of soldiers for war. Of the 30% Vaishyas, 20% could be agriculturists, the remaining 10% being traders. Thus, in addition to the Kshatriyas, it was the Vaishya who may have provided the sinews of war.

Shudras

[handwritten note: artisans? where?]

The Shudras were the recipients of the shabbiest treatment by the Hindu society. For all practical purposes, they were not even a part of society. The Shudras must have been simmering with discontent, and hatred for the higher castes. (Even presently, they are in the same frame of mind). In such a scenario, it would be reasonable to assume that the Shudras would not have had any stake in society. There was nothing in it for them, except humiliation and gross injustice. Therefore, if foreign elements were to take over such a society, the Shudras could not have cared less; their lot could not get any worse. The new masters may even give them a bit of a better deal. That reasoning would lead us to believe that Shudras might have had no stake in defending the country. Even if they were physically present in the battlefield, their heart was not in fighting. And, mental attitudes have a major influence in determining the outcome of war, i.e. victory or defeat.

[handwritten marginal note: ? kept Hinduism alive!]

From the above, we may be perhaps justified in stating that a significant part of Hindu society may have had very little stake in, or contribution to war effort. That has been often cited as a factor for poor military performance of the Hindus over the ages. However, that may not have been so in actual practice, as we shall see in the following paragraphs.

Actual Performance by the Castes

What we stated above is based on the conventional role of the castes in Hindu society. However, study of Hindu mythology and early history reveals a substantially different picture. We start our study with mythology.

Caste Factor in Mythology

In actual practice, we find a number of Brahmins performing Kshatriya functions, with a vengeance; some examples:

- Parshurama: He was a Brahmin, son of a *dev-rishi* and the sixth *avatar* of Vishnu. He wiped out the entire Kshatriya clans 21 times. For that purpose, Parshurama used the *parshu* (axe) provided by Lord Shiva. It would appear from this episode that the two great gods Vishnu and Shiva may not have been favorably inclined towards the Kshatriyas.

- Ravan: He was a blue-blooded Brahmin being the grandson of a *dev-rishi*. He was one of the greatest warriors of all times.
- Dronacharya: He was the Brahmin guru of the Kauravas and the Pandavas of the *Mahabharta*; he taught martial arts to the princes. After Bhisham Pitama, he was perhaps the greatest fighter of those days. Dronacharya could be defeated only by an ingenious plan of Lord Krishna.
- Chandravanshi line of the *Mahabharta*: The Kshatriya blood in this line was replaced by Brahmin blood, when the great Pauava King Bharat adopted Rishi Bhardwaj as his son. A son of the Rishi ascended the throne after King Bharat (see *Mahabharta* in chapter 5.8).

It appears that the great war of the *Ramayana* had just two Kshatriyas, i.e. the great Lord himself and Lakshmana. Ravan, his brother Kumbhkaran and his son Meghnad were blue-blooded Brahmins. Ravan's army of rakshasas could be presumed to be casteless. So must have been the monkey army of Lord Rama. The great warrior god Hanuman may also have been without any caste appellation.

In our chapters on the *Mahabharta* and the Kshatriyas, we have already brought out that by the time of the *Mahabharta* war, Kshatriyas were highly weakened. Dhritrashtra and Pandu were the biological sons of Brahmin Ved Vyas, who was himself born out of the union of a Brahmin *rishi* and a low-caste fisherwoman. In addition, the five Pandava brothers and their half-brother Karan were biological sons of gods, who could be considered casteless. Both Arjun and Karan were the greatest warriors of all times.

Caste Factor in History

At the very start of recorded history in India in the 7th–6th century BC, we find the Sisunaga dynasty ruling the state of Magadh in eastern India; it soon developed into the dominant power of India under Bimbisara in the 6th century BC. The Sisunaga dynasty is believed to have been low-caste. It went on to rule till the end of the 5th century BC.

In the early 4th century BC, an upstart Mahapadma Nanda overthrew the Sisunagas ruling in Magadh; he is believed to belong to the low caste barber class and founded the Nanda dynasty. At the time of Alexander's invasion in 326 BC, the Nandas were ruling major parts of India.

In 320 BC, the Mauryas replaced the Nanda dynasty; the Mauryas set up the first great pan-Indian empire, which ruled over the entire North India. Its founder, the great Chandragupta Maurya is believed to be of low caste origin, if not an actual Shudra. Chandragupta is generally considered to be the son of a later Nanda king through a low caste woman named Mauri, or of the Mauri tribe. The Mauryan dynasty ruled for 140 years, and included Emperor Ashoka, selected as the emblem of modern India.

The Arthshastra is the world's very first and one of the most comprehensive treatises on warfare. Its author was not a Kshatriya, but a Brahmin, i.e. Kautilya Chanakya, who lived in the 4th century BC. It was because of Chanakya's military acumen that Chandragupta Maurya could establish the first pan Indian Hindu Empire in 320 BC.

On the decline of the Mauryan Empire around 180 BC, power in Eastern India was taken over by the Shunga dynasty which in turn, was replaced by the Kanva dynasty. Both Shunga and Kanva were Brahmin dynasties, which ruled in the Eastern parts of India in the 2nd and 1st centuries BC. In North West India, the (foreign) barbarian tribes snatched power from the crumbling Mauryan dynasty, around 180 BC. These barbarian tribes ruled over North West India for some 500 yeas; Hindus designated them *mllechha*, i.e. lowest of the low caste.

In 320 AD, Chandragupta I, a Vaishya, established his rule over North India. He founded the Gupta dynasty, which after the Mauryas, were the second pan-Indian empire to rule over greater part of North India. They were the longest ruling Hindu dynasty, and ruled for about 200 years. The Guptas produced some great Hindu generals; the Gupta rule has been called the golden period of Indian history. From 606 to 647 AD, King Harshvardhan ruled over North West India; he is believed to be a non-Kshatriya.

It is most baffling to find Kshatriyas largely missing from the major ruling dynasties of India for about one thousand years from the 6th century BC to the 7th century AD. What is even more surprising is that no name of any great Kshatriya general or king emerges during that entire period; all the possible names are non-Kshatriyas:

- Bimbisara and Ajatsatru (Sisunaga)   — Low Caste
- Mahapadma Nanda   — Low Caste
- Chandragupta Maurya   — Low Caste
- Chandragupta I & II, Samundragupta — all Vaishyas

It is possible that some Kshatriya kings might have been ruling some smaller states in parts of India; Porus of Punjab may have been a Kshatriya. However, it is almost certain that Kshatriyas were not in the main line of the ruling classes, for any length of time.

In the 8th century AD, Rajputs appeared on the scene in a part of North West India, called Rajputana (present Rajasthan). Rajputs claimed that they were Kshatriyas, and of the highest class. We have discussed this issue in detail in chapter 23, where we have raised some serious doubts on the issue of the Rajputs being Kshatriyas, especially their claim of belonging to the Suryavanshi and Chandravanshi lines of the *Ramayana* and *Mahabharta* respectively.

In more recent times, we have the two sword-arms of the Hindus — the Marathas and the Sikhs; both are generally believed to be non-Kshatriya. Another martial race, the Jats, also appear to be non-Kshatriya. Still two other martial classes, the *majhbi* Sikhs, and Mahars of Maharashtra are scheduled castes. Yadavas could have been Kshatriya; but they want to be counted amongst the 'Other Backward Classes' (OBCs), which could hardly be considered a shining example of the (supposedly) ruling Kshatriya class.

We may note that we started our narrative with the conventional wisdom and belief that Kshatriyas were the warriors who fought wars on behalf of the Hindus; and that they were the natural and designated ruling classes. The actual recorded history, as outlined in the preceding paragraphs throws up a very different picture. Kshatriyas were nowhere near the fighting and ruling scenes, which was dominated by the low castes and Vaishyas. Thus, the caste scene in India, especially the one relating to the Kshatriyas, is most confusing, in fact, baffling. It would appear that non-Kshatriyas (Low Castes and Vaishyas) played major roles on the military scene, during the early history of India. That would call for a major revision of our conventionally held beliefs.

In view of the foregoing, it would not be right to attribute poor military performance of the Hindus to their caste system. We have seen that for major part of its history, it is the low castes and Vaishyas, who ruled India. They must have won their respective crowns in the battlefield. Thus, almost all castes must have contributed substantially to the war effort; there is nothing to suggest the predominance of the Kshatriyas.

Miscellaneous Factors

Geographical: From the military angle, India has a very secure location, and safe borders. In the South, the West, and major parts of the East, it is protected by the mighty seas. In the North are the formidable Himalayas that were considered impregnable until recently. The only small vulnerable point was an opening of the Khyber Pass in the North West. The Hindus could not defend that small opening; they offered very little resistance to the invaders coming through that route. There were many such invaders; not one of those could be beaten back. It would appear that the secure location of the country instead of being a positive point, proved a disadvantage and a drawback; it perhaps generated a sense of complacency.

Climate: Closely connected with the geographical location is the question of climate. India has hot and humid climate; part of the year, oppressively hot and humid. It is seen that vast majority of conquerors originated from cold or very cold regions. The barbarian tribes in 180 BC onwards were from the cold regions of Central Asia. So were the real tormentors of the Hindus, the Turko-Afghan Ghazni and Ghauri; as also were the Mongol/Tartar Tamberlane and Babur, and the Afghan Ahmed Shah Abdali. The English were from a very cold country. Does the cold strengthen the bones, and perhaps even the mind. There are no simple answers, and therefore we refrain from giving any. The reader may draw his/her own conclusions.

Food: Then there is the question of food. India has vast undulating plains with rich soil, well served by a vast network of river systems. That would appear to suggest that food should have been not only easily available, but also plentiful; though we do read reports of famines in India. The plentiful food could have had the following types of effects:

- With no worry about food, the mind of the king could have been engaged in making some foreign conquests.
- Adequacy of food brought in a sense of complacency, which diverted a king's mind to a life of ease and luxury.

It would appear that the Hindu rulers selected the second alternative, which proved very costly for the country and its people, and to the king himself.

In addition to quantity of food, there is the question of the type of food. All the invaders cum conquerors ate non-vegetarian food; meat was their staple diet. Hindus were largely vegetarian, which possibly reduces the ardor for warfare. In the Vedic times, sacrifice (slaughter) of animals in the honor of gods was a routine, almost daily activity. That accustomed those people to the sight and touch of blood. The vegetarian Puranic people abhorred the sight of blood, especially under influence of Buddhism and Jainism. It would not be unreasonable to assume that Hindus carried their distaste of blood to the battlefield, if not consciously, at least subconsciously. When the vegetarian food is combined with a hot and humid climate, the matters only worsen. It has been stated that this aspect may have even affected the Muslims in the long term, i.e. after they have spent say 200 years in this country. That is why even the Muslims, who had arrived earlier in India, could be defeated by the fresh arrivals, e.g. Babur could easily defeat Ibrahim Lodhi in 1526 AD.

Race Factor: Some races in history have been considered as naturally aggressive; some even bordering on the cruel who believed in raining murder and mayhem. Some examples are:

- The Goths — Visigoth and Ostogoth in Europe
- The Huns — Western (Atilla) and Eastern (Mihirkula)

- Mongol/Tartars — First Genghis Khan, then Tamberlane
- Islamic armies — from time to time

The Christians also displayed aggressive qualities during major parts of their history. Nations and even tribes with aggressive spirits went on to establish empires in various parts of the world, and rule over those for prolonged periods. Hindus appear to be the only civilization lacking a spirit of aggression; one can even identify a touch of pride in them for being non-aggressive. This could have been a significant factor in the military non-performance of the Hindus.

Naval Power: From the 15th century onwards, various world powers developed their naval fleets. Hindus were again found largely missing in action. They forgot to develop their naval power, except in a limited sense by the Marathas. The newly emerged Naval Powers, especially the European countries, had no difficulty in puncturing the safety of our seashores. There are also some references to the naval fleets of the Southern rulers. They are reported to have mounted naval expeditions to South East Asia, and recorded some victories in those parts. However, the details available are rather sketchy, and it is difficult to reach any specific conclusions.

42
SUMMARY OF CAUSES OF HINDU DEFEATS

At this stage, we will summarize the influence of the Hindu religion on the military value system of the Hindus, and go into the causes of Hindu military defeats. We will do that in light of the overall Hindu military performance over a 2,000 year period, and not restricted to any one particular event, decade, century, or period.

Causes — That Were Not

Before we go into the actual causes of Hindu defeats, we will spend a few minutes over the causes that were not responsible for the defeats, but are projected as being responsible. Like all defeated people, Hindus have shown great dexterity in inventing ingenious causes for their defeats. We list below such projected causes:

Projection 1: India of those days was divided into small states.

Fact 1: Most of the so-called 'small' Indian states were each bigger than Ghazni and Ghur, from which the invaders had originated. Some of the Indian states, including that of Prithviraj Chauhan, were of medium size, and individually many times the size of Ghazni and Ghur. Even otherwise, it was incumbent on one of the bigger Indian states to incorporate into itself, smaller states by persuasion, if possible; by the sword, if necessary. This is the route clearly chartered by Chanakya in his *Arthshastra*, and followed by every successful general of the world from Alexander, Caesar and Genghis Khan to Mahmud of Ghazni. No Hindu general after the 7th century AD considered that option. They had limited vision, and would often rejoice after recording a modest victory over a small neighbor. Bards in the service of the ruler did the rest; they magnified it as a great world-class victory. That was enough to boost the ego of the ruler.

In any case, the first period of Hindu slavery had started around 180 BC, when the barbarian tribes snatched power from the mighty Mauryan Empire (no excuse of small states there).

Projection 2: These Indian 'small' states were quarrelling with each other.

Fact 2: The referred quarrels were normal neighborly conflicts, which were a world norm those days; there was nothing unusual about it. These should not be given undue importance. At the time of the actual invasions by Mahmud of Ghazni (1009 AD) and Muhammad Ghauri (1192 AD), the Rajput rulers of North West India displayed exemplary unity. A large majority of them (almost all) sent their forces to fight the invasions under one command.

Projection 3: There were many collaborators among the Indian rulers. The often quoted names are of Jaichand of Kannauj (12th century), and Mir Jaffer of Bengal (18th century).

Fact 3: This is a highly exaggerated point. Some recent research has shown that there is no particular evidence to support the charge that Jaichand actually collaborated with Ghauri; the story is based largely on folklore. Anyway, collaborators have been known to exist in all civilizations. An odd collaborator in a few hundred years should not make an entire subcontinent go under with such ease. Surely, India of those days was not so fragile.

Projection 4: Ghazni and Ghauri are often blamed for descending on Bharat with a large fleet of swift-footed cavalry, with 'mounted archers' who could fire most accurately, even when at full gallop. (The Hindus armies had no answer for them.)

Fact 4: Hindus do not like to examine why no Hindu general could appreciate the central role of the horse as a 'weapon of war'. General after Hindu general continued to rely on the delusive strength of elephants, which let them down repeatedly, and at the most crucial moments of battle, as explained in chapter 39.

Projection 5: Hindus like to blame Babur for having come with gunpowder cannons. Tamberlane, Nadir Shah and others are often blamed for being barbaric.

Fact 5: Hindus were by far the most advanced civilization of those days. Why did they not consider inventing gunpowder? It appears that issues of war were no priority for them. War is a very dirty business; it is not a slugfest between ageing aunts. Hindus could have themselves displayed a bit of barbarity (meaning aggressive spirit).

Projection 6: Hindus often like to blame Muslim armies for not following the rules of war. In 2009, a TV serial showed Prithviraj Chauhan sitting on the top of Muhammad Ghauri, with his sword on the latter's throat. Just at that moment, sunset was announced, and Prithviraj let Ghauri go, to resume the fight the next day. Muhammad Ghauri could not believe his eyes, and his luck. As per this story, he attacked the same night and inflicted a crushing defeat on Prithviraj. That one defeat pushed the great Bharat varsha into 750 years of slavery.

Fact 6: The practice of only daylight fighting was prevalent in the *Ramayana/Mahabharta* days. The above story presumes that Ghauri would have been reading those epics before coming to Bharat. The episode is obviously not true; no sane person, leave alone a general of caliber, could have followed such a fatal practice in the 12th century AD. Still, the TV serial chose to put up this (imaginary) episode, presumably in the belief that the public would lap it up; some may even consider it as a high point of Hindu civilization. If our interpretation is even partly correct, things could not have got any worse.

Projection 7: It is fashionable for Hindus to blame the British for their policy of 'Divide and Rule'.

Fact 7: Instead of blaming the British, Hindus should ponder why they allowed themselves to be so easily divided. The cold fact is that we were hopelessly divided even before the coming of the British. The British followed their *dharma*; we forgot ours.

Projection 8: Whenever the question of prolonged slavery of the Hindus arises, many Hindus are often heard saying that in spite of all that, the Hindus civilization did survive.

Fact 8: It all depends upon as to how one defines 'survival'. For most part of the slavery period, Hindus had no control over their own destiny; their women were routinely dishonored, their mandirs demolished and gods humiliated. Every type and manner of atrocities was inflicted on them. If all that constitutes as 'survival', we concede the point.

Projection 9: Hindus love to project that they were always brave, and continue to be so now. They say that Bharatvarsha was always teeming with *shur-virs* (Bravest of Bravehearts), who would smash anyone casting a lustful eye on this most holy land. As an example, they often refer to the victory of Lord Rama over Ravan.

Fact 9: For the last about 1000 years, Hindus could not produce a '*Shur-vir*' who could teach a lesson or two, to the following types:
— Violators of Hindu hearth and homes, and tormentors of Hindus
— Molesters of Hindu women (in hundreds or tens of thousands)
— Demolishers of Hindu mandirs (in thousands). There was not one Hindu '*shur-vir*' who went to Ghazni and Ghur to avenge the honor of their women. The mighty and the powerful watched from the sidelines, waiting to be attacked. No provocation was enough to stir them for any type of offensive action.

Having covered the 'non-causes' of Hindu defeats, we analyze the possible cause of defeat in the following paragraphs.

Causes — That Were

Taking an overall and long-term view of issues and events, the military downfall of the Hindus during the last one thousand years, could be attributed to the following main factors, largely, due to the wrong interpretation of the Hindu religious scriptures:

* Too much stress on the individual, rather than on the nation. Hindus were obsessed with purifying their individual soul, and trying to merge it with the 'World Soul Brahman'. There was a fruitless search for an illusory entity called '*moksha*' (salvation). The Hindus' priority was to ensure a secure next life, rather than concentrate on the present one. They perhaps considered the present life as transitory, if not actually '*maya*' (illusion).

- Under the influence of the preceding factor, Hindus lost the distinction between 'mastery' and 'slavery'. They perhaps argued that due to the transitory nature of the present life, 'mastery and slavery' were some concepts largely in the mind. At the practical level, such types of issues did not make much of a difference, and were no reason for dispute, leave alone bloodshed. In the totality of circumstances and events (over 2000 years), it is difficult to avoid the impression that at some level, Hindus might have been even comfortable with their 'slavery'. Their efforts to get rid of slavery were few and far between, and mostly half-hearted. Even when opportunities for emancipation were presented to them, they failed to exploit these. We have covered those examples in our text earlier.

- At some stage, the Hindus locked on to the nation destroying concepts like *ahimsa* (non-violence), *shanti* (peace), *satya* (truth) — the '*ass*' syndrome. Whilst these issues might have some sort of a niche in an individual's life, these can possibly have no space in a nation's life. Hindus could not distinguish between the individual and the nation. They thought that what is good for the individual, must be good for the nation. The concept of *ahimsa* was entirely an import from Buddhism; the word does not even appear in the *Rig Veda*. If the word '*ahimsa*' appears in some Hindu texts, it does not have the meaning that we are trying to give it presently. We make these comments with due apologies to Mahatma Gandhi.

- In the Vedic times, animal sacrifice was the main means to please the Vedic gods. Animal slaughter was a daily affair. *Ashevamedha* (horse-sacrifice) was the most exalted ritual, which removed all sin. Under the influence of Buddhism, Hindus became averse to bloodshed (even in the service of the nation). It is possible that, gradually, they perhaps became averse to the very sight of blood.

- The hot and humid climate of India may have been a contributing factor. We may note that all invaders came from cold or very cold climates. Further, the invaders were all fiercely non-vegetarian. Now, nothing can be done about the climatic conditions. However, there may be an occasion to revisit the dietary habits.

- Hindus take pride in saying that their religion is tolerant, all-inclusive, assimilative, lacking assertiveness, etc. The net effect is that Hinduism gets projected as effete, meek and submissive.
- It has been sometimes expressed that a Hindu has the characteristics of *dáya, karuna* and *kshama* (compassion and forgiveness). From the military angle, these are self-destroying concepts.

The above interpretation of Hinduism emerges out of a gross mis-reading and wrong interpretation of the Hindu scriptures. Hindus also forgot an important principle enunciated in the following couplet:

Kshama sohati us bujangh (snake) ko, jis ke pas garal (poison) hai
Uska kya, jo dant-heen, vish-heen, vineet, saral hai.

(Only that snake can give forgiveness, which has poison in its fang;
What use is the one which is without fangs and poison, is humble
and simple.)

Chanakya in his *Arthshastra* says that a snake even without poison should behave as if he has poison in his fangs.

Aggressive Spirit Missing

In view of the above types of factors, Hindus lost their aggressive spirit; they were overtaken by a defensive mindset. Their central slogan became 'We will fight only when attacked'; and they stuck to it steadfastly. One irrefutable lesson of military history is that nations and generals without an offensive mindset can do no good even in the defensive mode. World military history proves the inviolability of this dictum. The only way to save 'Ajmer and Delhi' (and therefore Bharat) was for Prithviraj Chauhan to go and capture Ghazni and Ghur in Afghanistan. He had the capability and military muscle to do that; but the mindset was missing. But for that type of 'defensive mindset' Bharat would never have been a slave.

Closely allied with the aggressive spirit, is the question of attitude towards 'risk-taking'. There is a famous saying — 'No risk, no gain'; this dictum is particularly applicable to war situations. Only the bold and daring generals succeed. Fortune helps the brave, who will inherit the earth. In an earlier chapter, we have quoted the Sanskrit *shloka* '*Veera Bhoga Vasundhra* — the Brave will enjoy the Earth.' Lord Krishna in the *Bhagwad Gita*, effectively gave the same message; but, the Hindu antennas did not receive it. Thus, along with the loss of aggressive spirit, Hindus also became 'Risk-Averse'; 'Safety first' became their motto. Otherwise, there is no reason for Prithviraj Chauhan for not mounting a campaign to capture Ghazni and Ghur. If you dither at a crucial point in history, you are likely to be assigned to its dustbin. In view of the totality of the above factors —

- Ask not — 'Why the Hindus were defeated?'
- But ask — 'Why the Hindus were never on the OFFENSIVE?'

Unfortunately, this latter question has never been asked of the Hindus; neither by themselves, nor by anyone else. The impression that is sought to be created is that the 'Offensive' option was never available to the Hindus, and is not available now. Hindus themselves are at the forefront of creating this impression. Hindu apologists remark that 'Offensive' actions are not in Hindu culture. 'We are not that type of people' is a phrase often heard. This is a complete distortion of the Hindu religion.

The true Hindu scriptures are all for aggressive and offensive actions, for that one aim of achieving victory. Even the means adopted for that do not matter. We have covered this aspect in detail in our earlier chapters (40 and 41); here we just quote a *Rig Veda* Hymn (RV 6.75.2):

> *"With the bow let us win cows, with the bow let us win the contest and violent battles with the bow. The bow ruins the enemy's pleasure; with the bow let us conquer all the corners of the world."*

We must take note of the repeated use of the words 'Win' and 'Conquer' in the above short hymn; 'Violent' battles are recommended. 'World conquest' is a slogan given by the *Rig* at that stage of pre-history. All this establishes great stress of the *Rig* on 'Offensive Actions' and 'Victory' — always and under all circumstances.

The Hindus' problem lay in the fact that at some stage they got confused about their true scriptures. Most of the Hindu religious literature that emerged in the Christian era had a thick coating of Buddhism. That is true of the *Puranas* that dominate present day Hinduism. The ordinary folk are not able to discern that Buddhist coating; the learned perhaps are not interested.

We conclude this part by recording the following three broad reasons for the Hindu military defeats, especially during the 2nd millennium AD:

— Almost total absence of the 'Aggressive Spirit'
— General lack of enterprise and aversion to 'Risk Taking'
— The above two resulting in the absence of the 'Killer Spirit'

The above attitudes have arisen in the Hindus, out of misreading and wrong interpretation of Hindu scriptures, combined with their inability to identify their true scriptures. Some of that misreading may have been deliberate; it helped the Hindus explain away their prolonged slavery, in rather easy terms.

The overall conclusion that emerges is that Hindus like to blame everyone, except themselves for their woes; this is their trademark. Rather than facing the hard realities of life, they like to live in a cocoon of 'make-believe'; it helps their 'self-delusion'. Thus, the cause for Hindu defeats lay in their mind, rather than in their muscle. Only after Hindus accept and face this bitter truth that any recovery process can start. The issues involved are of such basic nature that the recovery process may extend over many decades, perhaps even a century. However, there is little probability that Hindus will accept this conclusion. Rather, they would attribute unholy motives to anyone talking along these lines, and call him ignorant, knave and prejudiced, who is not acquainted with the great Hindu culture.

43
SUMMARY AND CONCLUSIONS

This book on Hinduism starts by introducing the reader to other major religions of the world, viz.:

India Born Religions: Buddhism, Jainism and Sikhism
The Judaic Religions: Judaism, Christianity and Islam

The three Judaic religions are fiercely monotheist. Buddhism and Jainism have no need of God.

The above is followed by a study of the Hindu religion in all its splendor, variety and complexity. Keeping in view the myriad diversity of the Hindu religion, the book can do no better than to give broad contours of this religion, without in any way, taking a philosophical view. The aim is tell the story of Hinduism in a simple way.

The journey is started with Hindu literature, which falls in two broad categories;

Sruti: That which is of divine origin; it consists of the following;
Four Vedas (*Samhita*): Rig, Sam, Yajur & Atharva
13 Brahmanas: Expository prose portion of the Vedas
Aranyakas: Forest Books, four in number
Upanishads: 108 surviving ones

Smriti: That which is remembered. These include the two great epics Ramayana and Mahabharta (including Bhagwad Gita), 18 Puranas, Kalpa Sutras, Dharam Shastras and a host of other literature. In the two great epics lies the soul of Hindus.

Thereafter, the book goes on to explore the fascinating story of how the concept of 'godhood' evolved in Hinduism, perhaps from the very beginning of time. Summarized, the broad picture is on the following lines;

- In the very beginning was the concept of 33 gods in the Rig Veda, dominated by gods like Indra, *Surya* (Sun), *Vayu* (Wind), *Agni* (Fire), and Varuna. References to One Single Universal God in the Rig are largely by inference and rather at low key.
- In the second phase (around 800 to 600 BC), the Upanishads develop to its finality the concept of One Single Universal God called Brahman.
- In the third phase around 500 BC emerge two new religions – Buddhism and Jainism; these religions have no need of God. Buddhism flourished in Bharat.
- In the final phase, in the epics and the Puranas emerge a galaxy of new gods, headed by the Trimuti (Hindu Triad) of Brahma (different from Brahman), Vishnu and Shiva. The new gods include Rama and Krishna (the two *avatars* of Vishnu), Ganesha, Hanuman and many others. The Rig Veda gods were pushed to the margins. It is this concept of 'godhood' (called Puranic *Dharma*) which presently prevails in Bharat.

The book explores the Hindu concept of time in terms of the four *Yugas – Krita* (*Satya*), *Treta, Dvapar and Kali.* Theories about creation of the Universe are covered. Hinduism has the concept of 17 celestial seers (mind-born sons of Brahma); of these, 7 are generally called *dev-rishis* and 10 *prajapatis.* These include hallowed names like Vishvamitra, Vasishta, Bhrigu, Kashyap, Narad, and a prajapati called Daksha. Tradition primarily credits Kashyap and Daksha with start of humanity and populating the world. 13 daughters of Daksha were wives of Kashyap; one of them was Aditi. She gave birth to *adityas* (gods). One of the *adityas* was Vivaswat (*Surya*, the sun); he became father of Manu Vaivaswat who was progenitor of the human race. Tradition has it that in all there are 14 Manus who rule over the earth. Sons and a daughter of Manu established various ruling dynasties, which went on to rule various parts of Bharat, till the time of Ramayana and Mahabharta.

Hindu Military Value System – Past & Present

Of the recorded Hindu history of around 2300 years, Bharat was under jackboots of slavery for about 1300 years in two phases:

First Phase: 500 years – 200 BC to 300 AD
Second Phase: 800 years – 1200 AD to 1947 AD + 50 years

The 1st and 2nd Battles of Tarrain (near Karnal) in 1191 and 1192 AD respectively, both between Muhammad Ghauri and Prithviraj Chauhan were directly responsible for 800 years of Hindu slavery of the second phase. Prithviraj was the victor in the first battle; Ghauri was wounded and on the run. The most elementary military strategy required that Ghauri was pursued to his lair in Ghazni, and the Muslim power smashed at its very base; that was within the power of Prithviraj. But Prithviraj, rather than follow that route, took to premature celebrations. Ghauri returned within the year and smashed the Rajput and Hindu Power for all times to come, in the second battle of Tarrain. History does not excuse those who dither at a crucial moment in history. We are reminded of an Urdu couplet:

Lamhon ne khata ki thi, Saddiyon ne saza payi
(Lapse was momentary; resultant punishment was for centuries)

Something similar happened 750 years later. Pak irregulars (tribals) had invaded Kashmir in 1948 AD. The Indian army beat them back and the irregulars were on the run. Again, at a very crucial moment of history, Jawaharlal Nehru stopped the onward march of the victorious Indian army. We are still paying for that elementary mistake. Therefore, we may not be amiss in concluding that the mindsets of 12th century AD continued till the 20th century.

In our earlier chapters, we have traced the long bouts of Hindu slavery and their miseries to chinks in the Hindu Military Value System; we have termed that as 'Hindu mindset'. We have relied on the following factors/facts to reach that conclusion:

- Bharat was a huge country with humongous resources, including almost limitless manpower. Even then, Hindu forces never ventured out of their land borders to attack and capture foreign lands. Hindus considered (& still do) those 'non-invasions' as a high point of their civilization, though that was against all canons of military strategy; victory normally comes to armies which are on the offensive.

- Hindus in general and their sentinels Rajputs in particular, fought only when they were attacked; they displayed a singular lack of 'Offensive' spirit. There were innumerable occasions during the 600 years of Muslim rule when there were prolonged periods of power vacuum in Delhi and anarchy prevailed. But the next door Rajput rulers would not unsheathe their sword, mount their horse and march on Delhi. They would rather wait for the Muslim ruler of Delhi to gather muscle, and then come after them. *popularised by Gandhi*

- Hindus suffer from bouts of phony morality and bogus sense of self-righteousness. They have been wedded to body and soul destroying concepts of *ahimsa* (non-violence), *shanti* (peace) and *satya* (truth). Hindus also like to flaunt their (presumed) attributes of *daya, karuna* and *kshama* (compassion and forgiveness). All these are un-military like attributes, which must be shunned.

- It is astonishing to see the ease with which the great Hindu civilization went under. After the second battle of Tarrain (1192 AD), the whole of North India lay prostrate. Even the rulers of South India did not consider it prudent to challenge the intruders; they could see neither the actual nor potential threat; (Though we do hear of some Southern rulers having mounted expeditions to South East Asia).

- It is equally baffling to see the ease with which Hindus accepted their slavery. They adjusted to it with remarkable alacrity, almost as a duck takes to water. There was no great national upsurge, no fight back, even no major signs of resentment. Slavery appears to have been accepted as an inevitability to have happened naturally; it was not considered a affliction which had to be fought all the way.

- The over-all effort of the Hindus has been to convey the impression that the 'Hindu slavery' was really not their fault; someone else must have been responsible for that. They might have even reasoned that concepts like mastery and slavery were all in the mind only. These were not worth shedding blood; it is all *maya* (illusion). It is difficult to see any set of people other than Hindus, so comfortable with slavery.

The above state of affairs arose due to a false interpretation of Hindu scriptures. In early centuries of the Christian era, Buddhism started making major inroads in Hinduism; over time, it became difficult to distinguish between the two religions. Hindu scriptures got over-coated with Buddhist thought. Vedic *dharma* is as different from Buddhism as cheese from chalk. We repeat below a somewhat controversial statement which we have made earlier:

'Whatever is joyous and active (even aggressive) is from the Vedic *dharma*; whatever is pessimistic and passive is from Buddhism.'

We elucidate the above theory of ours by some examples:

- Rig Veda is the very first religious book of Hindus. It believes in joyous living, enjoying the bounties of nature. Animal sacrifice was a daily ritual. Rig is full of aggressive exploits of the great god Indra. He is shown repeatedly attacking demons and recording victories. One Rig *shloka* recommends even world conquest.
- Bhagwad Gita is the most holy book of Hindus. It says that not engaging in (righteous) war attracts sin. 'Kill, or be killed' has to be the motto on the battlefield. Holy book of no other religion glorifies war in this way.
- In Manusmriti (5.39, 44), Manu says as follows:
  "Killing in a sacrifice is not killing. ------*himsa* (violence) to those that move and those that do not move which is sanctioned by the Veda – that is known as *ahimsa* (non-violence)."
- The soul of Hindus lies in the two great epics, Ramayana and Mahabhata, which are essentially tales of war, and glorify it.
- In Mahabharta, at one time Yudhishtra expressed his desire to renounce the world. Arjuna dissuaded him from doing that by informing him as follows:

'People honor most the gods who are killers. Rudra

(Shiva) is a killer, and so are Skanda, Indra, Agni, Varuna, and Yama. I do not see anyone living in this world with *ahimsa* (non-violence); even great renunciats cannot stay alive without killing.'   MB2.15.16, 20, 21, 24
After hearing the above, Yudhistra gave up his plans.

- The great lord Vishnu took *avatar* as Parshuram. He carried out mass slaughter of the whole clans of Kshatriyas 21 times. Shiva gave his *parshu* (axe) to Parshuram for the purpose. Thus, both the great gods were involved in that wanton exercise of bloodshed on an unprecedented scale.
- Hindu philosophy resolves around '*matsya-nyaya*' (Law of the fish) i.e. the bigger fish gobbles up the smaller fish. In other words, the bigger state must take over the small or weak state. Chanakya's Arthshastra recommends it unashamedly. Chanakya goes on to say that a snake even without poison, must behave as if it has poison.
- The ancient Hindu military theory rested on the concept of 'Circle' i.e. the country on your border is your enemy, and your enemy's neighbor is your ally. Such a theory was nothing but an invitation for relentless aggression; kings were always on the prowl for a 'kill'. In ancient times, the name for the Hindu king was '*vijigishu*' – meaning, the one who wishes to conquer.
- Manusmriti 5.29 states as follows:

"Those that do not move are food for those that move; and those that have no fangs are food for those that have fangs. Those that have no hands are food for those that have hands; and cowards are food for the brave."

In the above, words 'fang' and 'hands' are not to be taken literally; these words represent 'power'; in other words, the powerful rules, should rule and has to rule. Ancient Hindus had the necessary amount of poison in their fang; however, at some stage, they decided to put that in wraps, and pretend that they have neither fang, nor poison. Over time, Hindus even came to project that as a high point of their civilization.

In an overall analysis, there is no doubt that Hindus of the yore were a set of proud people, who believed in living life to the full. They had an offensive mindset, and believed in fighting all forms of injustices; that is the central message of their holy book, the Bhagwad Gita. Hindus of those days knew how to defend their freedom, and enjoy its fruits. There is neither cause nor justification to push such vibrant set of people in the *bhul-bhuliyan* (dark alleys) of *ahimsa* (non-violence), *shanti* (peace) and *satya* (truth); but they did get pushed. People without power and those who are cowards, have no place in the Hindu scheme of things. It is indeed a great irony that Hindus who have their texts rooted around 'power' and 'militarism', shy away from these very concepts. The train of Hindu religion got derailed by hitting the Buddhist boulder lying on the track. That is what led to the deluded sense of Hindu *dharma* that presently prevails in Bharat. That is what constitutes the 'Hindu mindset', which resulted in prolonged bouts of slavery and misery for the Hindus.

Position in Present-Day Bharat

The million dollar question is whether traces of 'Hindu mindset' continue to prevail in independent India. The obvious and simple answer is, "No, No way. We are a vibrant secular democracy. So, where is the scope for Hindu thinking to prevail? The very question is irrelevant." But many a time, things are not what they appear to be on the surface. Let us examine the issue in some depth.

Without any shadow of doubt, India is not a Hindu country. But, it cannot be denied that it is a Hindu-majority state; about 85 % are Hindus. However, the more crucial factor is that since 1947, more than 95 % of all important decision-making posts have always been occupied by Hindus We can have a look at the Central Cabinets since independence, starting with that of the first Prime Minister Jawaharlal Nehru. Important Cabinet posts of PM, Home, Defense, Finance and External Affairs have been always with Hindus (an odd Christian/ Parsi once in a while does not make much of a difference). Persons of the dominant minority i.e. Muslims were and are entrusted with mostly innocuous portfolios like Education, Minority Affairs, Non-Conventional Energy, etc. Position is no different for top bureaucratic and military posts; most of these have been with Hindus. Since 1947, we have had some 60 Chiefs of Staff; only one has been a Muslim - an Air Chief. Yes, we have had three Muslim Presidents; but that is largely a ceremonial post with hardly any effective power.

In the present day 'secular' Bharat even official functions are normally started with 'Ganesh *puja*'; many other Hindu rituals like *saraswati puja*, *'diya'* lighting are also undertaken. Even the most committed secularists do not realize that those rituals are anathema to the dominant minority - the Muslims. The State has a Hindu *shloka* 'Satya mev Jayate' as its national motto. Another one *'Atithi Devo Bhava'* (with its negative implications as discussed in chapter 14.3) has been lately adopted. In TV debates, even official spokesmen often quote Hindu *shlokas* (e.g. *sarv dharma sambhav*) to prove their point.

From the totality of circumstances, it is difficult to avoid reaching the conclusion that the present-day Indian thinking is based predominantly on 'Hindu thinking', or at least is derived from that. Without anybody realizing it, a cold layer of Hindu mindset prevails at the sub-conscious level. Its direct effect has been on the military value system. The net effect was that Hindus cannot appreciate:

- Centrality of 'War' in affairs of nations
- Centrality of Armed Forces in affairs of war
- The crucial role of Generals in recording 'victory', which is the only aim of war.

Slide in the Indian military affairs started from the early period of Indian independence, from the days of the first Prime Minister Jawaharlal Nehru, a most prescient man of his time. India had inherited a very fine set of armed forces from the British. With that excellent base, India could have gone on to become a predominant, if not dominant, military power. But, that was not to be; there were several reasons for the same.

Nehru was the high priest of secularism. But in an irony of fate, this de-hard secularist locked on to typical Hindu concepts of *ahimsa* (non-violence), *shanti* (peace) and *satya* (truth). It is another matter that we have argued earlier that these concepts cropped up due to miss-reading and miss-interpretation of Hindu scriptures i.e. these emerged out of a deluded sense of *dharma*. *Ahimsa* was the flavor of the season in 1947, pushed relentlessly by Mahatma Gandhi; it was mandatory for all politicians of those days to have deep faith in *ahimsa*. Nehru also locked on the concept of *shanti*; so much so that he started lecturing the world about it, with some very unwelcome results. *Ahimsa* and *shanti* became Nehru's obsession. A typical Hindu *shloka* 'Satya mev Jayate' was adopted as the State motto.

Nehru selected (Emperor) Ashoka as the icon of free India. In chapter 17.3, we have established that Ashoka was responsible for pushing Bharat of the closing centuries BC, into slavery. In his later years, Ashoka almost outlawed war; he left written instructions to his heirs to do likewise. The passive mindset advocated by Ashoka was constructively responsible even for the second phase of Hindu slavery, starting 1200 AD. It was Ashoka's grandfather Chandragupta Maurya who had set up the great Mauryan Empire, from scratch. He is the only Hindu general who defeated a foreign general i.e. Selecus (Greek), and extracted territory from him. Chandragupta has all the credentials to be free India's icon; but that was not to be. Military exploits of Ashoka, if any, compared to Chandragupta were quite modest, in fact insignificant; yet, Ashoka occupies the prime place and Chandragupta is nowhere. This is an example of the topsy-turvy view of the military value system that prevails in this woeful land.

Military value system does not come easily to people who are wedded to (rather wallow in) types of concepts and attributes listed in the preceding paragraphs. Such concepts are not conducive to nurturing the tender sapling of the military value system; that requires a different type of mindset. The slide in the fortune and culture of the armed forces was inevitable; and that was what happened.

On India gaining independence, the political class showed a marked degree of disinterest in military affairs; for the politician, military was an avoidable distraction. Armed forces came to be considered a type of unavoidable non-necessity. Bureaucrats moved in to fill the vacuum, as the generals were slow in taking stock of the situation. The new

Defense management structure concentrated all powers in the hands of bureaucrats in the Ministry of Defense. Generals were effectively marginalized; an impenetrable wall came to be built between the politicians and generals. The concept of 'Civil control over military' was distorted out of shape. Over time, Chiefs of Staff were pushed down the Protocol List. A policy decision was taken not to allow any General near the Pay Commissions that are periodically set-up. That facilitated gradual down-grading of the military ranks and military value system.

The overall effect of various acts of commission and omission was that the armed forces were gradually eased out of the center of national consciousness, and pushed to the sidelines. This is well illustrated by the puny little size of the 'Tomb of the unknown Soldier' built for soldiers of independent India; it is grandly called *Amar Jawan Jyoti*. It is in the form of a tiny, insignificant, almost unnoticeable *chabutra* (platform) under the very shadow of gigantic structure (India Gate) made by the British for the 'slave soldiers' of Indian origin, who fought for the British Empire in the First World War. No soldier, dead or alive, can be proud of that *chabutra* like structure; question of drawing any inspiration from the same does not only arise. Stark contrast between the British and Indian (read Hindu) attitudes towards military issues is well illustrated by the comparative sizes of the two structures.

We give another small example. It is customary to hear nationalistic songs blaring out of colleges, schools and TV channels on days of national importance e.g. Republic day. These songs are meant to inspire the youth and arouse nationalistic feelings in them. One of those songs has the following line:

Duniya ke zulum sehna, aur munh se kuchh na kehna

Loosely translated, the couplet means 'It is a great tradition of ours to bear all type and manner of atrocities, without ever complaining.'

Now what future can a nation have which sends the above type of message to its youth? It even goes against the central message of the Bhagwad Gita, wherein the great Lord exhorts Arjuna to fight against every form of injustice (*zulum*), even at the cost of one's life.

The general and overall neglect of the 'Military Value System' and resultant Hindu slavery leads us to the following conclusions:

- The most prized attribute of humankind is civilization, which requires a delicate balance of liberty and order, militarism and culture. Civilization will be overthrown by invaders from abroad unless it is jealously guarded by people from within. Eternal vigilance is the price of civilization and liberty, to be paid in human blood.
- Nations may love peace; but they must keep their powder dry. Nations wanting to live with honor must have steel in their soul.
- When the chips are down and balloon is up (Army speak for war), it is only a General or two of caliber who will save the *Izzat* (honor) of the nation. Such Generals have to be groomed over the long term i.e. over decades; they cannot be produced on the eve of war.

Effect of general all-round benign neglect of the armed forces was reflected on the battle-field. As true nationalists, we all like to hope and believe that our armed forces would always perform gallantly on the battle-field, and bring glory to the nation. But, in our enthusiasm, we should not over-look the actual ground reality. In chapter 29, we have analyzed the four major military engagements of free India; the over-all performance of the Indian armed forces falls in the 'Highly Deficient' category. That should been very worrisome; but no one cares.

That was not the end of the story. In August 2009, the incumbent Chairman, Chiefs of Staff Committee (highest ranking Defense Official) made a public statement that India was no (military) match for China, and that the gap was unbridgeable. Such a grave statement should have caused all-round turmoil; but not a leaf stirred. There was no response from the government; it tried to give the impression that things were as these should be, and there was nothing to worry. The government appeared to be saying, "Our plate is already full of a number of serious problems like terrorism, Maoists, Telangana, price rise, etc. Why are some people trying to push a low-level issue like Defense to the fore?

Inshah-Allah (God willing) we will attend to Defense as soon as we find some breathing space." That, in a nutshell reflects Hindu attitude to issues of war.

The whole Defense culture in India is in the wrong lane; the Defense issues in this woeful land are at the bottom of the heap. Presently, we are spending some 70 to 80 % of our waking hours on terrorism; 'Defense' gets, if at all, less than 5 % of political attention. No doubt, terrorism is an important issue; but it is still a short term one. Sooner or later (say in 10 years), terrorism will sort itself out, as it did in Punjab. But the Defense issues are here to stay for the long term (for decades and centuries), and affect the very life-blood of the nation, the very concept of mastery and slavery. The Defense culture in this land needs some fundamental changes. The process can only start if and when the Indian politician starts taking genuine interest in Defense issues.

The question of Hindu military mindset and resultant prolonged bouts of slavery are of basic nature. We have raised the issue, but may not have been able to provide all the answers due to the extremely complex nature of the problem. The issues involved occupy the sub-conscious space, which is not easy to reach, leave alone analyze; highly prescient minds are required to do that. We hope that at some stage in the future the military issues will come to occupy their rightful place in the minds of people inhabiting this ancient and holy land. But, we are not sanguine that that will actually happen. So, we can do no better than rest our case with the words:

Que Sera Sera – Whatever will be, will be

Appendix
SOME RANDOM THOUGHTS
Quotes, Definitions, Views & General Comments

A: Quotes of Famous Men

1. Truth
Man will occasionally stumble over the truth; but, most of the time, he will pick himself up and continue on — Churchill

In war, every truth has to have an escort of lies — Churchill

2. Justice
That which is in the interest of the stronger party — Socrates

3. Rumor
That which should not be believed till it has been officially denied — Churchill

4. Rationality
Men and nations will act rationally when all other possibilities have been exhausted — Kartz's Law

5. War
Four things greater than all things are

Women and Horses, Power and War — Kipling

6. Rule of God
There can be no rule of God in the present state of inequities, in which a few roll in riches and the masses do not get enough to eat. — Mahatma Gandhi

7. Foolishness

Better to remain silent and be thought to be a fool, than to speak and remove all doubts — Abraham Lincoln

8. Expert

An expert is a person who avoids small errors while sweeping on to the great fallacy — Weinberg's Corollary

9. Intelligence

The sum total of Intelligence on this planet is constant; and the population is growing — Colin's Axiom

10. *Bhagwad Gita*

Karmanye evadhikarste, ma phalasu kadachin — BG. 2. 47
(translated)
Your right is only to perform action, and not to its fruit

B: Anonymous

1. Law

Howsoever high you may be; law is above you.

2. Selfishness

Adam, Adam, Adam Smith
Listen, what I charge you with
Didn't you say?

In the class one day
That Selfishness was bound to pay
Of all doctrines that was the pith
Wasn't it, wasn't it, Adam Smith.

Note: Adam Smith (1723–1790), a Scottish philosopher, is considered the father of modern Economics, and a believer in 'Free Enterprise'.

3. System Failure

A favorite alibi for sloth, indolence and non-performance

It is the table at which the bunch of corruption, non-accountability, negligence and inefficiency rests.

4. Government

At the best of times, government is a necessary evil;

And at the worst, it is more evil than necessary.

5. Bureaucracy

The only way to root out bureaucratic inefficiency and corruption is not to have bureaucracy.

6. Love/Heart (in French)

Le Coeur a ses raisons, que le raison ne connait pas —

The heart has its reasons, which the reason knows not

7. Democracy

That borrowed concept by which Indian politicians, acting in the name of people, have accumulated unbridled power and vast illegal wealth and reduced the country to a position where moral fiber is torn, national feeling gone and social justice buried.

Intellectuals have not been able to summon a roar of protest, a battlecry against riggers of our democracy.

Some distinguishing features of the present Indian democracy:

— Politics without Accountability

— Economics without Efficiency

— Media without Integrity

— Crime, Corruption and Contempt for Norms

— Law-breaking, Smuggling, Drug trafficking

— Self-aggrandisement and Unbridled Greed — Mortgaging of the future in pursuit of endless present Gratification

— Some ½ billion illiterates, famished or half-famished

— Top Defence Man declaring Bharat as 'No match' for China

8. Pay-Back Time

Everything that the rich and powerful have is taken from society;

Some by hook, but mostly by crook

How about some Pay-Back?

9. The Rich and the Righteous
The rich and righteous who

> — consider themselves to be the embodiment of civilization,

> — but consider the despoiled and the deprived as the embodiment of evil,

Are destined for hubris; in other words, the arrogance which leads to downfall, defeat and self-destruction.

10. To Whom You Love
Too often, too much remains unsaid

And then, suddenly, the time is up

You must find time and opportunity to tell people around you of your feelings for them

11. One's Own People (An Urdu Couplet)

Mujhe apno ne hi mara	I was tricked by my own people
Gairon mein kahan dum tha	Others dared not do that
Meri kashti wahan doobi	My boat was drowned
Jahan pani bahut kum tha	Where water was shallow
	(Basically, I was over-confident)

12. Might & Right
Might is always Right

Right emerges out of Might

The 'high class' criminals think that they will not get caught. If caught, they know they will not be convicted. If convicted by chance, they are sure they will get a light sentence, and immediate bail. We the non-criminals, have contributed a great deal to that way of thinking.

13. Civilization
Civilization is a precarious thing whose delicate balance of liberty and order, peace and culture may at any time be overthrown by enemies invading from abroad, and sometimes even arising from

within. Eternal vigilance is the price of civilization, to be paid in human blood.

A great civilization is not conquered from without unless it has first destroyed itself from within; that is a settled law of nature and history.

14. A Vile Man
A vile man is not the one who lies, murders, betrays or drinks too much;
But the one who is cowardly and weak; and does not win in war.

15. Austerity
Austerity, and not endless gratification of wishes and wants, is the hallmark of an advanced civilization.

16. Success
If you are successful, you will have false friends and true enemies.

Succeed anyway!

C: By the Author

1. Freedom and Bloodshed
Freedom is not for free
There is a price to pay
The price is in human blood
Sheep's blood has no say

It is only when men die that nations survive. We have to get used to bloodshed in service of the nation.

2. Slavery
Slavery is the worst affliction of humankind.

Hindus must find time to reflect on 750 years of their slavery; presently, they are not in a mood to spare even 750 seconds for that. Hindus must learn to face and confront unpleasant facts and bitter truths of their past.

3. Generals and Armed Forces
When the chips are down, and the balloon is up (army speak for war), only a General or two of caliber (if we can produce them) will

save the honor and *izzat* of the country.

Bharat must start nurturing its generals, here and now; tomorrow may be too late.

Armed forces require *Samman* (Honor), Status and Salary, in that order. It is for civil society to give these; not for the armed forces to covet. The day when the armed forces have to clamour for these, is indeed a sad day for the nation.

For all its tall talk, the Indian (read Hindu) society has never understood the concept of honoring its soldier; all their strindent claims to the contrary are insincere, hollow and bogus.

4. Winning and Victory

In War,

a) Winning/Victory is not everything, it is the ONLY thing.
b) Defeat is a disgrace — always and under all circumstances;
c) No excuses are good enough to justify defeat — never and under no circumstances.
d) It is better to be a foot-soldier of a victorious army, than being the General of a defeated one.

From the battlefield, one must walk out Victorious; or not walk out at all, i.e. head for heaven, where Lord Krishna ensures guaranteed entry.

A Hindu enters and should enter war only for two reasons — to attain heaven, if slain; to rule over the earth, if victorious.

5. Offensive Mindset

All world-class generals had an Offensive mindset, and always acted in the Offensive mode. That is how they could record victory after victory.

History records that the party on the Offensive won 8 or 9 times out of 10.

6. Readiness for War

Nations may love peace, but they must keep their 'powder dry' (i.e. be ever prepared & ready for war).

A sharp sword is the best (and the only) guarantee for *shanti* (peace) and honor.

A rusted sword is a standing invitation for invasion, defeat and slavery.

7. Poverty

The second worst affliction (after slavery) of humankind

The people who have not experienced it personally, cannot even start to realize its dehumanizing power

It should be a strict 'no-no' in a society claiming to be civilized

8. Time

That un-understood entity which has flowed timelessly

From the beginning which there was not;

And which will flow to the end, which there will not be

Time is the only entity which no one has been able to fool or evade.

9. Death

The final (unreal) Reality

The ultimate (uncertain) Certainty

The most feared, but the least avoidable

Other than God, the only real truth

10. God

The only Absolute Truth and Reality

Lives in the poor and the destitute,

Whose service is the only true path to God;

Rest is all humbug

11. Religion

That by which civilized man lives

But, also the cause for maximum human conflict

12. *Ahimsa* (Non-violence), *Shanti* (Peace), *Satya* (Truth)

The 'A-S-S' syndrome

Most Hindus believe that *Ahimsa, Shanti, Satya* are in their DNA; that is a grave misconception

Vedic *dharma* has no place for these attributes at the State level, even if these could be tolerated at the individual level.

We must learn to distinguish between the State and the Individual.

13. *Ahimsa* (Non-Violence)

Vedic *dharma* requires a Hindu to be strong. To be truly Vedic, a Hindu has to be powerful. If he is weak, his *dharma* is dubious, delirious and distasteful

Hindus were worshippers of *Shakti* (Power); they could not have been votaries of *Ahimsa*, which has been dreamt up by those who want to enjoy cowardice (Mahatma Gandhi was an honorable exception; his *Ahimsa* was in a specific context, which is above the understanding of us, the normal mortals)

Ahimsa is a concept imposed on Hinduism by Buddhism. Hindus are unable to see this crucial difference.

14. *Satya* (Truth): In matters of the State

Satya has seldom succeeded;

Asatya (Untruth) has rarely failed

Satya has no place in matters of the State.

15. *Shanti* (Peace)

The concept of *Shanti* is strictly for application at the individual level.

The State has nothing to do with it, except as a ruse.

16. Arrogance, Ego, Ignorance (AEI syndrome)

These are a part of human DNA; man being a bundle of these.

17. Modesty (Humility)

Modesty (Humility) is best when not put too much on display.

18. Consistency

Only fools are consistent (largely in their foolishness)

The wise always change and adapt

19. Mediocrity and Excellence

Mediocrity is as common, as Excellence is rare;

Be assured, you have lot of company.

20 Dignity

There is a need to impart a modicum of dignity to the poor and the deprived of this woeful land, by giving them at least a small measure of control over their lives, and self-reliance for their basic needs (say toilets).

21. Marriage

Marriage as an institution

Is an unexciting proposition

Meant mainly for procreation

It is a stage,

Where life takes a bend

And towards dull routine tends

Where duties and obligations predominate

And all romance ends.

D: Political Scene in India (2009)

The Indian Politician

Indian politicians now appear to be like foxes sent to watch the chickens, simply because they have had a lot of experience in the hen-house.

The nation is now in the hands of men who will stop at nothing, including ruse and cunning, to effectively wage the struggle to grab power. They are the inept merchants of dubious virtue, spending all their energy in safeguarding their own edifice.

The *aam admi* (common man), with his feudalistic servitude and fatalistic resignation, has no choice but to look upon the political scene in sorrow, and upon the politicians in (feigned) anger. His future appears to be enveloped in uncertainty, despair, gloom and chaos.

Note: There are some honorable exceptions to the above description of politicians.

BJP Politicians

After the electoral defeat of 2009, the BJP politicians are much in the manner of bucking horses in a rodeo ring dashing helter-skelter, trying to make sense of the disaster that has hit them. They are running around in circles defending the indefensible, and trying to locate a peg on which to hang the reasons for their (shattering) defeat. They take an impasse for a breakthrough, a handshake for a rapprochement, and a (feigned) smile for unbounded love. They want to convince themselves (and no one else) that the storm has blown over.

By their frequent trips (both joint and single) to pay obeisance at Jhandewalan (RSS Delhi HQs), they wish to finally convince the skeptics and the cynics that the RSS and BJP are indeed independent organizations — no *choli-daman ka sath* (no oneness, only two-ness).

After the BJP's electoral defeat, its senior-most leader of 'Jinnah praise fame', with the supreme confidence of a man dialing his own number:

— Sits through party conclaves in enigmatic silence.

— Steadfastly refuses to answer the grave charges flung at him, including by a senior (expelled) member (another Jinnah acolyte) of his own party.

— Offers to resign citing some quaint reasons; then withdraws the offer for reasons even more quaint and bizarre.

— Offers to opt out of politics, but waits to be pushed out (gracefully).

The Secular Elite

The so-called forward-looking secularist elite of India is

fundamentally intolerant and illiberal — the same as they accuse others of. It has crafted an image of itself, which mirrors its own fads and predilections.

Liberalization

Liberalization is a word for the process which sets in motion the prospects of enrichment of those who are already wealthy. In this system the already rich will become spectacularly rich, and the already poor will become abjectly poor.

On the one hand, capitalism requires the engine of self-interest, otherwise called greed. On the other hand, society requires attention to the general interest of the masses, otherwise called 'taming of greed'. We are forever pulled back and forth between these two poles. We need a politician who can see the big picture, and has the courage of conviction to act decisively and ruthlessly according to his/her beliefs.

Liberalization is also a word to tell the deprived and the poor that in any scheme of things, they will have to finally foot the bill. So, they should tighten their belt further (all for the good of the country).

Each billionaire added in the metropolitan cities implies pushing down the poverty slope of about a million in the countryside. But then, who cares; the media is obsessed with billionaires.

Liberalization works by converting our Desires into Needs. Of all the exploitative mechanisms invented by man, there is perhaps nothing more sinister and intractable than Liberalization and Globalization.

Law's Own Course

'Law's Own Course' does not provide the deprived and the poor of this land the same rights, as the rich and the elegant have in the pursuit of Depravity.

In our society, poverty by itself is a cognizable offence. The crimes of the poor, since these arise out of unfulfilled basic necessities, are legally actionable, with speed and promptitude.

The crimes of the rich, committed in wantonness, are subject to stay orders and endless bails (even at night); these are amenable to legal wrangling by affluent lawyers, expert at hair-splitting for the sole benefit of the rich and the super rich. Granted such absolute immunity from law (which takes its own course relentlessly and resolutely, but aimlessly and selectively), it is inevitable that most of us are tempted to yield to our worst instincts.

BIBLIOGRAPHY

1. *The Indian Theology — Brahma, Visnu and Siva* by Sukumari Bhattacharya; Penguin Books

2. *The 'A' to 'Z' of Hinduism* by Bruce M Sulluvan; Vision Books

3. *The Book of Shiva* by Namita Gokhale; Viking, Penguin Books

4. *The Rig Veda — an Anthology* by Wendy Doniger O'Flaherty, Penguin Books

5. *Hindu History* by Ashoka K Majumdar; Rupa & Co

6. *Questioning Ramayana* edited by Paula Richman; Oxford University Press

7. *Religion as Knowledge — The Hindu Concept* by Janaki Abhesheki; Akshaya Prakashan

8. *Ancient Indian Historical Tradition* by FE Pargiter; Motilal Banarsidas Publishers

9. *Hindu Myths — Penguin Classics Introduction* by Wendy Doniger O' Flaherty

10. *Myths and Legend Series India* by Donald A Mackenzie; The Mustic Press, London

11. *The Gods of India* by Rev E Osborn Martin; Cosmo Publication, Delhi

12. *A Classical Dictionary of Hindu Mythology & Religion* by John Dowson; Oriental Books, New Delhi

13. *The Mind of India* edited by William Gerber

14. *India — A History* by John Keay; Harper Collins, India

15, *The 1965 War — The Inside Story* by RD Pradhan; Atlantic Publishers, New Delhi

16. *Hinduism — An Introduction* by Shakuntla Jagadnathan Publication

17. *Soul of India* by Amury De Rencourt Publication

18. *Ramayana* by Griffith Publication

19. *Wonder that was India* by AL Basham; Sidgwich & Jackson, London

20. *History of the Indian People* by DP Singhal; Methlen London Ltd.

21. *The Bhagwad Gita* by Nataraja Guru; Vikas Publishing House

22. *Srimad Bhagwad Gita* by Pujya Ma (translated); Arpana Publications

23. *The Essential Teachings of Hinduism,* Edited by Kerry Brown; Arrow Books

24. *The Hindu Religious Tradition* by Pratima Bowes; Allied Publishers

25. *The Vedas* by Fredrich Max Muller; Susil Gupta (India) Ltd.

26. *Myths of the Hindus and Buddhists* by Ananda Coomarswamy; Dover Publications, New York

27. *Indian Legends* by Nagendra Kr Singh; APH Publishing Corporation

28. *Hinduism* by Karan Singh; Ratna Sagar (Pvt) Ltd

29. *No Substitute for Victory* by Theodore Kini and Donna Kini Publication

30. *History of India* by Monstaurt Elphinstine Publication

31. *War Despatches* by Lt Gen Harbaksh Singh Publication

32. *The Laws of Manu* — Penguin Classics

INDEX